# The Cell Cycle

# The Practical Approach Series

**SERIES EDITORS**
**D. RICKWOOD**
*Department of Biology, University of Essex*
*Wivenhoe Park, Colchester, Essex CO4 3SQ, UK*

**B. D. HAMES**
*Department of Biochemistry and Molecular Biology*
*University of Leeds, Leeds LS2 9JT, UK*

Affinity Chromatography
Anaerobic Microbiology
Animal Cell Culture (2nd edition)
Animal Virus Pathogenesis
Antibodies I and II
Behavioural Neuroscience
Biochemical Toxicology
Biological Data Analysis
Biological Membranes
Biomechanics—Materials
Biomechanics—Structures and Systems
Biosensors
Carbohydrate Analysis
Cell–Cell Interactions
The Cell Cycle
Cell Growth and Division
Cellular Calcium
Cellular Interactions in Development
Cellular Neurobiology
Centrifugation (2nd edition)
Clinical Immunology
Computers in Microbiology
Crystallization of Nucleic Acids and Proteins
Cytokines
The Cytoskeleton
Diagnostic Molecular Pathology I and II
Directed Mutagenesis
DNA Cloning I, II, and III
Drosophila
Electron Microscopy in Biology
Electron Microscopy in Molecular Biology

Electrophysiology
Enzyme Assays
Essential Developmental Biology
Essential Molecular Biology I and II
Experimental Neuroanatomy
Extracellular Matrix
Fermentation
Flow Cytometry
Gas Chromatography
Gel Electrophoresis of Nucleic Acids (2nd edition)
Gel Electrophoresis of Proteins (2nd edition)
Gene Targeting
Gene Transcription
Genome Analysis
Glycobiology
Growth Factors
Haemopoiesis
Histocompatibility Testing
HPLC of Macromolecules
HPLC of Small Molecules
Human Cytogenetics I and II (2nd edition)
Human Genetic Disease Analysis
Immobilised Cells and Enzymes
Immunocytochemistry
In Situ Hybridization
Iodinated Density Gradient Media
Light Microscopy in Biology
Lipid Analysis
Lipid Modification of Proteins
Lipoprotein Analysis
Liposomes
Lymphocytes
Mammalian Cell Biotechnology
Mammalian Development
Medical Bacteriology
Medical Mycology
Microcomputers in Biochemistry
Microcomputers in Biology
Microcomputers in Physiology
Mitochondria
Molecular Genetic Analysis of Populations

Molecular Imaging in Neuroscience
Molecular Neurobiology
Molecular Plant Pathology I and II
Molecular Virology
Monitoring Neuronal Activity
Mutagenicity Testing
Neural Transplantation
Neurochemistry
Neuronal Cell Lines
NMR of Biological Macromolecules
Nucleic Acid and Protein Sequence Analysis
Nucleic Acid Hybridisation
Nucleic Acids Sequencing
Oligonucleotides and Analogues
Oligonucleotide Synthesis
PCR
Peptide Hormone Action
Peptide Hormone Secretion
Photosynthesis: Energy Transduction
Plant Cell Biology
Plant Cell Culture
Plant Molecular Biology
Plasmids (2nd edition)
Pollination Ecology
Postimplantation Mammalian Embryos
Preparative Centrifugation
Prostaglandins and Related Substances
Protein Architecture
Protein Blotting
Protein Engineering
Protein Function
Protein Phosphorylation
Protein Purification Applications
Protein Purification Methods
Protein Sequencing
Protein Structure
Protein Targeting
Proteolytic Enzymes
Radioisotopes in Biology
Receptor Biochemistry
Receptor–Effector Coupling
Receptor–Ligand Interactions
Ribosomes and Protein Synthesis

RNA Processing I and II

Signal Transduction

Solid Phase Peptide Synthesis

Spectrophotometry and Spectrofluorimetry

Steroid Hormones

Teratocarcinomas and Embryonic Stem Cells

Transcription Factors

Transcription and Translation

Tumour Immunobiology

Virology

Yeast

# The Cell Cycle
## A Practical Approach

Edited by

PETER FANTES
*Institute of Cell and Molecular Biology*
*University of Edinburgh*

and

ROBERT BROOKS
*Anatomy and Human Biology Group*
*Biomedical Sciences Division*
*King's College, Strand*
*London WC2R 2LS*

—at—
OXFORD UNIVERSITY PRESS
Oxford  New York  Tokyo

Oxford University Press, Walton Street, Oxford OX2 6DP
Oxford New York Toronto
Delhi Bombay Calcutta Madras Karachi
Kuala Lumpur Singapore Hong Kong Tokyo
Nairobi Dar es Salaam Cape Town
Melbourne Auckland Madrid
and associated companies in
Berlin Ibadan

Oxford is a trade mark of Oxford University Press

A Practical Approach ⓘ is a registered trade mark
of the Chancellor, Masters, and Scholars of the University of Oxford
trading as Oxford University Press

Published in the United States
by Oxford University Press Inc., New York

© Oxford University Press, 1993

All rights reserved. No part of this publication may be
reproduced, stored in a retrieval system, or transmitted, in any
form or by any means, without the prior permission in writing of Oxford
University Press. Within the UK, exceptions are allowed in respect of any
fair dealing for the purpose of research or private study, or criticism or
review, as permitted under the Copyright, Designs and Patents Act, 1988, or
in the case of reprographic reproduction in accordance with the terms of
licences issued by the Copyright Licensing Agency. Enquiries concerning
reproduction outside those terms and in other countries should be sent to
the Rights Department, Oxford University Press, at the address above.

This book is sold subject to the condition that it shall not,
by way of trade or otherwise, be lent, re-sold, hired out or otherwise
circulated without the publisher's prior consent in any form of binding
or cover other than that in which it is published and without a similar
condition including this condition being imposed
on the subsequent purchaser.

Users of books in the Practical Approach series are advised that prudent
laboratory safety procedures should be followed at all times. Oxford
University Press makes no representation, express or implied, in respect of
the accuracy of the material set forth in books in this series and cannot
accept any legal responsibility or liability for errors or omissions
that may be made.

A catalogue record for this book is available from the British Library

Library of Congress Cataloging in Publication Data
The Cell cycle: a practical approach/edited by Peter Fantes and
Robert Brooks.
p. cm.—(The Practical approach series)
Includes bibliographical references and index.
1. Cell cycle—Research—Methodology. 2. Cellular control
mechanisms—Research—Methodology. I. Fantes, Peter. II. Brooks,
Robert, 1949–. III. Series.
QH605.C423    1994    574.87'623—dc20    93-26200
ISBN 0–19–963395–9 (h/b)
ISBN 0–19–963394–0 (p/b)

Typeset by Cambrian Typesetters, Frimley, Surrey
Printed in Great Britain by
Information Press Ltd, Eynsham, Oxford

# Preface

THE ability to combine genetical research in yeasts and other simple eukaryotes with biochemical investigations in higher cells has led in the past few years to an explosive expansion in research into the eukaryotic cell cycle. The events triggering this were the isolation in 1987 of a human $p34^{cdc2}$ homologue, and the discovery in the following year that maturation-promoting factor (MPF), which induces *Xenopus* oocytes to enter meiotic metaphase, contains $p34^{cdc2}$ and cyclin. Genetical investigations in fission yeast had previously identified the $cdc2^+$ gene as a key element in regulating entry into mitosis, and its product, $p34^{cdc2}$, had been shown to have protein kinase activity. At the same time cyclins had been identified in marine invertebrate eggs as proteins whose levels varied dramatically through the cell cycle, and were proposed as regulatory elements controlling entry into mitosis. These very different approaches led to the realization that the mitotic control system was highly conserved, a view that has subsequently been supported by the identification of $p34^{cdc2}$ and cyclin homologues in every eukaryotic cell examined.

Alongside these exciting findings came the development of cell-free systems to investigate the regulation of S phase and mitosis, *in vitro* systems able to initiate DNA replication, biochemical assays for MPF activity, and many other important advances. DNA manipulation techniques have allowed functional homologues of yeast cell-cycle genes to be isolated and characterized from evolutionarily diverse higher eukaryotes, while antibodies raised against either native or recombinant proteins have been used for biochemical and cytological analysis, and also for isolating genes by reverse genetics.

The opening up of the cell-cycle field in this way has allowed researchers well-versed in one system to extend their studies to other systems better suited to the problem under investigation. Although bringing great rewards, this has imposed great demands on investigators: while in the past it was sufficient to have expertise in a relatively narrow area, many studies now require the use of a far wider range of techniques. This diversification has not always been easy to achieve, because of a lack of familiarity with the experimental possibilities of other systems as much as technical ignorance. Consequently, when David Hames approached us in 1991 suggesting putting together a *Practical Approach* book on the cell cycle, we were enthusiastic about the idea, since it seemed to us that there was a real need for a compendium of cell biological, biochemical, and genetical methods used in cell-cycle research. The problem we faced was the very breadth and diversity

## Preface

of the material to be included. We clearly needed to include methods for handling cells and generating synchronous populations, ways of analysing the cell cycle by modern methods such as flow microfluorimetry and fluorescence labelling of macromolecules, while at the same time providing sufficient guidance about more traditional methods which are often still the most appropriate. We needed to include enough information about genetical methods, particularly in yeasts, to give the reader an idea of what techniques are available, in addition to providing detailed experimental protocols. Other specialized methods for the analysis of DNA replication and its initiation, for preparing and assaying the activity and components of MPF, and for isolating nuclear-associated organelles, also needed to be included. The amount of information we might include seemed limitless, and compromises clearly had to be made. Perhaps the most difficult decision was whether chapters should be arranged by cell type, or by technique. After extensive discussions we decided that no unitary way of dividing the material would work well, and so this volume is a hybrid: some chapters deal with the handling of specific organisms, while others cross phylogenetic boundaries and describe instead biochemical or cell biological procedures applicable in principle to many cell types.

Inevitably in a book of this size it has not been possible to include every conceivable technique that might be needed. Where space has not allowed a comprehensive list of methods to be presented, our aim has been to inform the reader of the range of techniques available, and to provide guidance about their suitability for particular purposes. We hope that the book will introduce many relevant techniques to those new to the cell cycle, and will help established workers to transcend traditional boundaries.

Our thanks are extended to all the authors for agreeing to undertake the time-consuming job of presenting specialized techniques for a wide readership, and for succeeding in doing so.

| | |
|---|---:|
| *Edinburgh* | P. F. |
| *London* | R. B. |
| March, 1993 | |

# Contents

*List of contributors*     xix

*Abbreviations*     xxi

**1. The synchronization of mammalian cells**     1
R. T. Johnson, C. S. Downes, and R. E. Meyn

   **1.** Introduction     1
   **2.** The production of non-cycling $G_0$ cells     2
       Quiescent human diploid fibroblasts     2
   **3.** Production of synchronized $G_1$ populations     4
       Very early $G_1$     4
       $G_0/G_1$ transition     4
       Isoleucine-depletion arrest point     4
       Late $G_1$ arrest by mimosine of HTDCT     5
   **4.** S-phase synchronization     6
       Early S-phase arrest of normal human fibroblasts     7
       Early S-phase arrest of permanent human and rodent cell lines     8
       Caveat: problems of extended S-phase inhibition     8
   **5.** Synchronization in $G_2$     9
   **6.** Detachment and collection of mitotic cells     11
       Selection of mitotic cells without drug use     11
       Freezing and recovery of mitotic cells     13
   **7.** The collection of mitotic cells by arresting agents     15
       Peculiarities of the mitotic checkpoint     15
       Collection of mitotic cells by arrest followed by shake-off     16
       Automatic nitrous oxide synchrony in mitosis     16
   **8.** Cautionary note     19
   **9.** Separation of cell-cycle phases by centrifugal elutriation     20
   Acknowledgements     22
   References     22

## 2. Preparation of synchronous cultures of the yeasts, *Saccharomyces cerevisiae* and *Schizosaccharomyces pombe* — 25

*J. Creanor and J. Toyne*

1. Introduction — 25
2. Division profiles of synchronous cultures — 26
   - *S. cerevisiae* — 26
   - *S. pombe* — 28
3. Selection synchrony — 29
   - Gradient separation in tubes — 29
   - Density separation — 31
   - Preparation of synchronous cultures by elutriation — 31
   - Preparation of synchronous cultures using a zonal rotor — 34
   - Conclusions — 36
4. Induction methods of synchronization — 36
   - Block and release methods — 36
5. Age fractionation — 42
6. Concluding remarks — 43

References — 43

## 3. Mammalian cell-cycle analysis — 45

*Zbigniew Darzynkiewicz*

1. Introduction — 45
2. Autoradiographic methods — 46
   - [$^3$H]TdR labelling index — 46
   - Fraction of labelled mitoses (FLM) analysis — 47
   - Tumour growth fraction estimation — 49
3. Flow cytometric methods of cell-cycle analysis — 50
   - Univariate cellular DNA content distribution — 51
   - Multiparameter analysis — 53
4. Flow cytometric kinetic methods — 60
   - Stathmokinetic methods — 60
   - Methods based on BrdU incorporation — 62

Acknowledgements — 66

References — 67

## 4. Analysis of the cell cycle in *Saccharomyces cerevisiae*     69

*Bruce Futcher*

1. Introduction     69
2. The cell cycle of S. cerevisiae     70
3. Monitoring the cell cycle     71
   - Introduction     71
   - Preparation of cells arrested at cell-cycle blocks     72
   - Measurement of cell size     74
   - Budding     74
   - DNA content and FACS analysis     75
   - Microtubule morphology     78
   - Staining DNA with DAPI     78
   - Reciprocal shift experiments     79
   - Histone H1 kinase assays     80
4. Analysis of heterologous genes in yeast     85
   - Cloning heterologous homologues of yeast genes by functional complementation     85
   - Testing known heterologous genes for their ability to complement mutations in yeast     89
   - Analysis of heterologous genes not having known homologues     90

Acknowledgements     91

References     91

## 5. Methods for analysis of the fission yeast cycle     93

*S. A. MacNeill and P. A. Fantes*

1. Introduction     93
2. Life cycle and genetics     93
3. Monitoring cell-cycle progress     95
   - Growing fission yeast cells     95
   - Determining cell number     96
   - Cell length and septation (cell plate) index     97
   - Measuring DNA content     97
   - Monitoring DNA replication     99
   - Flow cytometry     101
   - Measuring $p34^{cdc2}$ H1 kinase activity     104
4. Cell staining     106
   - Nuclear staining     106

| | |
|---|---|
| Staining the cell wall and septum | 108 |
| Staining actin structures | 108 |
| Immunofluorescence microscopy | 109 |
| Electron microscopy | 110 |
| 5. Synchronizing cells | 110 |
| 6. Cell-cycle inhibitors | 110 |
| Hydroxyurea | 110 |
| Benomyl and related compounds | 111 |
| Cycloheximide | 111 |
| 7. Cell-cycle genetics | 111 |
| Cell-cycle mutants | 111 |
| Characterization of new mutants | 112 |
| Cloning of cell-cycle genes | 113 |
| Manipulation of cloned genes | 118 |
| Isolation of higher eukaryotic homologues | 120 |
| 8. Summary | 122 |
| Acknowledgements | 122 |
| References | 122 |

## 6. Cell-cycle analysis using the filamentous fungus *Aspergillus nidulans* — 127

B. R. Oakley and S. A. Osmani

| | |
|---|---|
| 1. Introduction | 127 |
| 2. Microscopic examination of the cell cycle and scoring of mitosis | 128 |
| Determination of chromosome mitotic index | 129 |
| Determination of spindle mitotic index | 130 |
| 3. Methods based on transformation of *A. nidulans* | 135 |
| Transformation of *A. nidulans* | 135 |
| Generation of conditional mutations using the *alcA* promoter | 136 |
| Molecular disruption of essential genes and phenotypic analysis from heterokaryons | 137 |
| 4. Cell-cycle synchrony methods | 140 |
| Hydroxurea block–release | 140 |
| Cell-cycle synchrony using conditional mutations | 140 |
| Acknowledgements | 141 |
| References | 142 |

## 7. Techniques for studying mitosis in *Drosophila* — 143
*Cayetano Gonzalez and David M. Glover*

1. Introduction — 143
   - Cell-cycle control during *Drosophila* development — 143
   - General phenotypes of cell-division mutants — 144
2. Genetic approaches — 146
   - Generating *Drosophila* mutants — 146
   - Mapping mutants — 147
3. General approaches — 153
   - Localization of transcripts — 153
   - Immunolocalization studies — 156
4. Embryos — 158
   - Collection and dechorionation — 158
   - Introducing reagents into embryos — 159
   - Preparation of embryos for immunostaining — 160
5. Brains — 163
   - Procedures for making squashed preparations — 163
   - Procedures for immunostaining larval neuroblasts — 166
   - *In situ* hybridization to cells of the larval brain — 168
6. Mutations that also affect male meiosis — 169
   - Cytological studies on testes squashes — 169
   - Assays for non-disjunction in male meiosis — 173

Acknowledgements — 174

References — 174

## 8. The use of cell-free extracts of *Xenopus* eggs for studying DNA replication *in vitro* — 177
*C. J. Hutchison*

1. Introduction — 177
2. Maintaining a *Xenopus* colony — 177
3. Preparation of egg extracts — 178
4. Using egg extracts — 181
   - Fresh vs frozen — 181
   - Choice of DNA template — 182
   - Labelling protocols — 186

## Contents

| | |
|---|---|
| 5. Microscopic analysis of nuclear assembly | 191 |
|     Phase-contrast microscopy | 191 |
|     Immunofluorescence | 191 |
|     IgG exclusion assay | 193 |
| 6. Fractionation of extracts | 193 |
|     High-speed centrifugation | 193 |
|     Immunodepletion | 193 |
| References | 195 |

### 9. In vitro SV40 DNA replication — 197
*Mark K. Kenny*

| | |
|---|---|
| 1. Introduction | 197 |
| 2. T-antigen purification | 199 |
|     Column preparation | 199 |
|     Immunoaffinity purification | 201 |
| 3. Extract preparation | 204 |
| 4. Replication reaction | 205 |
|     The reagents | 206 |
|     The assay | 206 |
| 5. Further possibilities | 208 |
| Acknowledgements | 209 |
| References | 209 |

### 10. Analysis of DNA replication origins and directions by two-dimensional gel electrophoresis — 213
*Joel A. Huberman*

| | |
|---|---|
| 1. Introduction | 213 |
|     Importance of the two-dimensional (2D) gel methods | 213 |
|     Purpose of this chapter | 213 |
|     Historical antecedents | 213 |
| 2. How the 2D gel replicon mapping methods work | 214 |
|     Features common to both the N/N and N/A methods | 214 |
|     The N/N replicon mapping technique | 215 |
|     The N/A replicon mapping technique | 217 |
|     Advantages and disadvantages of the two techniques | 219 |
| 3. Methods | 220 |
|     DNA preparation | 220 |

|  |  |
|---|---|
| Incubation of purified DNA with a restriction enzyme | 220 |
| Enrichment for replicating molecules | 221 |
| First dimension electrophoresis | 224 |
| Second dimension electrophoresis | 226 |
| Modification of the N/N procedure to permit the determination of replication direction | 230 |
| Hybridization with specific probes | 233 |
| Additional information | 233 |

Acknowledgements 233

References 233

## 11. The isolation of functional mitotic organelles from tissue culture cells — 235

*Linda Wordeman*

| | |
|---|---|
| 1. Introduction | 235 |
| 2. Mitotic arrest in tissue culture cells | 236 |
|    Cell lines | 236 |
|    Methods | 237 |
| 3. The mitotic apparatus | 238 |
|    Structure and components | 238 |
|    Isolation of CHO mitotic spindles | 239 |
|    Other systems for mitotic apparatus isolation | 241 |
| 4. Centrosomes | 242 |
|    CHO cell centrosomes | 242 |
| 5. Metaphase chromosomes | 246 |
|    CHO cell chromosomes | 246 |

Acknowledgements 250

References 250

## 12. Xenopus egg extracts as a system for studying mitosis — 253

*Marie-Anne Félix, Paul R. Clarke, Julia Coleman, Fulvia Verde, and Eric Karsenti*

| | |
|---|---|
| 1. Introduction | 253 |
| 2. Preparation of frog egg extracts | 254 |
|    The different types of extracts | 255 |
|    The eggs | 258 |

| | | |
|---|---|---|
| | Preparation of the extracts | 259 |
| | Monitoring cdc2 H1 kinase activity as a cell-cycle marker in egg extracts | 267 |
| 3. | Assay of cyclin activity in egg extracts | 270 |
| | Production of cyclins as fusion proteins | 270 |
| | Analysis of cyclin activity in egg extracts | 273 |
| 4. | Assay of other cell-cycle regulatory molecules in egg extracts | 277 |
| 5. | Regulation of microtubule dynamics in egg extracts | 277 |
| | Analysis of microtubule dynamics using fixed timepoints and immunofluorescence microscopy | 278 |
| | Acknowledgements | 282 |
| | References | 282 |

## 13. Maturation-promoting factor and cyclin-dependent protein kinases  285

*Marcel Dorée, Thierry Lorca, and André Picard*

| | | |
|---|---|---|
| 1. | Introduction | 285 |
| 2. | Detection of MPF activity by direct cytoplasm transfer | 287 |
| | In *Xenopus* oocytes | 287 |
| | Detection of MPF activity by direct cytoplasm transfer in other species | 292 |
| 3. | Detection of MPF activity in cell-free extracts | 296 |
| | Extracts and assays | 296 |
| | Nature of MPF activities in cell-free extracts | 298 |
| 4. | Purification of cyclin B–cdc2 kinase (MPF) | 299 |
| 5. | Cyclin-dependent kinases | 300 |
| | Assay and properties of mitotic cdc2 kinases | 302 |
| | Properties of cdc2 kinases | 303 |
| | Production of mitotic cdc2 kinases *in vitro* | 303 |
| | Assay of cyclin degradation *in vitro* | 305 |
| | References | 307 |

## Appendix: Addresses of suppliers  311

## Index  317

# Contributors

PAUL R. CLARKE
EMBL, Postfach 10.2209, Meyerhofstrasse 1, 6900 Heidelberg, Germany.

JULIA COLEMAN
EMBL, Postfach 10.2209, Meyerhofstrasse 1, 6900 Heidelberg, Germany.

J. CREANOR
Institute of Cell, Animal and Population Biology, University of Edinburgh, West Mains Road, Edinburgh EH9 3JT, Scotland.

ZBIGNIEW DARZYNKIEWICZ
The Cancer Research Institute, New York Medical College, Valhalla, NY 10595, USA.

MARCEL DORÉE
Biochimie CNRS–INSERM, BP 5051, 34033 Montpellier cedex, France.

C. S. DOWNES
Cancer Research Campaign Mammalian Cell DNA Repair Research Group, Department of Zoology, University of Cambridge, Downing Street, Cambridge CB2 3EJ, UK.

P. A. FANTES
Institute of Cell and Molecular Biology, University of Edinburgh, Mayfield Road, Edinburgh EH9 3JR, Scotland.

MARIE-ANNE FÉLIX
EMBL, Postfach 10.2209, Meyerhofstrasse 1, 6900 Heidelberg, Germany. Present address: Institut Jacques Monod, Department of Developmental Biology, Tour 43, 2 Place Jussieu, 75251 Paris, France.

BRUCE FUTCHER
Cold Spring Harbor Laboratory, PO Box 100, Cold Spring Harbor, NY 11724, USA.

DAVID M. GLOVER
Cancer Research Campaign Laboratories, Cell Cycle Genetics Group, Department of Anatomy and Physiology, Medical Sciences Institute, University of Dundee, Dundee DD1 4HN, Scotland.

CAYETANO GONZALEZ
Cancer Research Campaign Laboratories, Cell Cycle Genetics Group, Department of Anatomy and Physiology, Medical Sciences Institute, University of Dundee, Dundee DD1 4HN, Scotland.

## Contributors

JOEL A. HUBERMAN
Department of Molecular and Cellular Biology, Roswell Park Cancer Institute, Buffalo, NY 14263, USA.

C. J. HUTCHISON
Department of Biological Sciences, University of Dundee, Dundee DD1 4HN, Scotland.

R. T. JOHNSON
Cancer Research Campaign Mammalian Cell DNA Repair Research Group, Department of Zoology, University of Cambridge, Downing Street, Cambridge CB2 3EJ, UK.

ERIC KARSENTI
EMBL, Postfach 10.2209, Meyerhofstrasse 1, 6900 Heidelberg, Germany.

MARK K. KENNY
The Picower Institute for Medical Research, 350 Community Drive, Manhasset, NY 11030, USA.

THIERRY LORCA
Biochimie CNRS–INSERM, BP 5051, 34033 Montpellier cedex, France.

S. A. MacNEILL
Institute of Cell and Molecular Biology, University of Edinburgh, Mayfield Road, Edinburgh EH9 3JR, Scotland.

R. E. MEYN
Department of Experimental Radiotherapy, The University of Texas, M. D. Anderson Cancer Center, 1515 Holcombe Boulevard, Houston, TX 77030, USA.

B. R. OAKLEY
Department of Molecular Genetics, Ohio State University, 484 West 12th Ave, Columbus, OH 43210, USA.

S. A. OSMANI
Weis Center for Research, Geisinger Clinic, 100 North Academy, Danville, PA 17822 26–17, USA.

ANDRÉ PICARD
Biochimie CNRS–INSERM, BP 5051, 34033 Montpellier cedex, France.

J. TOYNE
National Institute for Medical Research, The Ridgeway, Mill Hill, London NW7 1AA, UK.

FULVIA VERDE
EMBL, Postfach 10.2209, Meyerhofstrasse 1, 6900 Heidelberg, Germany.

LINDA WORDEMAN
Department of Pharmacology, University of California, San Francisco, CA 94143, USA.

# Abbreviations

| | |
|---|---|
| 1-MeAde | 1-methyladenine |
| AO | acridine orange |
| ATP | adenosine triphosphate |
| ATP-γS | adenosine-5'-$O$-(3-thiotriphosphate) |
| BAME | $N$-α-benzoyl-L-arginine methyl ester |
| BrdU | bromodeoxyuridine |
| BrdUTP | bromodeoxyuridine triphosphate |
| BSA | bovine serum albumin |
| C | unreplicated haploid DNA content |
| CSF | cytostatic factor |
| DAB | 3,3-diaminobenzidine |
| DAPI | 4',6'-diamidino-2-phenylindole dihydrochloride |
| dATP | deoxyadenosine triphosphate |
| dCTP | deoxycytosine triphosphate |
| dGTP | deoxyguanosine triphosphate |
| DMEM | Dulbecco's modified Eagle's medium |
| DMSO | dimethyl sulfoxide |
| DTT | dithiothreitol |
| dTTP | thymidine triphosphate |
| EB | ethidium bromide |
| EDTA | ethylenediaminetetraacetic acid |
| EGS | ethylene glycol bis-(succinic acid $N$-hydroxysuccinimide ester) |
| EGTA | ethylene glycol-bis($\beta$-aminoethyl ether) $N,N,N',N'$-tetraacetic acid |
| EM | electron microscope |
| FACS | fluorescence activated cell sorter |
| FCS | fetal calf serum |
| FLM | fraction of labelled mitoses |
| GST | glutathione-S-transferase |
| GTP | guanosine triphosphate |
| GTP-γS | guanosine-5'-$O$-(3-thiotriphosphate) |
| GVBD | germinal vesicle breakdown |
| HBSS | Hank's buffered (balanced) salt solution |
| HCG | human chorionic gonadotropin |
| HO258 | Hoechst 33258 |
| IPTG | isopropyl-thio-$\beta$-D-galactoside |
| m$^7$GpppG | $P^1$-5'-(7-methyl)-guanosine-$P^3$-5'-guanosine triphosphate |
| MAP | microtubule associated protein |
| MBC | methylbenzimidazol-2-yl carbamate |
| MBP | maltose-binding protein |
| MEM | Eagle's minimal essential medium |
| Mes | 2-[$N$-morphilino]ethanesulfonic acid |
| MPF | maturation promoting factor |
| NBT | nitro-blue tetrazolium salt |

## Abbreviations

| | |
|---|---|
| NCS | new-born calf serum |
| OD | optical density |
| ORF | open reading frame |
| $p34^{cdc2}$ | protein kinase product of $cdc2^+$ gene |
| PAGE | polyacrylamide gel electrophoresis |
| PBS | phosphate-buffered saline |
| PBT | 0.2% Triton in PBS |
| PCA | perchloric acid |
| PCR | polymerase chain reaction |
| PEG | polyethylene glycol |
| PI | propidium iodide |
| Pipes | Piperazine-$N,N'$-bis[2-ethanesulfonic acid] |
| PMSF | phenylmethylsulfonyl fluoride |
| PMSG | pregnant mare serum gonadotropin |
| PP | 4% paraformaldehyde in PBS |
| RI | refractive index |
| RNase | ribonuclease |
| RT | room temperature |
| SBTI | soybean trypsin inhibitor |
| SDS | sodium dodecyl sulfate |
| SSB | single-stranded DNA binding protein |
| SV40 | simian virus 40 |
| TBZ | thiabendazole |
| TCA | trichloroacetic acid |
| TdR | thymidine |
| TPCK | $N$-tosyl-L-phenylalanine chloromethyl ketone |
| X-gal | 5-bromo-4-chloro-3-indolyl-$\beta$-D-galactoside |
| X-phosphate | 5-bromo-4-chloro-3-indolyl-phosphate |

# 1

# The synchronization of mammalian cells

R. T. JOHNSON, C. S. DOWNES, and R. E. MEYN

## 1. Introduction

Once upon a time most cell biologists could synchronize mammalian cells in different stages of the cell cycle, albeit with different degrees of success. However, few molecular probes were then available. Now, with the great resurgence of interest in cell-cycle control and the beginnings of molecular understanding, many workers who have recently entered the field require tightly synchronized cells for their experiments. This chapter presents a selection of protocols which should provide synchronized cells in each cycle phase, and in most instances, cells from different parts of each phase. From the methods commonly used we have selected those which provide excellent synchrony with good reversibility, balanced growth, and high viability. We have also concentrated on methods which (except for centrifugal elutriation) can be carried out without expensive equipment, apart from the standard $CO_2$ incubators, bench centrifuges, inverted microscopes, Coulter counters, and −70 °C freezers. A 37 °C hot room is a useful resource for synchronization, and a good workshop is needed to build equipment for high-pressure nitrous oxide arrest. A Cytospin cytocentrifuge (Shandon Scientific) for the rapid production of cell monolayers for microscopic inspection is strongly recommended. Details of methods involving high technology—separation of viable cells on the basis of DNA content by fluorescence activated cell sorting, or automated mechanical synchronization systems—can be found in refs 1 and 2, respectively.

As yet, there are very few mammalian cell types for which synchronization schemes have been developed. HeLa (human papilloma-virus transformed) and CHO (Chinese hamster ovary) cells are well served, as are human primary fibroblasts, but much less is understood about synchronization of, for example, SV40 transformed fibroblasts or EB transformed lymphoblastoid cell lines. However, the methods that work for HeLa and CHO can be adapted for use with many permanent cell lines, though the concentrations of agents used may vary by at least one order of magnitude, and the duration of arrest will depend on the overall cycle time of the particular cultures. Many of the techniques that are now followed have been developed by Robert Tobey and his co-workers, whose careful work over a period of 25 years has laid the basis for the logical production of viable synchronized cells.

Methods of cell synchrony depend on two general strategies. One strategy exploits cycle-dependent changes in the physical properties of cells, such as size (elutriation) or differential adhesion (mitotic detachment). The other exploits cell-cycle checkpoint regulation; i.e. the systems that prevent cycle progression when any one essential process is left uncompleted. Drugs which inhibit cycle-specific activities, such as DNA synthesis or mitotic spindle formation, will therefore, cause all aspects of cycle progression to arrest at the affected stage. Note that this strategy will not work if the cells to be synchronized are defective in checkpoint controls. Transformed rodent cells can be difficult in this respect. Generally, however, cell-cycle checkpoint control can be used to arrest cells in each phase of the cycle. Because of the rapid decay of synchrony following release from arrest at any cell-cycle stage (which is a function of the large variation in rate of cycle passage among individual cells), release and resynchronization protocols may be necessary to generate adequately synchronized cultures.

## 2. The production of non-cycling $G_0$ cells

Cycling cells pass from mitosis through $G_1$ into S phase; transformed cells may be already committed to this trajectory as division is completed. The synchronization of cells at various points during $G_1$ is dealt with in Section 3; here we consider an additional pre-S phase stage, $G_0$, available to certain cells, of which skin-derived primary fibroblasts have been best studied. The $G_0$ state represents a condition of natural proliferative quiescence that is stable *in vivo*, in some cases for the lifespan of the organism. With suitable cells it is easy to achieve *in vitro* and, when reversed, provides a tightly synchronized cohort of early $G_1$ cells, which may be suitable for further synchronization, for example at the $G_1/S$ boundary.

### 2.1 Quiescent human diploid fibroblasts

The generation of stable $G_0$ human fibroblast arrest requires vigorous early passage cells (ideally passage 10 or less), and good nutritional conditions during arrest. While it is possible to accumulate cells in $G_0$ by drastically reducing the serum supplement, this is not the preferred method. Serum deprivation for 2 days or more causes irreversible damage, which restricts cell survival and may affect subsequent cycle characteristics; but lesser periods of serum deprivation may be inadequate to arrest all cells. We prefer synchrony based on growth to confluence, as described in *Protocol 1* (modified from ref. 3). According to the vigour and proliferation characteristics of individual fibroblast strains, this method can be adjusted to allow sufficient time for a given number of cells to reach confluence in a particular culture vessel.

Cultures held in quiescence can be characterized for diploid DNA content by flow analysis, or for absence of replication by DNA labelling ($[^3H]$-thymidine labelling and autoradiography, or BUdR labelling and antibody staining) (see Chapter 3). Release of cells from $G_0$ arrest by trypsin detachment and replating results in the coherent flow of cells towards S phase; synchrony deteriorates in late $G_1$.

It should be noted that this method is only suitable for fibroblast cultures of some mammalian species, especially human. Attempting to use it on primary mouse fibroblasts, except for those of very early passage number, will lead to a cell crisis, with massive killing and the emergence of transformed clones.

---

**Protocol 1.** Arrest of human fibroblasts in $G_0$

*Materials*

- Growth medium: Dulbecco's modified Eagle's medium (DMEM, Gibco BRL 041–01965) recommended for its enriched supply of vitamins and amino acids, supplemented with 10% (v/v) FCS and antibiotics; other normal growth medium will usually be satisfactory
- Low passage number fibroblasts (as far as possible)
- 80 or 75 cm$^2$ culture flasks (Falcon or Nunc)
- Trypsin (Gibco BRL, 10 × stock 043–05090, or pancreatin, (Gibco BRL, 4 × stock 043–05720)
- Coulter cell counter or haemocytometer
- Inverted microscope

*Method*

1. Remove exponentially growing fibroblasts from the substrate by trypsin or pancreatin treatment.
2. Count the cell suspension and add 3–4 × 10$^6$ cells to a culture flask in 20 ml of 37 °C prewarmed growth medium.
3. Allow 18 h for recovery from replating. Remove medium and add 20 ml 37 °C prewarmed growth medium. Early passage fibroblasts should reach confluency after 48 h and enter $G_0$, although an additional 24 h is recommended. Later passage fibroblasts (passages 10–20) will require extra time to reach a confluent state, possibly up to 2 weeks. Growth medium should be replaced twice during this period and cells used at least 4 days after the final medium change.

**Protocol 1.** *Continued*

4. Use the inverted microscope (ideally in a 37 °C warm room) to monitor the cultures for the presence of dividing cells, these are readily seen as rounded-up against a uniform background of flattened cells.

Transformed lines cannot normally be synchronized in $G_0$; serum starvation drives them into a rather unhealthy state. In the exceptional cases of hormone-dependent transformed lines, such as T-47D or MCF-A, efficient and reversible arrest in quiescence can be achieved by hormone deprivation (see refs 4 and 5 for full details).

## 3. Production of synchronized $G_1$ populations

### 3.1 Very early $G_1$

Large number of cells can be obtained within 2 h by reversal of a brief mitotic arrest, using nitrous oxide, Colcemid, or nocodazole, as described in Section 7. Suitable cells include most transformed human cells of epithelial origin and transformed rodent (Chinese hamster, Syrian hamster and mouse) cell lines.

### 3.2 $G_0/G_1$ transition

These cells are obtained by removing quiescent cells from the substrate using trypsin and seeding into new dishes with fresh medium at a subconfluent density.

### 3.3 Isoleucine-depletion arrest point

Deprivation of the amino acid isoleucine is most limiting to cells in mid-$G_1$, about 4 h from the $G_1/S$ boundary. (There is a tempting analogy here with $G_1$ control in *S. cerevisiae*, where the functioning of the cln3 $G_1$ cyclin is very sensitive to overall rates of protein synthesis.) Usually when cells are transferred to isoleucine-free medium, sufficient isoleucine remains to allow proliferation through the cycle up to mid-$G_1$ when arrest occurs. This excellent method, developed by Tobey and colleagues (6), is widely used with transformed hamster and mouse cell lines, but can also be applied to permanent cultures of human cells (HeLa, SV40-transformed fibroblasts). The use of the technique for a particular cell line requires careful monitoring as prolonged isoleucine depletion kills cells, some being more sensitive than others (SV40-transformed human cells much more than hamster cell lines). It may be necessary to adjust the period of growth in the selective medium as well as the degree of isoleucine depletion to ensure good viability and $G_1$ synchrony.

Isoleucine-free growth medium can be produced in the laboratory. We routinely prepare isoleucine-free Ham's F12 medium (7) for this purpose, since we use F12 for work with auxotrophic mutants and so have all the stock vitamin, nutrient, and salt solutions available; other media might equally be prepared in isoleucine-free form. The serum used to supplement any isoleucine-free medium must be dialysed over 4 days at 4 °C with GE buffer (20 × GE buffer stock: 740 g NaCl, 28.5 g KCl, 29 g $Na_2HPO_4.7H_2O$, in 5 litres of 2 × glass-distilled or RO cartridge-filtered water) to remove from it the bulk of low molecular weight components, including isoleucine.

For most cell cycles, 30–36 h isoleucine deprivation is sufficient time to allow the accumulation of 95% of cells at the $G_1$ restriction point. Chinese and Syrian hamster cell lines, mouse cell lines, Indian muntjac cell lines, and SV40 transformed, or otherwise transformed, human cell lines are all accumulated in $G_1$. In our experience human cells are less well synchronized by the standard schedule, with up to 10% in S phase after a 30 h isoleucine-free incubation. Better synchrony and better reversal and viability of the synchronized cells is achieved with transformed human cells by supplementing isoleucine-free F12 medium with 5 µM each deoxycytosine, deoxyguanosine, deoxyadenosine, and thymidine (Shoshana Squires, personal communication). For hamster cells, despite unbalanced growth in which the cells enlarge although the nucleus does not, isoleucine depletion produces excellent $G_1$ synchrony with high viability and rapid reversibility. Four hours after reversal the first cells move into S phase, but the natural decay of synchrony (50% of cells in S phase by 8 h) limits the usefulness of the reversed populations, unless a subsequent block is imposed at the $G_1/S$ boundary (see section 4.2).

## 3.4 Late $G_1$ arrest by mimosine of HTDCT

A useful addition to the repertoire of $G_1$ synchronizing agents, and one with apparent wide applicability, is the plant amino acid L-mimosine (Sigma, M-0253). Mimosine at 200 µM reversibly arrests human B cell-derived lymphoblastoid cells, HL60, a promyeloid leukaemia cell line, PHA stimulated T lymphocytes, and Chinese hamster ovary cells late in $G_1$, 15 min to 2 h before the $G_1/S$ boundary (8). Likewise, the Hoechst compound 768159 (alias HTDCT, a carboxythiazole derivative) acts on T lymphocytes, B cell lines and CHO cells at the same stage as mimosine (9, 10). Both inhibit the post-translational formation of the rare amino acid hypusine from lysine, and thereby diminish the activity of the protein synthesis initiation factor eIF-5A (Marc Lalande, personal communication). It is not clear why this produces cycle arrest.

Mimosine is proving particularly useful for the synchronization of cells just before the start of S phase. A human Epstein–Barr transformed cell line (LAZ463) grown for 16 h in 400 µM mimosine arrests approximately 15 min before S phase. CHO cells arrested in mid-$G_1$ by isoleucine deprivation and released into medium containing mimosine (400 µM) also appear to arrest just before the $G_1/S$ boundary. Release of mimosine arrest results in the rapid flow of LAZ463 cells into S phase within 15 min (11). Since mimosine also slows down the progress of cells through S phase (11), it is best used as a $G_1$ arresting agent to resynchronize cells released from an earlier mitotic, $G_1$ or $G_0$ arrest point.

Though mimosine clearly arrests cells in late $G_1$, some laboratories have found it difficult to reverse the cycle block. The reasons for the difficulties experienced are uncertain, but may reside in the particular strain of cells used since the precise point of arrest and, therefore, the nature of the disruption in $G_1$ are likely to vary among different cells. The optimum concentration of mimosine that accumulates $G_1$ cells and that can be rapidly reversed has to be determined for each cell strain. High concentrations (over 600 µM) are increasingly difficult to reverse. The potential wide applicability of mimosine, as well as HTDCT, for the synchronization of non-transformed, as well as permanent, cell lines from man and rodent makes these agents increasingly important. For example, mimosine may help in the isolation of replication origins from the tightly synchronized cohort of early S-phase cells.

## 4. S-phase synchronization

Arrest of cells in S phase is straightforward. Any inhibitors of DNA synthesis arrest replicating cells; those in common use are inhibitors of the synthesis of deoxyribonucleotide triphosphates—such as thymidine at high concentrations or, more drastically, hydroxyurea—or the more expensive direct DNA polymerase inhibitor aphidicolin. Cells that are outside S at the time of adding such inhibitors are arrested when they pass through the $G_1/S$ boundary. A double-arrest protocol, in which cells are accumulated anywhere in S during the first arrest, released, and given inhibitors again when the population has moved out of S, is useful for many transformed human cells, accumulating more than 90% of cells close to the $G_1/S$ boundary. For some purposes S-phase cells synchronized by these means may be satisfactory; for sturdy lines like HeLa, a double thymidine arrest (17 h 2 mM thymidine, rinse twice, 9 h growth medium, 15 h 2 mM thymidine) is a simple and cheap way of producing large numbers of S-phase cells.

However, prolonged treatment with thymidine and hydroxyurea can be toxic to S-phase cells, and do not always hold cells securely at the $G_1/S$ boundary. Also, the subsequent behaviour of cells so treated may be abnormal (12). Aphidicolin may equally upset subsequent growth if used for long. The thymidylate synthetase inhibitor fluorodeoxyuridine, and the leaky chain terminator cytosine arabinoside, arrest cells lethally in S phase and should never be used for synchronization. The methods that produce the tightest early S-phase synchrony and the most viable cells involve $G_1$ or $G_0$ arrest, followed by release and subsequent arrest at the $G_1/S$ boundary. These methods, described below, are applicable to a wide variety of human and rodent cells, both diploid and heteroploid.

It is a pity that all current S-phase arresting agents act indiscriminantly on every replicating cell. S phase is clearly structured, with the most active genes replicating early, and inactive genes late; moreover, there are clear peaks of replication that can be discerned in the best synchronous populations (13). Analysis of S phase would be assisted by protocols that ensure tight synchrony in middle or late S; unfortunately these do not, as yet, exist.

## 4.1. Early S-phase arrest of normal human fibroblasts (after ref. 3)

The method (*Protocol 2*) relies on the rapid accumulation of early passage fibroblasts in $G_0$ by a short period of subconfluent growth (to avoid later trypsin detachment) in medium supplemented with 0.1% serum and with ribo- and deoxyribonucleosides, followed by release in high-serum medium containing aphidicolin, an inhibitor of DNA polymerases α and δ, for long enough to arrest almost all cells at the $G_1$/S boundary. The use of excess thymidine or hydroxyurea as the S-phase blocking agent in these cells is not recommended, since viability is reduced. The protocol strikes a compromise by selecting a period of 24 h in aphidicolin, long enough to arrest almost all cells at the start of S, yet short enough to maintain reasonable viability of those cells. Eighty to ninety per cent of cells are usually found to move into S phase, as judged by an increase in DNA precursor incorporation during the first 60 min after release of the aphidicolin block.

---

**Protocol 2.** Synchronization of normal human fibroblasts in early S phase

*Materials*

- MEM alpha medium, supplemeneted with ribo- and deoxyribo-nucleosides (Gibco BRL 041-02571) plus fetal calf serum at 0.1% (v/v) or 10% (v/v)
- Low passage number fibroblasts
- Aphidicolin (Sigma, A0781) stock at 2 mg/ml (approx. 6 mM) dissolved in DMSO. **NB** aphidicolin is a possible carcinogen
- Dulbecco's phosphate-buffered saline (PBS; Gibco BRL 041-04040) at 37 °C

*Method*

1. Generate subconfluent $G_0$ arrested cultures by adding $3.7 \times 10^5$ trypsin- or pancreatin-detached cells from an exponential culture in

**Protocol 2.** *Continued*

 5 ml MEM 10% FCS to each of ten 58 mm culture dishes (Nunc 1-50288A).

2. After 18 h of recovery, wash the cultures twice with warm PBS to remove loosely attached and floating cells, and residual medium, then add 5 ml MEM 0.1% FCS to each dish. Leave for 48 h.
3. Change the medium to prewarmed MEM + 10% FCS containing aphidicolin at 5 μg/ml. Incubate for 24 h.
4. Release the cells from aphidicolin arrest by washing twice with warm PBS and then adding fresh MEM + 10% FCS.
5. Assay S-phase index within 30 min by method of choice (see Chapter 3).

## 4.2 Early S-phase arrest of permanent human and rodent cell lines

Methods of $G_1$ arrest outlined in Section 3 provide the initial synchronization step, followed by release and subsequent re-synchronization at the $G_1/S$ boundary. Isoleucine deprivation is particularly suitable for hamster cells. Twenty hours growth in isoleucine-free medium followed by release into complete medium plus aphidicolin (5 μg/ml) for 8 h will accumulate more than 95% of cells in S phase close to the $G_1/S$ boundary.

## 4.3 Caveat: problems of extended S-phase inhibition

Agents that arrest cells in S phase can cause time-dependent unbalanced growth, changes in chromatin including loss of histone H1, and loss of viability. Despite the apparent accumulation of cells at the $G_1/S$ border by the inhibitors listed above, it is clear that cells actually enter S phase and synthesize limited amounts of DNA (12, 14). Given the DNA precursor pool disturbance created by inhibition of DNA synthesis (15), the accuracy of DNA synthesized may be poor. In the presence of extended hydroxyurea (though not aphidicolin) arrest there is significant degradation of nascent DNA (12).

In transformed rodent cells, protocols involving extended arrests by hydroxyurea are likely to distort normal cycle progress. For example, CHO and BHK cells override the S-phase checkpoint, which normally does not permit progress into $G_2$ until replication is completed. Hydroxyurea treatment promotes a time-dependent loss of checkpoint control; human cell lines do not lose their S-phase checkpoint in this manner (16). For the purposes of biochemical analysis, therefore, rodent cell-cycle distortion by synchronizing agents is a significant problem.

Reversal of early S-phase arrest after excess thymidine or hydroxyurea is usually rapid, though the length of S phase may be shorter than usual. Release from aphidicolin can be surprisingly slow; 60–90 min are required to reach maximum DNA synthesis rate in Indian muntjac cells (Helen Strutt and Andy Ryan, personal communication). These considerations may be important in the design of experiments using cells in different regions of S phase.

## 5. Synchronization in $G_2$

$G_2$ is traditionally the most difficult of phases to collect. HeLa cells in $G_2$ have been routinely obtained by reversing a double thymidine arrest and waiting 6–8 h until the majority of cells are in $G_2$ (17). Addition of the microtubule antagonist nocodazole (methyl(5-[2-thienylcarbonyl]-1H-benzimidazol-2-yl)carbamate; Sigma, M 1404) at 0.04 µg/ml, after release from the second thymidine arrest causes the most rapidly cycling cells (which reach mitosis and would otherwise pass through into $G_1$) to arrest in mitosis; they can then be selectively removed by gentle shaking, as described in Section 6.1. This gives an improved yield of HeLa cells in $G_2$ (18).

A different approach to $G_2$ synchronization has been devised by Tobey and colleagues (19); it relies on the application of DNA topoisomerase II inhibitors. These drugs, which produce covalent complexes between the topoisomerase II molecules and DNA at the site of DNA strand breaks (20), arrest cells in $G_2$ by preventing the activation of $p34^{cdc2}$/cyclin B kinase which is essential for the onset of mitosis. Suppression of activation is an indirect effect of DNA breakage, by ionizing radiation or by topoisomerase II inhibitors (21).

To achieve efficient and reversible $G_2$ synchronization through DNA damage induced by topoisomerase II inhibitors, great care is needed to ensure maximum viability after reversal, and minimal replicative perturbation during drug treatment. Each cell line may differ considerably in the concentration of drug required. *Protocol 3* gives the method for $G_2$ synchronization of CHO cells and human diploid fibroblasts. The standard protocol generates a $G_2$ synchronization of 80–90% or more, with no gross imbalance of macromolecular synthesis in rodent or human cells. Both types recover rapidly from arrest, and flow into mitosis within 3 h. Viability, as estimated from colony formation, is said to be at control levels; but some chromosome damage is an inevitable consequence of exposure to topoisomerase II inhibitors, and subsequent mitoses may be abnormal. By carefully varying the drug concentration or duration of drug exposure populations of cells in different parts of $G_2$ can be generated (19). In *Protocol 3*, the Hoechst drug 33342 is used since it gives the best reversal, but VM26 (teniposide) at 0.5 µg/ml is suitable for use with CHO cells.

**Protocol 3.** Synchronization of human fibroblasts and CHO cells in $G_2$

*Materials*

- Growth medium: α-MEM (Gibco BRL 041-02751) for human fibroblasts with 10% or 0.1% (v/v) fetal calf serum (less expensive heat-inactivated calf serum (CS) can be substituted); F12 (Gibco BRL 041-01765) and F12 isoleucine-free medium for CHO cells, the former supplemented with 10% (v/v) FCS and the latter with 15% dialysed fetal calf (Gibco BRL 063-6300) or dialysed calf serum (Gibco BRL 063-06440); or use dialysed serum
- $10^{-1}$ M (100 × stock) hydroxyurea (Boehringer–Mannheim 106615) dissolved in PBS, filter sterilized, and used fresh
- 2 mg/ml (6 mM) stock aphidicolin (Sigma) dissolved in DMSO
- $1.3 \times 10^{-3}$ M (100 × stock for CHO) or $1.7 \times 10^{-5}$ M (100 × stock for human fibroblasts) of the DNA topoisomerase II inhibitor Hoechst 33342
- Tissue culture flasks (80 or 75 cm$^2$, Nunc or Falcon) or dishes (90 mm, Nunc)

*Method*

A. *CHO cells (monolayer)*

1. Establish a set of culture flasks with $10^6$ cells per flask from an exponential culture, using trypsin or pancreatin to detach the cells. Allow to recover for 18 h in F12 + 15% CS.
2. Remove the medium, wash twice in warm F12 isoleucine-free medium, and replace with F12 isoleucine-free medium supplemented with 15% dialysed calf serum. Incubate for 36 h to arrest cells in mid-$G_1$.
3. Resynchronize the cells in early S phase by removing the medium and replacing with prewarmed F12 medium + 15% CS containing $10^{-3}$ M hydroxyurea, incubate for 10 h at 37 °C.
4. Remove the hydroxyurea-containing medium, wash once with growth medium and replace with F12 + 15% CS and $1.3 \times 10^{-5}$ M (7.5 µg/ml) Hoechst 33342; incubate for 8 h to accumulate $G_2$ cells.

B. *Human fibroblasts (up to passage 10)*

1. Accumulate subconfluent cultures in $G_0$ by seeding $8 \times 10^5$ cells per culture flask (or $3-7 \times 10^5$ per 60 mm dish) in α-MEM + 15% CS. Allow to recover for 18 h before changing the medium to α-MEM + 0.1% CS. Incubate for 48 h.

2. Release the $G_0$ arrest by changing the medium to α-MEM + 10% CS plus aphidicolin (5 μg/ml) and resynchronize cells at the $G_1$/S boundary for 24 h.
3. Release the $G_1$/S arrest by removing the aphidicolin medium, washing the culture twice with α-MEM + 10% CS, and incubating in α-MEM + 10% CS containing 0.1 μg/ml Hoechst 33342. Ten hours later approximately 80% of the cells will be in $G_2$.
4. Reverse CHO or human fibroblast $G_2$ cultures by washing twice with drug-free medium and incubating in the appropriate growth medium. The flow of cells into mitosis is monitored by Colcemid arrest or by another method of choice.

Following release from a hydroxyurea block, treatment of rat 3Y1 fibroblasts with the protein kinase inhibitor staurosporine (100 ng/ml) has recently been reported to cause a reversible $G_2$ arrest (22). This presumably involves inhibition of $p34^{cdc2}$ kinase, which is a target of staurosporine (23). It is not clear how widely applicable this technique will be; staurosporine has several other targets, including kinase C, and in some cells release from a staurosporine-induced $G_2$ block is followed by re-replication and nuclear fragmentation (24).

# 6. Detachment and collection of mitotic cells

## 6.1 Selection of mitotic cells without drug use

At mitosis, most vertebrate cells round-up. In many cell lines, especially those of epithelial origin which grow in monolayers, attachment to the substrate also becomes much looser. this causes some mitotic cells to detach spontaneously; many more can be made to do so by gentle mechanical treatment: by hand or mechanical shaking, or by pipetting streams of medium over the cells. HeLa and CHO cells fall into this category. *Protocol 4* gives a method for mitotic selection of HeLa or CHO cells without the use of shaking equipment; this protocol is adapted from ref. 25. Shaking equipment and selection of CHO cells are described in ref 26.

Populations of cells detached in this way usually have a mitotic index in excess of 90%, although the yield from each culture vessel will be low, less than 2% of the cell population. A brief (30 sec) incubation at 4 °C or room temperature with a standard trypsin solution (as used for routine cell detachment at 37 °C) removes more of the mitotic cells but often reduces their purity. Many fibroblast cell lines or fibroblast-derived cells are much more difficult to detach from the substrate in mitosis. Several brief periods of trypsin treatment at room temperature may help. Looser cell attachment has been achieved by the use of low-calcium growth media, though these media may not provide satisfactory growth conditions for all cells. Selective detachment of mitotic cells has also been achieved by rapid washing of primary and transformed human cell monolayers with a hypotonic solution, though reversal may be slow (27).

Because a single episode of shaking yields few cells, it is often better to detach mitotic cells from the same culture at 10–30 min intervals over a period of 2–3 h, rapidly cooling each suspension to 0 °C, and then combining them. Cold storage for up to 8 h does not prevent the completion of mitosis when cells are returned to 37 °C; but entry into $G_1$ is slower if the period of cold storage exceeds 2 h. Longer periods of cold storage, or cold storage after treatment with mitotic inhibitors, induces apoptosis (28).

---

**Protocol 4.** Mitotic selection from HeLa or CHO cell monolayers

*Materials*
- Growth medium: MEM (Gibco BRL 0410-2571) or F10 (Gibco BRL 041-01550) supplemented with 5% (v/v) fetal calf serum
- 80 or 75 cm$^2$ tissue culture flasks (Nunc 1-53732A or Falcon 3824)
- 37 °C walk-in warm room (ideally)
- Ice-water bath at 0 °C
- Cytospin (Shandon Scientific)
- Crystal violet (Sigma C3886)[a]

*Method*
1. *Day 1*: inoculate $10^6$ HeLa cells into each of several T75 flasks, incubate and six hours later replace the growth medium with 20 ml prewarmed (37 °C) fresh medium.
2. *Day 2*: Transfer flasks to a 37 °C warm room. Remove loosely attached and floating cells by washing each flask three times, 15 min between washes, with 5 ml prewarmed (37 °C) growth medium each time. Keep medium gassed with $CO_2$ to hold the pH down (or use Hepes-buffered medium).
3. Thirty minutes after the last wash shake each flask with sharp lateral movements. Hold the flask loosely by one end and tap sharply on the other with a finger; 30–60 sec of shaking/tapping is required. Check the production of floating cells using the inverted microscope.
4. Remove the supernatant and if more mitotic cells are required replace with prewarmed medium. An aliquot of cells should be inspected for mitotic index. The easiest way is to spin cells directly out of

*1: The synchronization of mammalian cells*

suspension on to a microscope slide by means of a Cytospin, 0.5 ml cell suspension/chamber, 500 r.p.m. for 5 min, followed by fixation in absolute methanol:glacial acetic acid (3:1), and staining in crystal violet. Alternatively, a smear can be prepared by drawing a drop of cell suspension over a microscope slide with the edge of another slide. When dry the slide is fixed and stained as described above. A reasonable technique should ensure a mitotic index of 80% plus, of which the great majority will be in metaphase.

5. Pool the populations detached by repeated treatments at 0 °C immediately after harvest to obtain a greater number of mitotic cells. Shake the flasks at 15–30 min intervals, remove the cells, and rapidly cool in suspension in an ice bath. The yield of mitotic cells from flasks for one period of shaking is about $1-2 \times 10^4$ per $10^6$ cells in culture.

*NB* Reversal: in the absence of cooling, detached mitotic cells pass rapidly and synchronously into $G_1$ (CHO, 20–30 min; HeLa, 60 min). Up to 2 h storage at 0 °C does not delay the subsequent passage into $G_1$, but longer cold storage causes slower and, therefore, less sychronous passage into $G_1$.

[a] Crystal violet stock solution: *clastogenic, wear protective gloves*. Mix the following two solutions in equal parts. Solution 1, 10 g crystal violet and 100 ml 95% ethanol; solution 2, 4 g ammonium oxalate crystals and 400 ml distilled or cartridge-filtered water. Filter through Whatman No. 1 paper and store in a dark brown bottle. Use at 1 : 25 dilution in distilled water for staining slides.

## 6.2 Freezing and recovery of mitotic cells

For many biochemical purposes it would be desirable to accumulate very large numbers of mitotic cells and to have these available on demand. One solution which has been devised for CHO cells (29) is to freeze synchronized mitotic cells to −70 °C. Thawed cells have a clonability similar to that found with control cells, and a normal rate of passage into $G_1$ and through the first cycle. The key factor in the freezing protocol is the use of low-sodium, high-potassium, and high lactic acid medium immediately before the addition of DMSO and freezing. Very large numbers of mitotic (or $G_1$) cells can be preserved in this way for subsequent use. The method is given in *Protocol 5*. Its suitability for cells in other parts of the cycle has not been determined.

**Protocol 5.** Freezing and recovery of mitotic cells

*Materials*

- AnalaR grade $KH_2PO_4$, KOH, NaCl, $MgCl_2.6H_2O$ and dimethyl sulfoxide (DMSO)

**Protocol 5.** Continued

- L-lactic acid, (Sigma L1750), dextrose (Sigma G8270), sorbitol (Sigma S1876)
- Tissue culture grade water (Sigma W3500) or water purified by reverse osmosis (RO) or cartridge-filtration system (e.g. Milli-Q, Millipore Inc.)
- 0.22 μm filter (e.g. Gelman 0.22 μm Acrodiscs)
- Freezing vials (Cryotubes, Nunc 3–40711)
- −70 °C refrigerator

*Method*

1. Make up 100 ml of solution in a beaker by adding all the reagents except DMSO:

    - $KH_2PO_4$      68 mg
    - KOH      140 mg
    - NaCl      175 mg
    - lactic acid      180 mg
    - dextrose      90 mg
    - $MgCl_2.6H_2O$      10 mg
    - sorbitol      4 g
    - water      to 80 ml

    Filter sterilize this solution and divide into two equal parts of 40 ml each. To one aliquot add 10 ml sterile water; to the other add 10 ml DMSO, i.e. this will give two solutions: one which is DMSO-free and another which contains 20% DMSO.

2. Prepare mitotic cells for freezing by centrifuging the cell suspension 150 *g* for 10 min at 4 °C) and thoroughly aspirating the supernatant. Resuspend in 250 μl of DMSO-free freezing solution (ice-cold. *Most important*, allow the solution to rest for 7 min on ice. To maintain rapid passage out of mitosis when the cells are eventually thawed the period of time spent at 4 °C or on ice should not exceed 30–40 min.
40 min.

3. Add an equal (250 μl) volume of 20% DMSO-freezing solution (to give a final 10% DMSO concentration) and mix thoroughly. Aliquot or transfer cell suspension into cryotubes, wrap these with insulating materal (e.g. expanded polystyrene), and transfer immediately to the −70 °C freezer. Cell concentrations in the range of $10^6$–$10^7$/ml are suitable.

4. Recover the frozen cells by rapidly thawing the vials at 37 °C and resuspending the cells before adding to prewarmed growth medium (aproximately 100 × volume), transfer to a spinner vessel or to Petri dishes.

## 7. The collection of mitotic cells by arresting agents

### 7.1 Peculiarities of the mitotic checkpoint

Exit from metaphase normally requires the operation of a microtubular spindle. Agents that disrupt tubulin assembly inhibit spindle formation, while allowing the interphase cell cycle to progress with normal kinetics, so that cells accumulate in metaphase. They can then be selectively detached, as described in Section 6.1. Colcemid and the vinca alkaloids (vinblastine, vincristine) are the traditional arresting agents of this type. Nitrous oxide, delivered at a pressure of 80 p.s.i., causes much less microtubule disruption and results in the assembly of a spindle which fails to engage the condensed chromosomes correctly (30). Nocodazole, an increasingly widely used inhibitor, causes selective loss of non-kinetochore microtubules and prohibits centrosome separation (31). Both these agents allow unhindered passage through interphase, giving excellent and reversible metaphase accumulations provided the duration of arrest is short (2–8 h).

The efficiency of mitotic arrest agents varies, depending on the drug concentration and the duration of treatment, and even more on the cell type and its species of origin. It has been known for decades that Colcemid does not produce stable mitotic arrest in Chinese hamster cell lines (32); after some hours, they pass into a tetraploid interphase without normal chromatid segregation (33). In contrast, HeLa cells arrest permanently in metaphase. In modern terms, this indicates that the mitotic checkpoint control (which prevents cells moving into interphase without passing through a normal anaphase–telophase) is firm in HeLa but not in Chinese hamster cell lines.

Further studies have shown that mammalian cells fall into two classes with regard to mitotic checkpoint control. In one class are all known transformed rodent cells, the adenovirus-transformed human line 293 (34), human cells from patients with the radiosensitive disease ataxia telangiectasia (35), and the Indian muntjac cell line SVM (S. R. R. Musk and R. T. Johnson, unpublished results); these have loose control, less loose in SVM. The other class contains everything else, including non-transformed (primary or 3T3-style) rodent cells, which have strict control. This division applies to arrest with Colcemid, nitrous oxide, or nocodazole. It is intriguingly reminiscent of the discrepancies between transformed rodent cells and others in the loss of checkpoint control in prolonged S-phase arrest by hydroxyurea (see Section 4.3). However, both 293 cells (C. S. Downes and R. T. Johnson, unpublished results) and SVM (16) are normal in their response to hydroxyurea.

## 7.2 Collection of mitotic cells by arrest followed by shake-off

The consequence of this great divide between cell types is that, for synchrony of transformed rodent lines in mitosis by arrest and shake-off, the mitotic arrest (by Colcemid, nitrous oxide, or nocodazole) must be short, usually 2–4 h; longer arrests do not proportionately increase mitotic yield. Short arrests also ensure that 100% reversal of the arrest takes place, soon after the drug is removed. For CHO cells arrested for 4 h by 40 ng/ml nocodazole, reversal is almost complete by 1 h; for HeLa 15% remain mitotic at the same time.

For human cells arrested with Colcemid, the duration of arrest is even more critical if reversal is desired. This drug is (for unknown reasons) much more toxic in human cells than it is for hamster cells, and more than 2 h arrest causes increasing irreversible blockage, abnormal patterns of segregation, and slow passage into $G_1$. Similarly, extended periods in nitrous oxide (or nocodazole), beyond 9 h arrest, result in slow reversal and abnormal segregation or multipolar spindles (30).

For cytogenetic studies, extended prophase or prometaphase chromosomes are ideal and can be obtained by application of a 10 min Colcemid (70 ng/ml) arrest given 5–6 h after release of a 16–20 h amethopterin (methotrexate) arrest of S-phase cells (36).

*Protocol 6* gives a method (after refs 28, 37, 38) for arresting human cells in mitosis, from monolayer cultures, by nitrous oxide following release of S-phase arrest by excess thymidine. The use of an S-phase synchronization is optional, but it does increase the subsequent mitotic yield. The manufacture and testing of the pressure vessel to house the cells can be carried out in any laboratory workshop with welding experience. The choice of vessel dimensions depends on the number of synchronized cells required, but for greatest use a vessel with an internal diameter able to accommodate 150 mm diameter tissue culture dishes in a stack of 10 is desirable. Details of pressure vessel construction are given in Johnson *et al.* (28).

## 7.3 Automatic nitrous oxide synchrony in mitosis

*Figure 1* shows a plan of the equipment used in the automatic version of the nitrous oxide pressure arrest system, full details of which are given in ref. 38. Automatic delivery of nitrous oxide, around or after midnight, reduces the strain of subsequent long cell-cycle experiments. Since nitrous oxide is narcotic, and in some proportions explosive when mixed with air, and since commercial pressure cylinders of nitrous oxide contain a liquid at about 800 p.s.i., an automatic direct flow of gas from the high-pressure cylinder to the cell-contining pressure vessel is potentially hazardous. To limit this hazard, the equipment includes a second gas cylinder interposed between the main cylinder and the pressure vessel, containing only sufficient nitrous oxide to charge the pressure vessel to 80 p.s.i. This container is charged by hand

# 1: The synchronization of mammalian cells

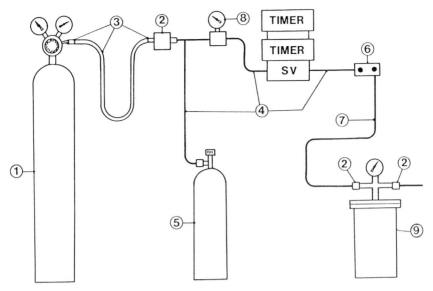

**Figure 1.** Automatic synchronization apparatus. (1) $N_2O$ high-pressure cylinder with outlet nozzle modified to accept screw-on female high-pressure hose with bayonet fitting. (2) Valves and diecast bodies modified to accept either pressure hose fittings or 7 mm copper tubing. (3) High-pressure hose with male bayonet fitting at either end. (4) Section of 7 mm o.d. copper tubing. (5) A 7 lb $CO_2$ cylinder with fitting made to accept 7 mm copper tube. (6) Brass manifold with provision for two or more chambers to be connected. (7) Brake pipe, 500 p.s.i. (Austin Rover). (8) Pressure gauge, 0–200/p.s.i. (9) Steel pressure chamber for cells. (Reproduced with kind permission from Academic Press.)

toward the end of the working day and discharged into the pressure chamber automatically at night; even if the discharge were to leak, the quantities involved would not be hazardous.

---

**Protocol 6.** Arrest of mitotic HeLa cells from monolayer cultures by high-pressure nitrous oxide

*Materials*
- Steel pressure vessel and nitrous oxide delivery system (manual or automatic), with pressure vessel kept in a 37 °C warm room, or warmed for the duration of experiment in a 37 °C waterbath
- Nitrous oxide cylinder and reducing valve
- $CO_2$ cylinder fitted with a reducing valve
- 100 ml plastic syringe plus tubing to fit $CO_2$ outlet and pressure vessel inlet, for charging pressure vessel with $CO_2$

**Protocol 6.** *Continued*

- 150 × 25 mm sterile plastic culture dishes (Falcon 3824) or 140 × 20 mm culture dishes (Nunc, 1-68381)
- 10 × stock of 25 mM thymidine (Sigma, T9250), made up in growth medium, sterilized by filtration, and kept frozen at −20 °C
- Shandon Cytospin (see *Protocol 1*)

*Method* (using automatic nitrous oxide delivery system)

1. *Day 1.* Add $2 \times 10^6$ HeLa cells from a growing culture to each of 10 140 mm Petri dishes in a total volume of 20 ml of growth medium per dish.
2. *Day 2* (around 8 p.m.). Remove the medium from the dishes and add 18 ml fresh medium prewarmed to 37 °C to each dish. Add 2 ml of prewarmed 25 mM thymidine stock to give a final thymidine concentration of 2.5 mM.
3. *Day 3* (around 8 p.m.). Remove cells from the thymidine block and rinse gently with two changes of prewarmed medium (total 10 ml per dish); add 20 ml prewarmed medium to each dish and quickly stack the plates, tape the stack together leaving two loops of tape attached to the top dish in stack (ideally in a 37 °C warm room). Place the stacked dishes into the prewarmed pressure vessel, lower the stack by holding on to the loops of tape. Bolt on the lid, and inject $CO_2$ with a syringe, through the outlet valve to 5% of final volume in each vessel; close the outlet valve. Check that the inlet valve on the vessel is open. Open the regulator on the high-pressure cylinder and charge the intermediate cylinder. Close the high-pressure regulator and set the timeswitch for 4–5 h (i.e. sufficient time for the thymidine-released S-phase cells to reach $G_2$ or late S phase but not to pass through division). Four to five hours later the pressure in the vessel will rise to 80 p.s.i.
4. *Day 4* (around 8 a.m.). Open the outlet valve on the pressure vessel and reduce the pressure gradually over 15 min, to avoid giving the cells the 'bends'. Void the gas into the open air by a tube attached to the outlet valve, leading to the outside of the building. Remove the dishes and inspect them by means of an inverted microscope.
5. The majority of cells should be seen to be floating, or loosely attached, rounded-up and mitotic. Swirl gently to release residual nitrous oxide and detach loosely attached cells; collect the suspension by pipette. Cytospin preparations reveal mitotic indices usually greater than 90%, frequently > 95%.

This method is applicable to a wide range of cells, especially transformed human lines, but the duration of high-pressure treatment may need to be adjusted, usually shortened, to increase viability and speed of exit from mitosis. With a 4 h arrest, HeLa cells exit rapidly; at 2 h after release, only 10% remain in mitosis. Diploid human cells are also arrested by nitrous oxide, but despite release into $G_1$ they do not make a satisfactory entry into S phase.

Rodent cells are also arrested in mitosis by nitrous oxide, but the period of arrest should not exceed 5–7 h, since beyond that time the block is overridden and cells pass abnormally into tetraploid interphase. With a 4 h arrest, exit is rapid; within 1 h of release, less than 5% of CHO mitoses remain.

We have found that this protocol also works for WAK fibroblasts derived from the 'honorary mammal', *Xenopus laevis*, with the following modifications. WAK cells are grown in 3 : 1 MEM : distilled water, supplemented with glutamine, antibiotics, and 10% fetal calf serum, in a $CO_2$ incubator at 25 °C. Nitrous oxide synchrony is performed in a pressure cylinder immersed in a 25 °C waterbath. Mitotic synchrony is excellent (routinely > 95%) and readily reversible.

In the absence of cooling, detached mitotic cells pass rapidly and synchronously into $G_1$ (CHO, 20–30 min; HeLa, 60 min) (39). Up to 2 h storage at 0 °C does not delay the subsequent passage into $G_1$, but longer cold storage causes slower and, therefore, less synchronous passage into $G_1$.

## 8. Cautionary note

Mammalian cell-cycle control systems are damnably intricate. All the methods described above, except for $G_0$ accumulation (Section 2) and shake-off of unarrested mitotic cells (Section 6.1), depend on inhibiting one normal cellular process that occurs at a specific stage of the cycle, and relying on checkpoint controls to arrest the cycle at this point. But checkpoint controls are not inflexible. *Even when cells' progress through the cycle is efficiently arrested, and they return to the cycle after release from arrest, they may well be arrested in a state where the cycle controls are distorted, and do not correspond to anything that occurs in the normal uninterrupted cycle.* For example, cells arrested in late $G_1$ with mimosine have high levels of cyclin A–complexed $p34^{cdc2}$ kinase, normally not seen till well into S phase (40). BHK cells arrested with hydroxyurea at the $G_1/S$ boundary have high levels of cyclin B, normally characteristic of $G_2$ (41, 42). HeLa cells arrested in metaphase with nocodazole lose cyclin A, which is normally undegraded up to the metaphase/anaphase transition (43). It would be unwise to suppose that such alterations in control systems as a result of cycle arrest are not more widespread. Consequently, the use of chemically arrested cell populations to study cell-cycle controls may well produce significant changes in the systems it is intended to study.

## 9. Separation of cell-cycle phases by centrifugal elutriation

The other synchrony method, in addition to mitotic shake-off (see Section 6) in which all of the cells in a particular cell-cycle phase are physically separated from the asynchronous population, is centrifugal elutriation. Centrifugal elutriation is a general purpose technique for separating cells from a population on the basis of their relative size and/or density and is widely used in biology for a number of different applications (44). The method requires a specially designed centrifuge and rotor that contains a separation chamber in which cells of the same size are held in position when their sedimentation velocity is exactly counterbalanced by medium flow in the opposite direction. When used to generate synchronous subpopulations, centrifugal elutriation takes advantage of the fact that the volume of a cell approximately doubles as it progresses from one division to the next and therefore its size increases accordingly. Since the density of cells remains constant the cells in the various stages of the cell cycle can be separated on the basis of this size change (45).

The advantages of centrifugal elutriation are several. It is relatively quick—cells being separated in times as short as 20 min. The cells are always kept in their normal growth medium at room temperature so perturbations in growth conditions are minimized and the physical forces applied to the cells, a maximum of about 400 $g$, are no greater than those used to pellet cells in a standard centrifuge during a wash step. Up to $2 \times 10^8$ cells can be separated into 8–10 fractions, representing subpopulations enriched in $G_1$; early, mid, and late S; and $G_2 + M$. However, the purity of these fractions is not uniform. For example, the purity of fractions of CHO cells separated by centrifugal elutriation has been shown to be 96% for $G_1$, 68% for S, and 65% for $G_2 + M$ (46). On the other hand, the purity of even the $G_2 + M$ fraction may be comparable to that achievable by methods using blocking agents. Another advantage for centrifugal elutriation is that the various fractions of cells separated into the cell-cycle phases are generated essentially at the same time so that they can be directly compared for biochemical analysis. If required, cells from these fractions (especially $G_1$) can be placed back into culture where they will progress synchronously (47).

Many different lines of mammalian cells of both human (e.g. HL-60) and rodent (e.g. CHO) origin have been subjected to synchronization by centrifugal elutriation (48). The major limitation appears to be the requirement that for proper separation the technique is highly dependent on having the cells in single-cell suspension (i.e. less than 1–2% of the cells in clumps). Sometimes, this can be difficult to achieve for certain lines growing attached

# 1: The synchronization of mammalian cells

to a substrate, although, general methods for avoiding clumps have been developed (49). The main disadvantages of centrifugal elutriation are cost and space, the elutriation rotor and special centrifuge are relatively expensive and take up valuable laboratory space.

---

**Protocol 7.** Separation of cell-cycle phases by centrifugal elutriation

*Materials*

- Normal growth medium, appropriate for the cells in question (1.5 litres, serum concentration reduced to 5%) should be used for the elutriation medium
- Dulbecco's phosphate-buffered saline (2 litres)
- 70% ethanol (2 litres) for system sterilization
- Sterile plastic disposable 50 ml sample tubes (one for each fraction to be taken, typically 12, Falcon no. 2098).
- Trypsin (0.025%, Gibco) plus DNase (20 mg/ml, Sigma)
- Elutriation rotor (Beckman, JE-6B), centrifuge (Beckman, J-21 series), and a variable speed peristaltic tubing pump (Cole–Parmer MasterFlex pump) and pump-head for 1/4" o.d. Silastic (Dow–Corning) tubing to yield flow-rates of 5–100 ml/min
- Nylon cloth (25 μm pore size, Fisher Scientific)
- Coulter Counter with Coulter Channelyzer attachment (calibrated with polystyrene beads of known diameter from Coulter)

*Method*

1. Determine the minimum and maximum cell volumes present in the asynchronous population using the Coulter Counter and Channelyzer. Then, calculate the rotor speed and flow-rates required to separate cells having that range of volumes, use the nomogram supplied with the rotor or see ref. 49 for a derivation of the calculations required.
2. Assemble the rotor and calibrate the rotor speed according to the rotor instruction manual. Determine the pump speed settings for each flow-rate required using a stopwatch and graduated cylinder.
3. Sterilize the system with 70% ethanol, rinse with PBS, fill the system with medium and remove all bubbles from the rotor chamber and pump tubing.
4. Prepare a cell suspension from exponentially growing monolayers using trypsin plus DNase, assay clumps microscopically, break up

**Protocol 7.** *Continued*

   clumps by pipetting, remove any remaining clumps by filtering the suspension through nylon cloth, determine cell count, pellet cells, resuspend cells ($2 \times 10^8$) in 20 ml of elutriation medium.
5. Adjust the rotor speed and flow-rate to the lowest setting (typically 1600 r.p.m. (247 $g$) and 12.6 ml/min for CHO cells (45) ) and then load the cells into the elutriation rotor through a 20 ml syringe connected to input tubing.
6. Collect 50 ml fractions by leaving the rotor speed constant and increasing the flow-rate in increments *or* leaving the flow-rate constant and decreasing rotor speed in increments. The appropriate increments are determined from the range of flow-rates or rotor speeds necessary to separate the cells, determined from step 1 above, by dividing the range by the number of fractions needed (typically for CHO cells increment flow-rate by 1.8 ml/min per fraction and collect 13–14 fractions).
7. Analyse cells in each fraction with the Coulter Channelyzer to verify separation efficiency by matching the cell volume in a given fraction against the cell volume predicted for that fraction (determined from the flow-rate and rotor speed used to collect that fraction, as calculated in step 1 above).

# Acknowledgements

We are grateful to the Cancer Research Campaign (UK) and to grant CA23270 awarded by the National Cancer Institute for support; and to M. Cornforth, E. A. Musgrove, A. Ryan, K. Sperling, and S. Squires for valuable discussions.

# References

1. Crissman, H. A., Hofland, M. H., Stevenson, A. P., Wilder, M. E., and Tobey, R. A. (1990). In *Methods in cell biology*, Vol. 33 (ed. Z. Darzynkiewicz and H. A. Crissman), pp. 59–95. Academic Press, New York.
2. Klevecz, R. R. (1975). In *Methods in cell biology*, Vol. 10 (ed. D. M. Prescott), pp. 157–172. Academic Press, New York.
3. Tobey, R. A., Valdez, J. G., and Crissmann, H. A. (1988). *Exp. Cell Res.*, **179**, 400.
4. Reddel, R. R., Murphy, L. C., and Sutherland, R. L. (1984). *Cancer Res.*, **44**, 2398.
5. Musgrove, E. A., Lee, C. S. L., and Sutherland, R. L. (1991). *Mol. Cell. Biol.*, **11**, 5032.

6. Tobey, R. A. (1973). In *Methods in cell biology*, Vol. 6 (ed. D. M. Prescott) pp. 67–112. Academic Press, New York.
7. Ham, R. G. (1965). *Proc. Natl. Acad. Sci. USA*, **53**, 288.
8. Lalande, M. (1990) *Exp. Cell Res.*, **186**, 332.
9. Lalande, M. and Hanauske-Abel, H. H. (1990). *Exp. Cell Res.*, **188**, 117.
10. Hoffman, B. D., Hanauske-Abel, H. H., Flint, A., and Lalande, M. (1991). *Cytometry*, **12**, 26.
11. Watson, P. A., Hanauske-Abel, H. H., Flint, A., and Lalande, M. (1991). *Cytometry*, **12**, 242.
12. D'Anna, J. A., Crissman, H. A., Jackson, P. J., and Tobey, R. A. (1988). *Biochemistry*, **24**, 5020.
13. Kapp, L. N. and Painter, R. B. (1977). *Exp. Cell Res.*, **107**, 429.
14. D'Anna, J. A. and Tobey, R. A. (1984). *Biochemistry*, **23**, 5024.
15. Nicander, B. and Reichard, P. (1985). *J. Biol. Chem.*, **260**, 5375.
16. Downes, C. S., Musk, S. R. R., Watson, J. V., and Johnson, R. T. (1990). *J. Cell Biol.*, **110**, 1855.
17. Rao, P. N. and Johnson, R. T. (1970). *Nature*, **225**, 159.
18. Leno, G. H., Downes, C. S., and Laskey, R. A. (1992). *Cell* **69**, 151.
19. Tobey, R. A., Oishi, N., and Crissman, H. A. (1990). *Proc. Natl Acad. Sci. USA*, **87**, 5104.
20. Smith, P. J., Bell, S. M., Dee, A., and Sykes, H. (1990). *Carcinogenesis*, **11**, 659.
21. Lock, R. B. and Ross, W. E. (1990). *Cancer Res.*, **50**, 3761.
22. Abe, K., Yoshida, M., Usui, T., Horinouchi, S., and Beppu, T. (1991). *Exp. Cell Res.*, **192**, 1922.
23. Gadbois, D. M., Hamaguchi, J. R., Swank, R. A., and Bradbury, E. M. (1992). *Biochem. Biophys. Res. Comm.*, **184**, 80.
24. Bruno, S., Ardelt, B., Skierski, J. S., Traganos, F., and Darzynkiewicz, Z. (1992). *Cancer Res.*, **51**, 470.
25. Lesser, B. and Brent, T. P. (1970). *Exp. Cell Res.*, **62**, 470.
26. Tobey, R. A., Andersen, E. C., and Petersen, D. F. (1970). *J. Cell Physiol.*, **70**, 63.
27. Gaffney, E. V. (1975). In *Methods in cell biology*, Vol. 9 (ed. D. M. Prescott), pp. 71–84. Academic Press, New York.
28. Johnson, R. T., Mullinger, A. M., and Downes, C. S. (1978). In *Methods in cell biology*, Vol. 20 (ed. D. M. Prescott), pp. 255–314. Academic Press, New York.
29. Borrelli, M. J., Mackey, M. A., and Dewey, W. C. (1987). *Exp. Cell Res.*, **170**, 363.
30. Brinkley, B. R. and Rao, P. N. (1973). *J. Cell Biol.*, **58**, 96.
31. Sweet, S. C., Rogers, C. M., and Welsh, M. J. (1988) *J. Cell Biol.*, **107**, 2243.
32. Puck, T. T., Sanders, P., and Petersen, D. (1964). *Biophys. J.*, **4**, 441.
33. Stubblefield, E. (1964). In *Symp. Int. Soc. Cell Biol.*, Vol. 3 (ed. R. J. C. Harris), pp. 223–248. Academic Press, New York.
34. Kung, A. L., Sherwood, S. W., and Schimke, R. T. (1990). *Proc. Natl Acad. Sci. USA*, **87**, 9553.
35. Schimke, R. T., Kung, A. L., Rush, D. F., and Sherwood, S. W. (1991). *Cold Spring Harbor Symp. Quant. Biol.*, **56**, 417.
36. Barch, M. J. (1991). *The ACT cytogenetic laboratory manual*, (2nd edn), pp. 58–65, Raven Press, New York.

37. Rao, P. N. (1968). *Science*, **140**, 802.
38. Downes, C. S., Unwin, D. M., Northfield, R. G. W., and Berry, M. J. (1987). *Analyt. Biochem.*, **165**, 56.
39. Jost, E. and Johnson, R. T. (1981). *J. Cell Sci.*, **47**, 25.
40. Marraccino, R. L., Firpo, E. J., and Roberts, J. M. (1992). *Mol. Biol. Cell*, **3**, 389.
41. Yamashita, K., Yasuda, H., Pines, J., Yasumoto, K., Nishitani, H., Ohtsubo, M., Hunter, T., Sugimura, T., and Nishimoto, T. (1990). *EMBO J.* **9**, 4331–8.
42. Steinmann, K. E., Belinsky, G. S., Lee, D., and Schlegel, R. T. (1991). *Proc. Natl Acad. Sci. USA*, **88**, 6843.
43. Pines, J. and Hunter, T. (1990). *Nature*, **346**, 760.
44. Grabske, R. J. (1978). In *Fractions* Vol 1. pp. 1–8. Spinco Division of Beckman Instruments, Inc., Palo Alto, CA.
45. Anderson, E. C., Peterson, D. F., and Tobey, R. A. (1970). *Biophys. J.*, **10**, 630.
46. Meyn, R. E., Meistrich, M. L., and White, R. A. (1980). *J. Natl Cancer Inst.*, **64**, 1215.
47. Grdina, D. J., Meistrich, M. L., Meyn, R. E., Johnson, T. S., and White, R. A. (1984). *Cell Tissue Kinet.*, **17**, 223.
48. Applications data, centrifugal elutriation of living cells, (1989). Publication no. DS-534D. Spinco Division of Beckman Instruments, Inc., Palo Alto, CA.
49. Meistrich, M. L., Meyn, R. E., and Barlogie, B. (1977). *Exp. Cell Res.*, **105**, 169.

# 2

# Preparation of synchronous cultures of the yeasts, *Saccharomyces cerevisiae* and *Schizosaccharomyces pombe*

J. CREANOR and J. TOYNE

## 1. Introduction

During the last 25 years a great deal of information has come from cell-cycle studies of yeast, and one of the essential tools has been the production of synchronous cultures. These cultures have been produced by a variety of techniques and the purpose of this chapter is to describe their use in the most frequently studied yeasts, *Saccharomyces cerevisiae* and *Schizosaccharomyces pombe*. The choice of method is usually determined by the available apparatus, by the cycle event or cellular component one is investigating, by the degree and duration of synchrony required, and, particularly in *S. cerevisiae*, by the compatibility of the methods available with particular strains.

There are, in principle, two ways in which a synchronous culture can be made: a subpopulation of the culture, which contains cells of the same age, can be selected and grown as a synchronous culture; alternatively, the entire culture can be induced to divide at the same time by the use of chemical inhibitors, by blocking and releasing cells which contain temperature-sensitive cell-cycle defects, or by feeding and starvation.

One other method of studying the cell cycle relies not on growing a selected fraction but on separating an entire culture into discrete populations of cells, within each of which the cells are at a similar cycle stage. This method has been used with some success with *S. cerevisiae*, but it does not seem to work so well with *S. pombe*.

In the budding yeast *Saccharomyces cerevisiae*, cell division involves the formation of a bud which increases in size before it is released, giving rise to mother and daughter cells without buds. The exact timing of bud formation is strain dependent, being initiated at the beginning of S phase in some strains and as late as the middle of S phase for others. It is, therefore, a rough indication of a cell's position in the cell cycle (1, 2). Except in the fastest growing cultures, newly-born daughter cells are usually smaller than the mother cells from which they are released and, as a result, daughters take longer than mothers to complete the subsequent division cycle. This places a limitation on the duration for which a synchronous culture of budding yeast can be maintained. Thus, the synchrony inevitably decreases with each cell division and, for routine methods, will have decayed noticeably after two division cycles. Another reason why synchrony decays is because of intrinsic variation in cycle times between cells, which is a feature of most cell types. A simple way to assess the degree of synchrony is to count the proportion of budded cells in each sample using phase-contrast microscopy or by cell counting using a Coulter counter.

*Schizosaccharomyces pombe* grows in length only during the cycle (3) and divides by binary fission. Division is more or less symmetrical and similar sized daughter cells are obtained. This results in synchrony being maintained for longer than is the case with budding yeast, and up to four cycles showing a good degree of synchrony can be obtained.

Other approaches to investigating the cell cycles of yeasts are described in Chapters 4 and 5.

## 2. Division profiles of synchronous cultures

### 2.1 *S. cerevisiae*

Counting buds as a means of measuring the degree of synchrony has the advantage of requiring only a sonicator probe and a phase-contrast microscope. Because the initiation of bud formation coincides approximately with the beginning of the S phase, unbudded cells correspond to the $G_1$/early S phase of the cell cycle. Thus, for synchronously dividing cultures, the proportion of budded cells reaches a minimum when the number of cells in $G_1$ and early S phase reaches a maximum. However, in rapidly growing yeast cultures, release of the bud does not occur until the bud has itself started to form a bud, giving rise to small chains and clumps of cells. Therefore, in order to observe unbudded cells, it is necessary to assist the process of bud release by sonicating a sample of the culture immediately before making observations. By using phase-contrast microscopy (500 × magnification), in which living cells appear bright, the degree of sonication can be adjusted empirically to eliminate all multiple budded clumps, but to cause the least amount of damage to the individual cells. Usually a random sample of at least 100 cells is scored for this purpose. When the 'budding profile' from a synchronously dividing culture is displayed graphically, the minima, corresponding to $G_1$/early S phase, stand out as sharp dips. For rapidly growing cultures the unbudded phase is very short so it may be necessary to take samples at least every five minutes during this period. In addition to observing the periodic dips in the budding profile, the average size of buds increases uniformly between divisions, which is readily detectable in a well-synchronized culture. This contrasts with the situation is asynchronous cultures in which buds of all possible sizes are encountered simultaneously and in which the proportion of budded cells remains constant at around 60–70%. As the synchrony decays in a synchronous culture, there is a flattening and broadening of the dips in the budding profile (see *Figure 1* for budding profiles obtained by each of the synchrony methods described here).

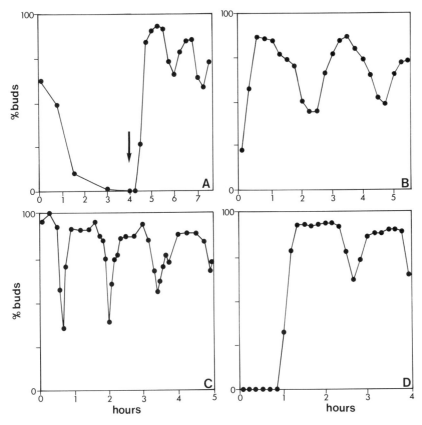

**Figure 1.** Budding profiles resulting from four different synchronous cultures of *S. cerevisiae*. (A) Synchrony was induced by the use of α-factor which was added at time 0 and removed after 4 h as described in *Protocol 7*. Samples were removed at various intervals to assess the proportion of budded cells. (B) A synchronous culture of strain NCYC239 was obtained by elutriation (see *Protocol 2*). (C) A synchronous culture was induced by means of a cell-cycle block using a diploid strain that was homozygous for *dbf*2 (22), a mutant with similar properties to *cdc*15 (see *Protocol 5*). (D) A synchronous culture was obtained by means of the feed–starve method (see *Protocol 8*).

An alternative method of scoring buds is to count only 'small buds'. This may be necessary if the cells are badly misshapen, as is the case for some mutants or in certain strains after release from α-factor arrest, making it difficult to distinguish between large budded cells and misshapen cells without buds. The resulting profile of small budded cells would then be a series of peaks, the position of which would approximate to the passage of the synchronous culture through S phase.

## 2.2 S. pombe

*S. pombe* grows in length only during the cycle, so a single measurement of length is sufficient to position the cell in the cycle with a fair degree of precision. For the first 75% of the cycle the cells increase in length in an approximately exponential manner. During the remaining 25% of the cycle

**Table 1.** Edinburgh Minimal Medium (EMM3) for growing *S. pombe*

| | | |
|---|---:|---|
| **Base** | | |
| Distilled water | 1 | litre |
| Glucose | 20 | g |
| Ammonium chloride | 5 | g |
| Potassium hydrogen phthalate | 3 | g |
| Disodium hydrogen phosphate | 1.8 | g |
| Magnesium chloride .6$H_2O$ | 1 | g |
| Sodium sulfate | 0.1 | g |
| Calcium chloride .2$H_2O$ | 0.015 | g |
| **Vitamins** | | |
| Inositol | 0.01 | g |
| Nicotinic acid | 0.01 | g |
| Calcium pantothenate | 0.001 | g |
| Biotin | 0.01 | mg |
| **Trace elements** | | |
| Boric acid | 0.05 | mg |
| Manganese sulfate | 0.4 | mg |
| Zinc sulfate | 0.4 | mg |
| Ferric chloride | 0.2 | mg |
| Molybdic acid | 0.16 | mg |
| Potassium iodide | 0.1 | mg |
| Copper sulfate | 0.04 | mg |
| Citric acid | 1 | mg |

Vitamins and trace elements can be made up as stock solutions at 1000 × concentrates: 1 ml is then used to make 1 litre of EMM3. All chemicals used are of AnalaR grade.

there is no change in length or volume; the nucleus divides at the beginning of this stage. Shortly afterwards, at about 85% of the way through the cycle, the septum appears transversely across the middle of the cell. The final stage is a splitting of the cell at the septum and the formation of two daughter cells of approximately the same size.

Synchrony can be monitored either by counting the percentage of dividing cells, i.e. those cells containing a septum, or by cell counting using either a haemocytometer or a Coulter Counter. The percentage of cells showing a septum can either be measured microscopically on live cells using phase contrast or, preferably, dark-ground illumination (300 × magnification) or on samples dried down and subsequently stained (4). *Figure 2* shows examples of the synchrony obtained in *S. pombe* using the techniques described in the following sections.

## 3. Selection synchrony

### 3.1 Gradient separation in tubes

The first widely used selection method for preparing synchronous cultures of yeast was by gradient separation in tubes (5). This method was originally described using *S. pombe*, *S. cerevisiae*, and *Escherichia coli*, but it will be described here for *S. pombe* (*Protocol 1*).

---

**Protocol 1.** Synchronization of yeast cultures by gradient separation

*Materials*
- Centrifuge with swing-out head, for example an MSE 41
- 4 litre culture of exponential phase cells in EMM3 (*Table 1*) at 3–4 × $10^6$ cells/ml
- Filter apparatus to take a 15 cm Whatman No. 50 filter paper
- 7.5–30% lactose gradient (w/w) or a 10–40% sucrose gradient (w/w), made up in EMM3
- 5 ml syringe with a long needle, 17 gauge
- 100–200 ml fresh or conditioned medium

*Method*
1. Collect the cells by filtration and scrape them off using a thin spatula.
2. Resuspend in 4 ml of medium and layer this on to the gradient.
3. Spin for approximately 5 min at room temperature at about 500 *g* by which time the top of the cell suspension will have moved about half-way down the tube.

**Protocol 1.** *Continued*

4. Use the syringe to remove the top layer of cells.
5. Inoculate cells into 100–150 ml filtered medium from the original culture, to give a final density of about $2 \times 10^6$ cells/ml. Fresh medium may be used but causes more perturbation to the cells during subsequent growth.

**Table 2.** Growth media for *S. cerevisiae*

YPD broth for growing S. cerevisiae
| | |
|---|---|
| Distilled water | 1 litre |
| Yeast extract (Difco) | 20 g |
| Peptone (Difco) | 10 g |
| Glucose | 20 g |

2 × MYGP medium for 'feed–starve' synchrony method
| | |
|---|---|
| Distilled water | 1 litre |
| Yeast extract (Difco) | 6 g |
| Malt extract (Difco) | 6 g |
| Peptone (Difco) | 10 g |
| Glucose | 20 g |

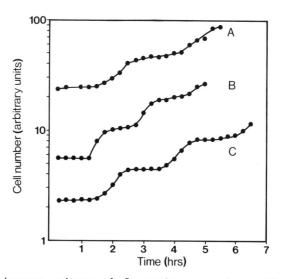

**Figure 2.** Synchronous cultures of *S. pombe* prepared by different methods. (A) Synchronous culture prepared by selection from a tube gradient (see *Protocol 1*). (B) Synchronous culture prepared by block and release of *cdc2–33* (see *Protocol 4*). (C) Synchronous culture prepared by elutriation (see *Protocol 2*).

In an attempt to measure the effect of the synchronization procedure on the subsequent growth of the culture, control cultures can be made. The same procedure is used except that after the centrifugation step, the gradient is mixed and a sample, which will contain cells at all stages of the cell cycle, inoculated into fresh or conditioned medium at the same cell density as was used in the synchronous culture. A culture of this nature is judged to be asynchronous from measurements of cell number increase. This method has proved to be satisfactory for measuring a number of components in *S. pombe* (reviewed in ref. 6), but its major problem was that the synchronization procedure itself has perturbing effects on a large number of the components measured (7). This might have been caused by the concentration of the cells during the collection procedure, by the high sugar content of the gradient, or by a temperature shock, but the significant fact was that the perturbations, in particular the activities of certain enzymes, persisted in asynchronous control cultures for some time after their preparation and occasionally with frequencies which were near to cell-cycle timing.

## 3.2 Density separation

Where there are changes in the buoyant density of cells during the cycle, it is possible to effect separation into cell-cycle stages by equilibrium centrifugation in a tube gradient. This has been used successfully with *S. cerevisiae* using Renografin (8) or dextrin gradients (9) and for *C. utilis* using dextrin (10), but it is not applicable to *S. pombe*, possibly because density changes during the cycle are small.

One advantage of this method is that it is possible to load large numbers of cells on the gradient, e.g. $5 \times 10^{10}$ cells have been loaded on to a 14 ml Renografin gradient. It is necessary to use high centrifugation speeds (up to 45 000 $g$), and during spinning the temperature has to be monitored. While this method is likely to give perturbations of the sort discussed above, this has not been examined in any of the published methods. Since the method has limited application and has largely been superseded by other techniques, it will not be described here.

## 3.3 Preparation of synchronous cultures by elutriation

In an effort to find a way of producing unperturbed synchronous cultures, various continuous flow rotors were tried, but the only really successful method in our experience was using the Beckman JE-5.0 Elutriator Rotor (*Protocol 2*). This works by a counter-current method in which yeast cells are kept suspended in the 40 ml rotor cell by an inward flow of warm growth medium. Separation of cells then takes place as a result of the opposing forces of the flow of medium and the centrifugation. By adjusting the flow rate and the centrifuge speed, a fraction of small cells at the beginning of the cycle can be pumped out of the top of the elutriation chamber while it is in operation.

However, for rapidly growing cultures of budding yeast, because of the time needed to collect the unbudded fraction, the earliest samples that can be taken for analysis may consist mainly of budded cells. These will divide synchronously for two generations before the synchrony decays.

---

**Protocol 2.** Synchronization of yeast using an elutriation rotor

*Materials*
- Beckman J-6M/E centrifuge with bubble trap and pressure gauge incorporated into the inflow line
- a suitable pulse-free peristaltic pump, e.g. Cole–Parmer Masterflex: the pump can be incorporated into either the inflow or the outflow lines, but in our experience there is less mixing in the chamber if the pump is connected in the outflow line
- Beckman JE-5.0 rotor, including a 40 ml elutriation chamber; a smaller chamber size (4 ml) may be sufficient but this will decrease the yield of synchronized cells
- 4 litres of log-phase cells growing in suitable medium (see *Tables 1* and *2*)
- several litres of fresh prewarmed medium

*Method*

NB Since the procedures which have been developed for *S. pombe* and *S. cerevisiae* are slightly different, these will be described in turn.

A. *S. pombe*
1. Pump the system full of culture medium and remove all air bubbles.
2. Increase the rotor speed to 4000 r.p.m.
3. Pump the cells into the rotor at about 150 ml/min from the exponential-phase culture which is kept in a waterbath adjacent to the centrifuge. It will take about 20 minutes to load $2 \times 10^{10}$ cells.
4. Watch the progress of the cells in the elutriation chamber using the stroboscope to ensure that all the cells are retained in the chamber.
5. Examine the effluent medium to ensure that it is free of cells; we normally retain this medium for pumping out fractions.
6. Make a synchronous culture by gradually increasing the pump speed until cells appear in the effluent. We have found that a more uniform cell population is obtained if the rotor speed is first reduced to about 3000 r.p.m. before increasing the pump speed, but it must be emphasized that alteration of the rotor speed causes a certain amount

of mixing in the chamber so one should wait for a few minutes after altering the rotor speed before starting to collect fractions.

7. Collect approximately 100 ml fractions and examine these under the microscope to check for a uniform population of small cells.

8. Pool the best samples and dilute if necessary with conditioned medium to give a culture of suitable density. A good degree of synchrony can be obtained using about 5% of the original culture, but obviously the smaller the fraction the better the synchrony.

B. S. cerevisiae

Steps 1 and 2 are identical to those described above for *S. pombe*.

3. Pump in the cells at an initial flow rate of 100 ml/min gradually increasing to 110 ml/min. Watch the progress of the cells in the elutriation chamber, using the stroboscope constantly.

4. Use as much of the culture as is necessary to concentrate $2 \times 10^{10}$ cells into the chamber, then continue pumping in fresh medium.

5. Slowly increase the flow rate to 130 ml/min until the cells reach the elutriation boundary located at the top of the chamber.

6. Collect the cells that are now pumped out in 300 ml fractions. This will take some 20–30 min.

7. Examine the elutriated cells by phase-contrast microscopy and pool the uniform fractions. The concentration of cells should be $1-2 \times 10^6$/ml.

This method has several obvious advantages. During collection the cells are maintained at growth temperature, since the temperature of the rotor is largely determined by the temperature of the medium flowing through the chamber. There is also a minimal risk of starvation since the medium in the chamber is changed about every 10 sec, and we have shown that a fraction containing $3 \times 10^9$ *S. pombe* cells continues to grow in the 4 ml chamber with a normal doubling time of 150 min for as long as medium is circulated through the system (33 ml/min).

The rotor chamber has to be cleaned after each use, and, because of the complex nature of the circulation system, it also has to be disinfected. We find Alcide Exspor (Life Science Laboratories) to be suitable for this, it is usually left overnight in the rotor and washed out with 1 litre of sterile distilled water just before cell collection.

To test the effect of the procedure on subsequent cell growth, we have prepared control cultures by collecting cells in the rotor, at a similar concentration as that used to prepare synchronous cultures, and then

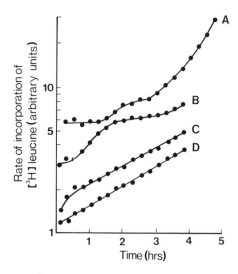

**Figure 3.** Incorporation of [$^3$H]leucine in control asynchronous cultures of *S. pombe*. Strain *NCYC*132 was used in these experiments. (A) and (B) Controls for tube gradients. Growing cells were concentrated by filtration, suspended for 4 min in medium containing 20% sucrose (w/w) and then diluted in the original medium to a cell density of $2 \times 10^6$ cells/ml at time 0. Successive samples were labelled for 14 min with [$^3$H]leucine (150 KBq/ml at 2.0 TBq/mM) and then treated with 5% trichloroacetic acid, filtered, washed, and counted. (C) and (D) Controls for the elutriator rotor. Growing cells were concentrated in the rotor, in (D) $6.5 \times 10^9$ cells were collected over a 40 min period. After collection, the cells were pumped out and diluted in the original medium to a density of $2 \times 10^6$ cells/ml. Samples were labelled as described above.

pumping all of them out into filtered medium by stopping the rotor. A sensitive test for possible perturbation of a control culture of this nature is to measure incorporation of amino acids into protein. *Figure 3* shows an experiment in which [$^3$H]leucine incorporation was measured in control cultures prepared by elutriation and by gradient centrifugation. Although these measurements do not show any perturbations, it cannot be assumed that they will always be absent and control cultures must be prepared for each parameter measured.

## 3.4 Preparation of synchronous cultures using a zonal rotor

If larger synchronous cultures are required, a zonal rotor Type A can be used, which has a capacity of 1500 ml (see *Protocol 3*). This has been used successfully with *S. pombe* and *S. cerevisiae*, and the method for each is essentially the same. This technique has been more or less superseded by the elutriation method, but it is described here since the apparatus is still widely available and is much less expensive that an elutriation system.

2: *Synchronous cultures of yeasts*

---

**Protocol 3.** Synchronization of *S. pombe* using a zonal rotor

*Materials*
- Zonal rotor Type A, MSE (Measuring Scientific Equipment)
- MSE 6L centrifuge
- Large gradient maker, capable of making a 1500 ml gradient
- Filter apparatus to hold a Whatman 32 cm No 50 filter paper
- Peristaltic pump
- 15–20 litres of an exponential-phase culture growing in EMM3 (see *Table 1*)
- 2.5 litres of fresh or conditioned EMM3
- 2.5 litres of 40% sucrose (w/w) made up in EMM3
- 1 litre of 15% sucrose (w/w) made up in EMM3

*Method*
1. Load a 15–40% sucrose gradient into the rotor using a peristaltic pump while the rotor is running at 600 r.p.m.
2. Harvest about $10^{11}$ cells by filtration as described in *Protocol 1* and resuspend in 40 ml medium
3. Load the cells into the centre of the rotor using a 50 ml syringe followed by about 30 ml of medium which displaces the cells into the top of the gradient.
4. Accelerate the rotor to 2000 r.p.m., at which time the cells begin to migrate into the gradient.
5. When the fastest moving cells have gone about 75% of the way to the edge of the rotor, reduce the speed to 600 r.p.m.
6. Remove the cells by pumping 40% sucrose into the outer edge of the rotor.
7. Collect fractions of about 50 ml displaced from the centre of the rotor.
8. Examine fractions microscopically and pool suitable fractions to give a synchronous culture, diluting if necessary with fresh or conditioned medium.

---

One advantage of this method is the large yield obtained compared with selection by a tube gradient. Cultures of *S. pombe* containing $5 \times 10^9$ cells and of *S. cerevisiae* containing $5 \times 10^{10}$ cells can easily be prepared. Indeed we have prepared larger cultures than this, but at the expense of the degree of synchrony.

Since it takes appreciably longer to prepare a synchronous culture by this method and cells are more concentrated during collection, perturbations are likely to be greater. It is, therefore, very important to prepare control cultures by pumping out all the cells and resuspending them in conditioned medium.

Materials other than sucrose can be used to prepare the gradient and there may be an advantage in using non-metabolized materials in order to reduce potential perturbations.

## 3.5 Conclusions

Preparation of synchronous cultures by velocity sedimentation in tube gradients is undoubtedly the most simple and inexpensive method and usually does not produce long-lasting perturbations. However, when the property of the culture under examination does not produce a smooth, exponential increase in control cultures, the method of choice for producing larger yields is the Beckman JE-5 system.

When high yield rather than an absence of perturbations is the main consideration, the zonal rotor can be used. However, experience has shown that in all cases it is important to measure whatever cell property is being examined in control cultures which have been subjected to the same procedures as the synchronous cultures.

# 4. Induction methods of synchronization

The aim here is not to separate a fraction of the population which contains cells of the same age, but to induce the entire population to divide at the same time. This can readily be achieved using large numbers of cells, but there is the problem of how far subsequent cell cycles are affected by the synchronizing procedure. However, this can often be difficult to measure since it is impossible to run control cultures using most of these methods as the very nature of the procedures is to make all of the cells divide at the same time.

## 4.1 Block and release methods

The earliest methods of block and release used with *S. pombe* were with chemical inhibitors. Here a block is imposed on the progression of cells through the DNA-division cycle for sufficient time for all the cells to reach the block point. This will usually be at least a generation. At this point the block is released and all the cells progress through the rest of the cycle for at least one subsequent cycle in a synchronous manner. Two chemicals that have been used are deoxyadenosine and hydroxyurea (11), but side effects of the inhibitors often made interpretation of the results difficult. The advent of *cdc* mutants has in general superseded this chemical technique so it will not be described here.

### 4.1.1 Induced synchrony by means of *cdc* mutants

In principle, any temperature-sensitive cell-cycle (*cdc*) mutant that causes a conditional, uniform and readily reversible arrest of the cell cycle can be used to induce synchrony, provided the conditions used do not result in cell death. In practice the majority of *cdc* mutants do not recover synchronously, even when the cell-cycle arrest they cause is fully reversible. See also Chapters 4 and 5.

*i. S. pombe*
Some of the temperature-sensitive cell-cycle mutants of *S. pombe* can be used to produce excellent synchronous cultures by the block and release method. The two mutants which have been used most successfully to produce induced synchronous cultures are *cdc* 2-33 and *cdc*25-22 both of which arrest at the $G_2$/M boundary. *Protocol 4* describes conditions appropriate for either.

---

**Protocol 4.** Induction of synchrony in *cdc* mutants of *S. pombe*

*Materials*

- A culture of a temperature-sensitive mutant, e.g. *cdc*2-33
- Two thermostatically-controlled waterbaths set for 27 °C and 35 °C

*Method*

1. Grow up a culture overnight at the permissive temperature (27 °C) in EMM3 (see *Table 1*).
2. Transfer the culture to the 35 °C bath and leave at this temperature for at least 3.5 h: longer blocks of 4–4.25 h have also been used.
3. Restore to the permissive temperature.

---

The culture will divide synchronously for at least one cycle, but the synchrony tends to decay rapidly. The reason for this is that during the time the cells are at the restrictive temperature, they continue to grow although they do not divide. Consequently, the cells are up to twice as large as normal but are as variable in size as in an asynchronous culture. This has the effect of shortening subsequent cycle times as the cells attempt to get back to normal size (12). It is possible to carry out a type of control for the induction method with *cdc* mutants by treating a wild-type culture to the same programme of temperature shifts. Other *cdc* mutants can be used to produce synchronous cultures of this type, but the degree of synchrony obtained will depend on the efficiency of the block and the ease with which cells recover from the high temperature.

### ii. S. cerevisiae

The mutants *cdc*14 and *cdc*15 are most frequently used to produce induced synchronous cultures. At the restrictive temperature, these arrest the cell cycle in mitosis (13). Thus, when cells from a log-phase culture are shifted to the restrictive temperature, they all stop dividing at the same point in M phase. This results in a population of uniform, large-budded cells. Synchronous rounds of division are initiated by shifting the culture back to the permissive temperature (see *Protocol 5*).

---

**Protocol 5.** Induction of synchrony in *S. cerevisiae cdc* mutants

*Materials*
- YPD broth (see *Table 2*)
- Two thermostatically-controlled shaking waterbaths at 25 °C and 37 °C

*Method*
1. Grow the mutant strain in YPD broth to a mid-log phase density of 1–5 × $10^6$ cells/ml in the 25 °C waterbath.
2. Transfer the culture flask to the 37 °C waterbath and incubate for about one generation time (typically 2 h). All cells should now have large buds.
3. Transfer the culture flask back to the 25 °C water bath to induce synchronous division.

---

### 4.1.2 Heat shocks

The technique of applying repetitive heat shocks to a culture in order to synchronize it was originally applied to *Tetrahymena* (14). However, the method can be used with *S. pombe* and results in a fair degree of synchrony.

---

**Protocol 6.** Preparation of a synchronous culture of *S. pombe* by heat shock

*Materials*
- A mid-log phase culture of *S. pombe* containing 1–5 × $10^6$ cells/ml
- Two thermostatically-controlled waterbaths at 32 °C and 41 °C

## 2: Synchronous cultures of yeasts

*Method*
1. Grow the culture overnight at 32 °C to the required cell density.
2. Transfer the culture to the 41 °C bath for 30 min.
3. Replace the culture in the 32 °C bath and incubate for 130 min.
4. Repeat steps 2 and 3 five or six times.

This is a very simple technique and large cultures can be made but there is a serious risk of cell-cycle perturbations occurring as a result of the temperature shocks. It is also made much simpler if an automatic device can be used for carrying out the temperature shifts.

### 4.1.3 Synchrony induced by α-factor

Budding yeast cells of *MATα* strain secrete a peptide pheromone, $α_1$-mating factor, that arrests *MATa* cells in $G_1$ (15). This property can be conveniently exploited (*Protocol 7*). Synthetic α-factor is added to log-phase *MATa* cells with the result that the entire culture becomes uniformly blocked as unbudded cells in $G_1$. Although division is blocked in the presence of α-factor, growth continues, resulting in slightly elongated, pear-shaped cells known as schmoos. Removal of α-factor allows the cells to proceed into the cell cycle and to divide synchronously.

**Protocol 7.** Synchronization of *S. cerevisiae* by α-factor

*Materials*
- YNB broth (Difco) or YPD broth (*Table 2*)
- 1 mg/ml α-factor stock solution (synthetic peptide, Sigma: catalogue number T 6901)
- Vacuum filtration apparatus
- Cellulose acetate membrane filters (0.45 μm pore size)
- 150 mM NaCl
- Sonicator with probe

*Method*
1. Grow the haploid *MATa* strain in YNB broth plus 2% glucose until mid-log phase, usually $5 \times 10^6$ to $1 \times 10^7$ cells/ml. Alternatively, YPD broth can be used but does not usually give as good synchrony.
2. Add α-factor to a final concentration of between 2–5 μg/ml. The minimum concentration that is required to arrest division of the particular strain should be used. Since α-factor is degraded by the

**Protocol 7.** *Continued*

cells, the appropriate concentration may depend on the length of time for which cell division is to be arrested.

3. Incubate the culture until less than 2% of the cells are budded. The optimum incubation time for cells in the presence of α-factor is strain dependent. Incubation for slightly more than one doubling time is often sufficient.
4. Harvest the culture by vacuum filtration.
5. Wash extensively with prewarmed medium.
6. The filter will now have a cake of yeast on it. Place the filter in a beaker and resuspend in fresh medium by gentle sonication.
7. Dilute the cells in fresh prewarmed medium to a density of $2 \times 10^6$/cells/ml. The cells should now bud synchronously.

Generally, this is the most readily applicable method since it can usually be carried out with the available laboratory yeast strains. However, *MAT*α cells and diploids cannot be treated in this way since they do not respond to α-factor. Even the *MATa* strains have a variable response to α-/factor, and it may be necessary to test both a number of appropriate strains and various concentrations of the α-factor in order to find optimal conditions.

### 4.1.4 Induction of synchrony in *S. cerevisiae* by the 'feed–starve' method

When a yeast culture runs out of nutrients, the cells enter stationary phase in which there are virtually no budded cells. In the feed–starve method (16), a stationary-phase culture is first obtained. The smaller cells, which cannot easily be synchronized, are then separated from the large cells by repeated cycles of centrifugal sedimentation. The remaining large cells are then exposed to several cycles of a treatment comprising suspension alternately in rich and in starvation medium. This method, described in *Protocol 8*, is limited to certain diploid strains.

**Protocol 8.** Synchronization of *S. cerevisiae* by the 'feed–starve' technique

*Materials*
- 500 ml of 2 × MYGP medium (see *Table 2*)
- Starvation medium (SM): 0.75 g/l KC1; 0.26 g/l $CaCl_2$; 0.5 g/l $MgCl_2.6H_2O$

- 15% mannitol
- Sterile filtered air supply

*Method*

1. Inoculate strain NCYC 239 into 500 ml 2 × MYGP medium.
2. Incubate at 26 °C with vigorous shaking for 10 days. The culture should now consist almost entirely of unbudded cells, of which some 40% are distinctly smaller than the rest. These smaller cells cannot readily be synchronized and are discarded as follows.
3. Harvest the culture and resuspend the pellet in 100 ml of 15% mannitol using 50 ml polypropylene tubes.
4. Centrifuge for 1–2 min at 1800–2000 $g$ in a swing-out rotor. The exact conditions used here need to be determined empirically since they will vary according to the rotor used.
5. Remove and discard the supernatant which contains the smaller cells.
6. Resuspend the pellet and repeat steps 4 and 5 until no more than 2–3% small cells remain.
7. Resuspend the pellet in 150 ml of 1 × MYGP, prewarmed at 26 °C, and incubate without agitation for 40 min.
8. Wash the yeast rapidly three times by centrifugation and re-suspension in starvation medium.
9. Resuspend the pellet in 150 ml of starvation medium using a round-bottomed flask and aerate vigorously with sterile filtered air for 5 hours.
10. Store the culture at 5 °C overnight (without aeration).
11. Repeat steps 7–10 at least twice after centrifuging the cells. After 2–3 cycles of treatment, inoculate the synchronous cells as described in step 7.
12. Initiate a synchronously dividing culture by resuspending a sample of the cells (the final 4 °C overnight incubation is not necessary at this stage) at a density of 5 × 10$^6$ cells/ml in YPD at 26 °C. The cells can be stored at 4 °C for a week or more before they are used to initiate the synchronous culture.
13. Plot a budding profile for the synchronous culture that is obtained. If the degree of synchrony is poor, carry out another feed–starve cycle (step 7 onwards).

This was the first synchronization technique to be developed in yeast, and this method has three main advantages; it is simple, the yield is high, and the synchrony is good. However, it does have the disadvantages that it only works with some strains and cultures take a long time to prepare. There is also some evidence that the first cycle is unbalanced and should be disregarded when the relationship between growth and division is critical.

## 5. Age fractionation

In principle, cells in a gradient or an elutriator are separated according to size (which is equivalent to age), therefore, fractions taken down the gradient should represent cells of increasing age.

The method of cell-cycle analysis has been used with *S. cerevisiae* using an elutriator rotor (17) or a zonal rotor (18) but has not been used successfully with *S. pombe*. The technique is similar to that described for selection synchrony using the elutriator (*Protocol 2*) except that fractions of the whole population are taken by increasing the pump speed in stages and collecting all the cells that come out at that particular setting. The rotor is kept cold during the separation so that cells do not grow during the procedure and it is also possible to add cycloheximide (0.1 mg/ml) to stop protein synthesis. Use of the zonal rotor to fractionate the cell cycle is similar to that described in *Protocol 3* and fractions are collected until all the cells have been removed from the rotor.

There are several advantages to this method of cell-cycle analysis. The yield is greater since the entire culture is used. Since the culture is not grown after separation there will be no distortions of cell-cycle patterns caused by the selection procedure. The cells are separated at 4 °C so there is a much smaller risk of perturbations.

However, the analysis of the data and subsequent deduction of cell-cycle patterns does present a major problem (19). When a culture of *S. cerevisiae* is separated across a rotor, there is much more than a two-fold difference in size between the smallest and the largest cells, mainly because of the unequal division which is a feature of this cell type. So there is uncertainty in how to consider the small numbers of very small and very large cells. One criterion for dealing with this problem is suggested by Salmon and Poole (20) in which cells smaller than $\bar{V}$, the average birth volume, and larger than $2\bar{V}$ are rejected. If this is done, many patterns which at first sight appear to be stepped through the cycle become continuous increases.

Another major problem is that separation is much less efficient in the larger cells at the end of the cycle (21). The reason for this is not clear, but it is possible that cell adhesion may occur during the separation processes.

## 6. Concluding remarks

The most commonly used techniques for preparing synchronous cultures of yeast have been described. Obviously when one is considering a study of a particular cell-cycle event, a choice of cell type and strain has to be made and it is as well to make this choice bearing in mind the techniques available for the organism.

The best representation of the cell cycle of a single cell is probably obtained in synchronous cultures prepared by elutriation, but it has to be said that this apparatus is expensive and requires considerable skill in operation. However, its two main advantages are that large synchronous cultures can now be prepared using the JE 5.0 rotor and control cultures can easily be made to monitor the effect of the treatment on a culture. In very few cases, however, have control cultures been examined to monitor the effects of the selection procedure.

Since the introduction of the larger elutriation rotor system, the use of zonal rotors and age fractionation are no longer such useful systems, particularly as there is now considerable evidence as to their unreliabilty.

If it is the process of cell division that is being studied, then most of the induction methods are suitable for providing a large number of cells dividing at the same time. However, the cell cycle that follows the treatment necessary to produce an induced synchronous culture is almost certainly abnormal in many respects and this has to be borne in mind when extrapolating results from this situation to the cycle of a balanced cell.

It has been said that the counsel of perfection is to try more than one method of synchronization (19), and while this remains the best advice it is seldom followed.

## References

1. Rivin, C. J. and Fangman, W. L. (1980). *J. Cell Biol.*, **25**, 517.
2. Williamson, D. H. (1964). In *Synchrony in cell division and growth* (ed. E. Zeuthen), p. 589. Wiley, London.
3. Mitchison, J. M. (1957). *Exp. Cell Res.*, **76**, 99.
4. Mitchison, J. M. (1970). In *Methods in cell physiology*, Vol. 4 (ed. D. M. Prescott), p. 131. Academic Press, New York.
5. Mitchison, J. M. and Vincent, W. S. (1965). *Nature*, **205**, 987.
6. Mitchison, J. M. (1989). In *Molecular biology of the fission yeast* (ed. A. Nasim, P. Young, and B. F. Johnson), p. 205. Academic Press, London and New York.
7. Mitchison, J. M. (1977). In *Cell differentiation in microorganisms, plants and animals* (ed. L. Nover and K. Mothes), p. 377. Gustav Fischer Verlag, Jena.
8. Hartwell, L. H. (1970). *J. Bacteriol.*, **104**, 1280.
9. Wiemken, A., Matile, P., and Moore, H. (1970). *Arch. Microbiol.*, **70**, 89.

10. Nurse, P. and Wiemken, A. (1974). *J. Bacteriol.*, **117**, 1108.
11. Mitchison, J. M. and Creanor, J. (1971). *Exp. Cell Res.*, **67**, 368.
12. Fantes, P. A. (1977). *J. Cell Sci.*, **24**, 51.
13. Hartwell, L. H., Mortimer, R. K., Culotti, J., and Culotti, M. (1973). *Genetics*, **74**, 267.
14. Zeuthen, E. and Rasmussen, L. (1971). In *Research in protozoology*, Vol. 4, (ed. T. T. Chen), p. 11. Pergammon, Oxford.
15. Bucking-Throm, E., Duntz, W., Hartwell, L. H., and Manney, T. R. (1973). *Exp. Cell Res.*, **76**, 99.
16. Williamson, D. H. and Scopes, A. W. (1962). *Nature*, **193**, 256.
17. Gordon, C. N. and Elliott, S. G. (1977).*J. Gen. Microbiol.*, **112**, 385.
18. Sebastian, J., Carter, B. L. A., and Halvorson, H. O. (1971). *J. Bacteriol.*, **108**, 1045.
19. Mitchison, J. M. (1988). In *Yeast: a practical approach* (ed. I. Campbell, and J. H. Duffus), p. 51. IRL Press, Oxford and Washington, DC.
20. Salmon, I. and Poole, R. K. (1983). *J. Gen. Microbiol.*, **129**, 2129.
21. Ludwig, J. R., Foy, J. J., Elliott, S. G., and McLaughlin, C. S. (1982). *Mol. Cell Biol.*, **2**, 117.
22. Johnston, L. H., Eberly, S. L., Chapman, J. W., Araki, H., and Sugano, A. (1990). *Mol. Cell Biol.*, **10**, 1358.

# 3
# Mammalian cell-cycle analysis

ZBIGNIEW DARZYNKIEWICZ

## 1. Introduction

A plethora of methods designed to probe the cell cycle of mammalian cells has been developed in the past several decades. The scope of this chapter makes it impossible to fully describe them. Therefore, the essential features of the most useful techniques are presented. Outlines of particular protocols, virtues, drawbacks, and applications of the methods, and practical notes regarding critical steps, are given. To master these techniques, the reader is encouraged to consult the articles which describe them more extensively, and which are referenced throughout the text. The aim of this chapter, therefore, is to introduce diverse techniques, which may be of help in selecting the most appropriate approach for a particular task.

The early methods were based on the use of $^3$H or $^{14}$C-labelled thymidine (TdR) followed by detection of this precursor, upon its incorporation into cellular DNA, by autoradiography (1, 2). A wealth of information about cell proliferation in normal and tumour tissues was obtained with the use of this methodology. Kinetics of cell progression through the cell cycle was studied in great detail after the development of more advanced techniques, e.g. analysis of the fraction of labelled mitosis (FLM) (3–5). Autoradiography also provided the means to recognize the presence of resting, quiescent ($G_0$) cells and to enumerate cells in the reproductive pool in tumours, the growth fraction (GF).

Flow cytometry opened new possibilities for cell-cycle analysis (for reviews see refs 6, 7). The major advantage of flow cytometry stems from the fact that measurement of the cellular DNA content reveals the position of the cell in the cycle, while the correlated measurement of a second or third constituent followed by multivariate analysis of the data yields information about the cell-cycle phase specificity of expression of this constituent. The detection of BrdU incorporation by flow cytometry (8), which identifies the cells replicating DNA, now offers all the benefits of [$^3$H]TdR autoradiography but, compared with the latter, is less cumbersome and thus can be of wider application.

## 2. Autoradiographic methods

### 2.1 [³H]TdR labelling index

The labelling index (LI) is a measure of the cell population that incorporates the DNA precursor during pulse treatment, and is generally presumed to represent the fraction of cells progressing through S phase of the cycle.

---

**Protocol 1.** [³H]TdR labelling index (LI)

1. Expose live cells to a pulse of [³H]TdR. The precursor is usually given at a dose between 0.5–5.0 µCi (specific activity 3–100 Ci/mmol) per 1 g of animal body weight, or per 1 ml when applied *in vitro*. The pulse duration is 30 min to 1 h; during this short time the radiobiological effect of tritium can be neglected[a].

2. Fix cell samples (tissue sections, cell smears, or cytocentrifuge preparations) in a mixture of ethanol:acetic acid (9:1, v/v) for at least 2 h. After fixation, extensively rinse the slides in tap water to remove the unincorporated precursor.

3. In the darkroom, using the appropriate safelight, cover the slides with nuclear emulsion. Dry the autoradiographs thoroughly and expose them in lightproof boxes, in a dry atmosphere, at 4 °C, for a period of 1 day to several weeks[b,c].

4. Following exposure, process the autoradiographs through the photographic procedure (developer, rinse, 30% sodium thiosulphate, rinse).

5. Counterstain the cells (e.g. 0.1% toluidine blue, Giemsa, Wright, or haematoxylin) and count the percentage of labelled cells (LI). Use the Poisson distribution statistical evaluation of the results[d].

---

[a] When cells are pulse labelled with [³H]TdR but then grown for an extended time, e.g. as in studies designed to use the precursor as a marker of proliferating cells, or in the FLM method, the ³H-radiation may significantly perturb cell progression through S and/or $G_2$. Therefore, the lowest possible doses of [³H]TdR should be used. In separate experiments, by comparison of the cell growth rate in the presence and absence of [³H]TdR, the radiobiological effects of ³H should be evaluated.

[b] Description of the autoradiographic techniques is the subject of several books and chapters (e.g. ref. 9). Details are often provided with the emulsion by the supplier, especially pertaining to the use of particular types of nuclear emulsion. Follow the steps required for the particular emulsion type. Emulsion sensitizers can be used to shorten the exposure. Check different exposure times to find the optimal one, adequate for the discrimination of labelled from unlabelled cells.

[c] The intensity of cell labelling (which is generally in stoichiometric relation to dose of the precursor) should be adequate to discriminate between labelled and unlabelled cells within an exposure time of up to 2 weeks. With longer exposure

# 3: Mammalian cell-cycle analysis

> times fading of the latent image may occur. Control for negative chemography using unlabelled tissue and exposing emulsion to light (9).
> $^d$ Identification of [$^3$H]TdR labelled cells may be obscured by a high emulsion background. The latter may result from: inadequate rinsing of the sample to remove unincorporated precursor, positive chemography, defective emulsion, inappropriate conditions of exposure, and inappropriate safelight used during the photographic procedure. See reference 9 for solutions to such problems.

i. *Advantages*

(a) The [$^3$H]TdR-LI is a classical assay of cells progressing through S phase. The results, thus, can be compared directly with the extensive data in the literature.

(b) The assay is very sensitive, cells progressing through S even at very slow rates can be detected (high doses of [$^3$H]TdR are then required).

(c) The method is especially useful in studies of low cell number populations in tissue, when the spatial location of labelled cells with respect to other cells, vasculature, necrosis, tumour periphery, degree of differentiation, etc. is of importance for data interpretation.

ii. *Limitations*

(a) The technique is cumbersome and requires the use of a radioisotope. [$^3$H]TdR, however, can be substituted by BrdU, the latter may be detected immunocytochemically in tissue sections or in cells on smears, or by flow cytometry.

(b) Visual counting of labelled cells is tiring and subjective. The devices developed to automatically read the autoradiographs have not lived up to their promise.

(c) The method is based on the assumption that incorporation of [$^3$H]TdR by the S-phase cells is dose-dependent and that all cells have access to the precursor. Variability in endogenous TdR pool size, the possibility that the precursor is incorporated during DNA repair (unscheduled DNA synthesis), inaccessibility (*in vivo*) of some cells to [$^3$H]TdR, and other possible artefacts, all may complicate data interpretation (3, 10, 11).

## 2.2 Fraction of labelled mitoses (FLM) analysis

The FLM method is designed to estimate the duration of the cell cycle and of the individual phases of the cycle (3). It is based on pulse labelling of the cell population, generally with [$^3$H]TdR, followed by time-lapse analysis of the per cent of labelled cells that subsequently progress through mitosis. The duration of the phases of the cycle is determined by the curve representing the changes in the per cent of labelled mitotic cells (3–5, 12).

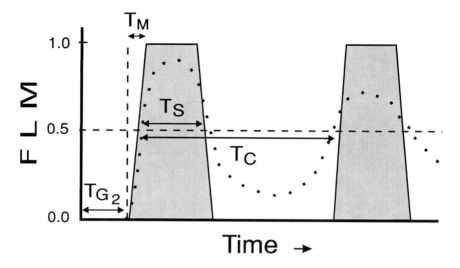

**Figure 1.** Estimation of the cell-cycle parameters by the FLM method. Following a pulse exposure of the cell population to [$^3$H]TdR, cells are periodically sampled to estimate the fraction of labelled M cells (FLM), which is then plotted as a function of time, starting with the moment of the addition of the precursor. The time elapsed until the appearance of the first labelled M cell reflects the minimal duration of $G_2$ ($T_{G2}$). The duration of mitosis ($T_M$) is estimated from that point on, until the maximal fraction of M cells is labelled. The length of S phase ($T_S$) is measured as the width of the FLM curve at 50% labelling level. The cell-cycle time ($T_C$) is the time between two equivalent points of the first and second curve (generally at 50% labelling level, on the ascending portion of the FLM curve). Due to variability in duration of particular phases of the cycle of individual cells and their transit through the stochastic compartment of the cycle (18), the actual curve (broken line) may vary considerably from the theoretical one. $^3$H-labelling may be substituted by labelling with BrdU.

---

**Protocol 2.** FLM analysis

1. Expose live cells to a short pulse of [$^3$H]TdR (as described in *Protocol 1*). Remove the non-incorporated precursor by rinsing the cells and transferring them to fresh, prewarmed medium. *In vivo*, a single injection of [$^3$H]TdR provides pulse labelling.

2. Harvest the cells at hourly intervals, for a period equivalent to approximately two generation times of these cells. Prepare, expose, develop, and fix the autoradiographs, as described in *Protocol 1*. Counterstain the cells.

3. Identify cells in mitosis on the autoradiographs and count at least 500 mitotic (M) cells per time point. Calculate the proportion of mitotic cells that are labelled.

*3: Mammalian cell-cycle analysis*

4. Plot the fraction of labelled M cells as a function of time after addition of [³H]TdR (see *Figure 1*).
5. Estimate the duration of the particular phases of the cycle, as shown in *Figure 1*.

i. *Advantages*
This is a classical technique which can provide several kinetic parameters simultaneously. Interpretation of the results is very straightforward.

ii. *Limitations*
(a) The technique is very cumbersome.
(b) In heterogenous cell populations, when there is a high variability of individual cells with respect to length of their cycles, or in the presence of slowly cycling or non-cycling cells (e.g. solid tumours), the elements of the curve needed to estimate the cell-cycle parameters, as shown in *Figure 1*, are poorly defined. Therefore, it is often difficult, if not impossible, to estimate cell kinetics in such situations.
(c) The radiobiological effects of incorporated [³H]TdR, in perturbing the cell cycle, cannot be disregarded.

## 2.3 Tumour growth fraction estimation

The term 'growth fraction' (GF) has been proposed to distinguish the cells progressing through the cycle (P cells) from the non-cycling, or quiescent (Q) cells in tumours (13):

$$GF = P/(P + Q)$$

Although flow cytometric methods to identify cycling and quiescent cells are now available (see below), the classical, autoradiographic techniques based on either pulse- or continuous-cell labelling with [³H]Tdr (13, 14) had often been used in the past: continuous labelling has proven to be of more practical value.

**Protocol 3.** Tumour growth fraction estimation

1. Expose the cells to [³H]TdR **continuously** for a duration approximately equivalent to the doubled median cell-cycle time of the cycling cells in the population. *In vivo*, thus, [³H]TdR should be repeatedly injected at intervals shorter than the minimal S-phase duration (generally 6 h) of the cells of interest.

> **Protocol 3.** *Continued*
> 2. Sample tumours sequentially in time (e.g. every 6 h), during the labelling period. Estimate the LI and plot it vs. time. The plot will show two phases: the initial, rather steep increase in LI, followed by a near plateau or slow increase in LI.
> 3. The per cent of labelled cells, on the plot, at the time point when the LI slope approaches the plateau, is an approximate estimate of GF.

i. *Advantages*

This is a classical approach and the results may be comparable with the data in the literature (13–15).

ii. *Limitations*

(a) Many ambiguities exist in the GF estimate (14). The plateau LI level can never be well defined because proliferating (labelled) cells continuously dilute the pool of unlabelled cells as they divide.

(b) The technique utilizes a radioisotope, is prone to the radiobiological effects of the precursor, and is time consuming.

## 3. Flow cytometric methods of cell-cycle analysis

The strategies of cell-cycle analysis by flow cytometry can be subdivided into three major categories. The first is the 'snapshot' or static analysis of the cell population. This analysis may be either univariate, based on a single measurement of DNA content alone, or multivariate (multiparameter). In the latter case, another cell feature is measured in addition to DNA, providing information about expression of a particular metabolic or molecular trait, which is generally known to correlate with progression through the cycle or with cell quiescence. Such a measurement, however, does not directly reveal whether the cell is progressing through the cycle or not. The kinetic information is inferred from the DNA content and from the cell metabolic or molecular profile.

The second strategy represents a combination of multiparameter flow cytometry with kinetic measurements. In this approach, the time lapse static observations are combined with the measurement of a rate of cell progression through the cycle, e.g. by using synchronized cell populations or the principle of stathmokinesis. As discussed later, the latter usually involves looking at the rate of accumulation of cells in mitosis following the addition of a metaphase-arresting agent such as colchicine or vinblastine.

The third strategy is based on analysis of DNA replication, by simultaneous measurement of BrdU incorporation and DNA content. Similar to [$^3$H]TdR, the cells that replicate DNA in the presence of BrdU can be recognized, making it possible to estimate their rate of entrance into S and progression through other phases. Examples of the methods utilizing these strategies are presented below.

## 3.1 Univariate cellular DNA content distribution

The cell's position in the cell cycle can be estimated based on the measurement of its DNA content (see *Figure 2*). Because the duration of S and $G_2$ and M phases is relatively constant while the duration of $G_1$ varies, and quiescent ($G_0$) cells have, in most cases, a DNA content equivalent to that of $G_1$ cells, the sum of cells in S, $G_2$, and M phases reflects the proliferative potential of the cell population.

All cells in $G_1$ have a uniform DNA content, as do cells in $G_2$. Under ideal conditions of DNA staining, the fluorescence intensities of all $G_1$ and $G_2$ cells are expected to be uniform, and after digitization of the electronic signal from the photomultiplier (representing their fluorescence intensity), to have uniform numerical values, respectively. This, however, is never the case, and on frequency histograms the $G_1$ and $G_2$ cell populations are represented by peaks of various widths. The coefficient of variation ($CV$) of the mean value of DNA-associated fluorescence of the $G_1$ population is a measure of the width of these peaks, which most often results from inaccuracy of DNA

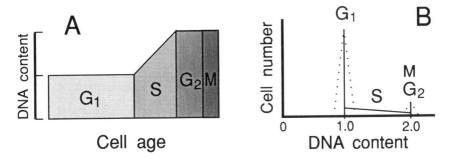

**Figure 2.** Estimation of cell position in the cell cycle based on DNA content measurement. Nuclear DNA content doubles during S phase and therefore the cell age during S can be estimated based on the amount of replicated DNA (increase in DNA content). In contrast, cells in $G_1$, as well as in $G_2$ + M, have identical DNA content, equivalent to the DNA ploidy index 1.0 and 2.0, respectively. Under ideal conditions of DNA measurements based on DNA-specific fluorescence, the $G_1$ and $G_2$ + M cells, therefore, would have uniform DNA values and be represented on the frequency histograms (after their fluorescence values are transformed to electronic signals, which are then digitized) as the bars of a single channel width (B). Due to the inaccuracies in DNA measurements, the actual data are in the form of $G_1$ and $G_2$ + M peaks (broken lines). Proportions of cells represented by these peaks, and in S phase, are estimated by deconvolution of the histograms, using a variety of mathematical models.

measurements. Several mathematical methods and computer programs are available to deconvolute the histograms, and to obtain the percentage of cells in $G_1$, S, and $G_2 + M$ (6).

The DNA frequency histograms do not provide any direct information on cell kinetics. However, when the cell proliferation rate (e.g. the cell doubling time) is estimated in parallel, it is possible, with certain assumptions and approximations, to estimate the duration of the individual phases of the cell cycle, as is illustrated in *Figure 3*.

Numerous fluorochromes can be used to stain cellular DNA and a variety of methods applicable to flow cytometry have been developed (6, 7). One of the simplest is detailed in *Protocol 4*.

---

**Protocol 4.** Univariate cellular DNA content distribution

1. Mix 0.2 ml of cell suspension ($10^5$–$10^6$ cells, taken either directly from the tissue culture, or prefixed in 70% ethanol and then resuspended in buffered saline), with 2 ml of a solution containing:

   - 0.1% Triton X-100
   - 2 mM $MgCl_2$
   - 0.1 M NaCl
   - 10 mM Pipes buffer (pH 6.8)
   - 1 μg/ml DAPI

2. Transfer the sample to the flow cytometer. Measure blue cell fluorescence within the wavelength band of 460–520 nm, under excitation with UV light (laser tuned to UV light emission, or, when a mercury lamp is the excitation source, use a UG 1 excitation filter).

3. Record the fluorescence intensities (the integrated area of the electronic pulse signal) of $10^4$ or more cells. Deconvolute the frequency histograms, e.g. using commercially-available software, to obtain the per cent of cells in $G_0 + G_1$ (the first peak on the histograms), S, and $G_2 + M$ (the second peak).

4. Estimate the approximate duration of individual phases of the cell cycle, by calculating the cell doubling time (time needed for cells to double in number) in the culture, from the cell growth curve (number of cells vs. time of culturing. Plot (as shown in *Figure 3*), the fraction of cells (*f*) in particular phases of the cycle [ln (1 + *f*) ] vs. cell doubling time (linear scale). This estimate is carried out under the assumptions that no significant cell death occurs, cells grow exponentially and asynchronously, and all cells divide (16).

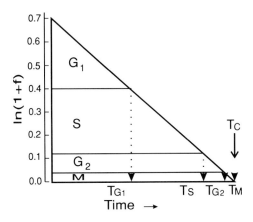

**Figure 3.** Graphic method of estimation of duration of particular cell-cycle phases. The method is based on the assumption that cells are in the exponential phase of proliferation and that the GF is 1.0. The proportions of cells in various phases of the cycle are obtained by flow cytometry, and the value of $T_C$ from the growth curves as the cell-doubling time in culture ($T_C$ equals the cell-doubling time when GF = 1.0). Fractions of cells in particular phases of the cycle (f) are plotted exponentially, ln (1 + f) and the respective points on the time coordinate represent the duration of these phases (16).

i. *Advantages*
(a) The method is rapid, simple, and applicable to large cell populations. A variety of other staining protocols utilizing different fluorochromes are available (6, 7). The univariate DNA content analysis can even be performed on archival samples, namely on the nuclei isolated from tissues embedded in paraffin blocks (17).
(b) List-mode data acquisition and analysis are not required. Thus, there is no need for a large computer for data storage, or complex software.

ii. *Limitations*
(a) The method does not discriminate between $G_0$ vs. $G_1$, and $G_2$ vs. M cells.
(b) Unless combined with a cell proliferation measurement (as shown in *Figure 3*), DNA content analysis alone provides no kinetic information.

## 3.2 Multiparameter analysis

### 3.2.1 DNA vs RNA content

Due to the low number of ribosomes, the quiescent cells ($G_0$), compared to their cycling counterparts, are often characterized by a much lower total cellular RNA content (18). The differences in cellular RNA content, therefore, discriminate between $G_0$ and $G_1$ cells, and in some cell systems, identify the quiescent cells that remain with an S or $G_2$ DNA content ($S_Q$, $G_{2Q}$) (19). The protocol presented below is based on the use of the metachromatic fluorochrome, acridine orange (AO), which can differentially stain DNA and RNA under the appropriate conditions (AO concentration, ionic strength, pH) (see *Figure 4*). Namely, this dye intercalates into double-stranded DNA, and in this form fluoresces green when excited with blue light. On the other hand, all cellular RNA, under these staining conditions, is transformed into the single-stranded form which reacts with AO, forming complexes which luminesce in red. The intensity of green and red luminescence of cells stained in this manner is proportional to their DNA and RNA content, respectively (18, 19).

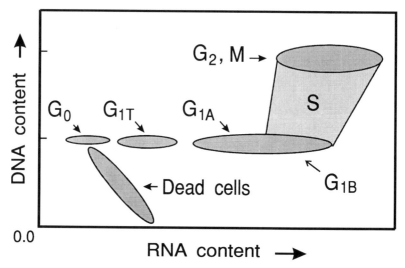

**Figure 4.** Schematic representation of cell populations distinguished by bivariate analysis of DNA and RNA content. Non-cycling ($G_0$) cells (e.g. such as peripheral blood lymphocytes) have minimal RNA content. Among the cycling cells, the threshold RNA content prior to entrance to S allows one to discriminate the early ($G_{1A}$) and late ($G_{1B}$) $G_1$ cells; cell progression through early $G_1$ has a stochastic component (18). The cells in transition from $G_0$ to the cycle ($G_{1T}$) have intermediate RNA values. Loss of DNA characterizes dead cells, especially when they die by apoptosis.

---

**Protocol 5.** RNA vs DNA content

1. Gently admix 0.2 ml of cell suspension, taken directly from tissue culture (at most $2 \times 10^5$ cells), to 0.4 ml of the ice-cold solution containing[a]:
   - 0.1% Triton X-100
   - 0.08 M HCl
   - 0.15 M NaCl

2. After 15 sec add 1.4 ml of the ice-cold solution containing:
   - 20 μM (6 μg/ml) AO
   - 1 mM EDTA–Na
   - 0.15 M NaCl
   - $Na_2HPO_4$ citric acid buffer, pH 6 (the buffer is prepared by mixing 37 parts of 0.1 M citric acid with 63 parts of 0.2 M $Na_2HPO_4$)

3. Transfer the sample to the flow cytometer. Optimal excitation of AO is with blue light (488 nm laser line or BG 12 filter). The DNA-associated green fluorescence is measured at a bandwidth between 515–545 nm, and the red luminescence, representing RNA, optimally above 650 nm (long-pass filter). Measure the cell luminescence between 2–10 min after addition of the AO solution.

[a] Cells prefixed in 70% ethanol and then resuspended in saline can also be used in this assay.

i. *Advantages*
(a) This is a rapid, simple method which can discriminate between quiescent and cycling cells in several cell types. It is ideally suited, e.g. to measure the $G_0$ to $G_1$ transition during mitogenic stimulation of lymphocytes (*Figure 4*) (11, 19).
(b) The method can be used to assay the degree of unbalanced growth (altered RNA/DNA ratio), when the cells are arrested in the cell cycle, e.g. by cytostatic drugs.

ii. *Limitations*
(a) *Protocol 5* requires very stringent conditions of dye concentration, pH, and ionic strength. The cell number (DNA:AO ratio) cannot exceed $2 \times 10^5$ cells per 1 ml of the staining solution. Some instruments require adjustment of the AO concentration in the staining solution to compensate for dye diffusion into the sheath fluid (7).
(b) The assay cannot be used in those cells that contain large amounts of glycosaminoglycans or keratins, such as primary fibroblasts, mast cells, or differentiated keratinocytes, because of the stainability of these compounds with AO.
(c) The strongly fluorescing dye AO binds to the tubing of flow cytometers, which may adversely affect measurements of weak immunofluorescence of subsequent samples. Rinsing with bleaching solution followed by saline between measurements is required (7).

### 3.2.2 DNA vs chromatin changes during the cycle

Chromatin undergoes condensation during mitosis and is also more condensed in the nuclei of $G_0$ cells compared to their cycling counterparts in interphase. The chromatin condensation is associated with the increased sensitivity of DNA *in situ* to denaturation: this marker, therefore, can be used to identify mitotic and quiescent cells (6, 19, 20).

Analysis of DNA denaturability *in situ* is simple and also involves the use of AO. After cell fixation and incubation with RNase, the cells are exposed to 0.1 M HCl, to partially denature DNA. The denatured and double-stranded DNA sections are then differentially stained with AO. The staining is performed at low pH to prevent DNA renaturation.

---

**Protocol 6.** DNA vs chromatin changes during the cycle

1. Fix cells in 80% ethanol for at least 2 h.
2. Centrifuge fixed cells for 10 min at 300 *g*. Resuspend the cell pellet in 1 ml of Hanks' buffered salt solution (HBSS) and add 100 units of RNase A.
3. Incubate at 37 °C for 1 h. Centrifuge as before and resuspend in 1 ml of HBSS.
4. Take 0.2 ml aliquot of this supension and add to 0.5 ml of 0.1 M HCl, at room temperature.
5. After 30 sec add 2 ml of a solution containing:
   - 20 µM (6 µg/ml) AO in:
   - $Na_2HPO_4$ – citric acid buffer at pH 2.6; (the buffer is prepared by mixing 9 parts of 0.1 M citric acid with 1 part of 0.2 M $Na_2HPO_4$). Use this solution at room temperature.
6. Measure cell fluorescence. Excitation and emission is as described in *Protocol 5* for DNA and RNA staining. Identification of cells in particular phases of the cell cycle is as shown in *Figure 5*.

---

i. *Advantages*

(a) This is a unique, rather simple procedure that allows, in a single measurement, discrimination of $G_0$, $G_1$, S, $G_2$, and M cells.

(b) Because M cells can be quantified, the method can be used to score mitotic indices and in stathmokinetic experiments that employ mitotic blockers such as vinblastine or Colcemid.

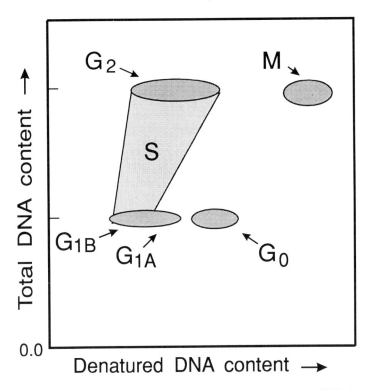

**Figure 5.** Discrimination of the cell-cycle phases based on partial DNA denaturation followed by differential staining of denatured and non-denatured DNA with AO. Sensitivity of DNA *in situ* to denaturation correlates with the degree of chromatin condensation. The most sensitive is DNA in condensed chromatin of mitotic and $G_0$ cells, the least sensitive of the cells entering S phase. The product of the interaction of AO with denatured DNA fluoresces red, whereas AO bound to double-stranded DNA fluoresces green. The sum of red plus green fluorescence intensities, properly standardized, reflects the total DNA content, whereas the proportion of denatured DNA is measured as an $\alpha_t$ index, representing the ratio of red to red plus green fluorescence intensities (7).

ii. *Limitations*
(a) Since the technique employs AO, similar technical limitations apply as those described above for the RNA/DNA procedure.
(b) DNA in apoptotic and necrotic cells is excessively sensitive to denaturation. Their presence in the sample, thus, may complicate identification of M and $G_0$ cells.

### 3.2.3 DNA vs proliferation associated antigens
Numerous proteins are selectively expressed either in cycling or non-cycling cells, or during particular phases of the cell cycle. Antibodies against many of

these proteins, often monoclonal, are commercially available. Immunocytochemical detection of these proteins, therefore, provides information on the proliferative status of the cell. Many techniques have been developed which combine simultaneous, correlated measurements of one or two of these proteins with the DNA content of the cell. Bivariate or multivariate analysis of such data can relate the expression of the particular protein with the cell position in the cycle.

Unfortunately, the optimal conditions (cell fixation, permeabilization, counterstaining, etc.) for detection of the particular antigen vary, depending on the primary structure, conformation, and milieu of the epitope. The scope of this chapter does not allow presentation of the different protocols tailored for individual antigens. The indirect immunofluorescence approach described below is based on mild cell fixation with formaldehyde and their permeabilization with a non-ionic detergent (21, 22). With minor modifications, this protocol has been applied to numerous antigens (23). The most useful markers of proliferating cells are:

- the antigen detected by Ki-67 antibody (24)
- the Proliferating Cells Nuclear Antigen (PCNA), which is a cofactor of DNA polymerase δ, specifically expressed in S-phase cells (25, 26)
- p120 nucleolar antigen (27)

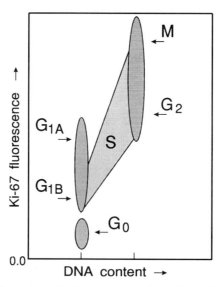

**Figure 6.** Schematic illustration of cell populations discriminated by bivariate analysis of Ki-67 antigen and DNA content. The Ki-67 antigen is expressed only in proliferating cells, reaching a maximum and $G_2$ and mitosis. It is degraded in the post-mitotic, early $G_1$ cells ($G_{1A}$) and accumulates mostly during S phase (24).

# 3: Mammalian cell-cycle analysis

- dihydrofolate reductase (dihydrofolate reductase can also be detected by binding the fluoresceinated inhibitor, methotrexate, see ref. 28)
- a 170 kDa isoform of DNA topoisomerase II (29)

The nuclear protein statin (30), on the other hand, can be used as a marker of non-cycling cells. An example of cell reactivity with the Ki-67 antibody, in relation to its position in the cycle, is presented in *Figure 6*.

---

**Protocol 7.** DNA vs proliferation-associated antigens

1. Fix $1 \times 10^6$–$2 \times 10^6$ cells in 1 ml of 0.5% formaldehyde solution prepared in HBSS, at pH 7.4, on ice, for 15 min. Centrifuge at 400 $g$ for 5 min.
2. Resuspend cell pellet in 1 ml of HBSS, containing 0.1% Triton X-100 and 1% bovine serum albumin (BSA). Keep on ice for 5 min, centrifuge.
3. Resuspend cell pellet in 100 µl of HBSS containing the primary antibody, 1% BSA and 0.1% sodium azide. Follow the instructions, usually supplied by the vendor, regarding the final dilution of the particular antibody (generally, the final dilution is 1:20 to 1:50). Use an isotype immunoglobulin as a negative control for non-specific staining, i.e. an irrelevant monoclonal (not reactive with the cells used) of the same isotype as the primary antibody. Incubate for 30–60 min at room temperature with gentle agitation. Add 2 ml of HBSS with 1% BSA, centrifuge.
4. Resuspend cells in 100 µl of HBSS solution containing the fluorescein-conjugated secondary antibody (generally, the final dilution is 1:20 to 1:40; if the primary reagent is a mouse monoclonal antibody, a fluorescein-labelled goat anti-mouse immunoglobulin is often used as the secondary antibody), 1% BSA and 0.1% sodium azide. Incubate at room temperature for 1 h with gentle agitation. Centrifuge.
5. Resuspend cells in 1 ml of a HBSS solution containing 5 µg/ml of propidium iodide and 100 units of RNase A. Keep at room temperature 1 h. Measure cell fluorescence using excitation with blue light (488 nm laser line) and collect fluorescence emission in the green (fluorescein, 530 + 15 nm) and red (propidium, > 630 nm) wavelengths.

---

i. *Advantages*
(a) By detecting individual proteins and correlating their cellular content with the cell position in the cycle, this method yields well-defined results,

in terms of identification of the expression of particular genes at the translational level.

(b) A single measurement provides an estimate of the proliferative potential of the cell and, thus, indirect kinetic information.

ii. *Limitations*

(a) Immunocytochemical detection of intracellular protein cannot be truly quantitative. Accessibility of the epitope to the large immunoglobulin molecule, its possible conformational change during fixation or leakage from the cell during permeabilization, all affect stoichiometry of the measurement.

(b) Due to the different sensitivities of epitopes of the particular antigens, the fixation step cannot be standardized. Some proteins (e.g. PCNA) require fixation in alcohol rather than formaldehyde. Detection of every new antigen involves testing a variety of cell fixation/permeabilization procedures to ensure optimal preservation and accessibility of the epitope.

## 4. Flow cytometric kinetic methods

Whereas the cytometric methods presented above provide indirect kinetic information, e.g. based on the assumed relationship between the expression of a particular protein and cell-cycle progression, they do not reveal the actual kinetics. For example, when the cell is poisoned by a treatment with a high concentration of a particular drug, it may immediately stop in its cycle progression (regardless of the phase of the cycle), and remain 'frozen' in the cycle, with little change for a considerable period of time. Before it develops clear signs of apoptosis or necrosis, such a cell may be mistakenly assumed to be progressing through the cycle. Direct kinetic methods are needed to identify such cells. These methods also provide information about the duration of the particular phase or whole cycle.

### 4.1 Stathmokinetic methods

The stathmokinetic metaphase-arrest techniques were developed to estimate the rate of entry of cells into mitosis ('mitotic rate'), or the cell 'birth rate' (31). The approach is based on the use of agents such as colchicine or vinblastine, which affect the mitotic spindle of dividing cells. When added to exponentially growing cultures, these agents arrest cells in metaphase: the slope of the plot representing the accumulation of cells arrested in mitosis vs. time of stathmokinesis provides an estimate of the rate of cell entry to mitosis. Under certain conditions (asynchronous, exponential growth, known GF), this slope reveals the cell-cycle time (31, 32).

Flow cytometry allows one to quantify cells either in $G_2 + M$ (based on

DNA content alone) or in M (*Figure 5*; or by using M-phase specific antibodies). It is possible, therefore, to apply flow cytometry for analysis of the stathmokinetic assay. Because the percentage of cells in $G_1$, S, and $G_2$ can also be estimated, the analysis of the stathmokinesis by flow cytometry yields a plethora of kinetic parameters (31–33). *Protocol 8* and *Figure 7* illustrate such an application.

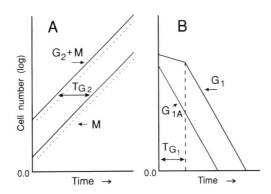

**Figure 7.** Graphic analysis of the stathmokinetic experiment. Numerous kinetic parameters can be estimated when an exponentially growing, asynchronous cell culture is treated with mitotic blocker and subsequently sampled to measure changes in the proportion of cells in M, and/or $G_2$ + M, $G_1$ and/or $G_{1A}$, at different times of the mitotic arrest The rate of entrance to M (or $G_2$ + M slope) reveals $T_C$ (31–33). The duration of $G_2$ ($T_{G2}$) is estimated based on the time-distance of these slopes. The rate of cell exit from $G_1$ and the early portion of $G_1$ ($G_{1A}$) has a typical stochastic component, characterized as a straight line a on log scale, revealing the half-time of cell residence in this compartment (18, 33). The point of inflection of the $G_1$ exit curve allows one to estimate the duration of $G_1$, excluding those cells with excessively long $G_1$ durations which represent the 'tail' of the stochastic distribution (33). The M arrest is usually delayed (A, broken line), as it takes time for the blocker to penetrate the cell and prevent the cells from completing mitosis.

**Protocol 8.** Stathmokinetic methods

1. Expose the exponentially, asynchronously growing cells continuously to 0.05 µg/ml of vinblastine (or Colcemid). Collect cells hourly, during a period equivalant to 30–40% of the cell-cycle time.

2. Fix the cells and process them for flow cytometry. Apply the flow cytometric methods which allow either identification of cells in $G_1$ vs. S vs. $G_2$ + M (univariate DNA content analysis), or multiparameter analysis, which allows discrimination between $G_2$ and M cells (e.g. as in *Figure 5*).

**Protocol 8.** *Continued*

3. Quantify the per cent of cells in the respective phases of the cycle, in each sample.
4. Plot the data as shown in *Figure 7*. Estimate the parameters of the cell cycle as described in the legend to this figure.

i. *Advantages*
(a) This short assay can provide several kinetic parameters simultaneously.
(b) The method is particularly suitable for analysis of the cell-cycle perturbations by cytostatic agents (33).

ii. *Limitations*
(a) The method, in its classical design (31), applies strictly to cells that grow exponentially, asynchronously and have GF = 1.0. It can be tailored, however, to accommodate exceptions (33).
(b) The effectiveness of the particular mitotic blocker may vary, depending on the cell type (e.g. in multidrug resistant cells) and the blocker may perturb the progression through phases other than M in some cell types. Pilot experiments, testing the choice of blocker and its optimal concentration, are often needed.

## 4.2 Methods based on BrdU incorporation

In principle, all of the approaches developed to study cell-cycle progression which are based on the use of [$^3$H]TdR can be adapted to BrdU. The latter is detected either immunocytochemically (18), or, cytochemically, as a result of its ability to quench fluorescence of several DNA fluorochromes following its incorporation to DNA (34).

Several approaches have been developed, based on either continuous or pulse-chase labelling with BrdU, to estimate various parameters of the cycle. Some are more applicable to *in vitro*, others to *in vivo* studies. The scope of this chapter does not permit the review of most of them.

The protocols for BrdU detection that are provided below, however, can be applied to different schemes of cell labelling. In the immunocytochemical method, DNA is partially denatured by acid or by heating to increase the accessibility of the incorporated BrdU to the antibody; the non-denatured sections of DNA are counterstained with PI. Cytochemical detection of the precursor is based on quenching of the fluorescence of Hoechst 33258 (HO 258) (34).

## 4.2.1 Immunocytochemical detection of BrdU

**Protocol 9.** Immunocytochemical detection of BrdU

1. After cell incubation with BrdU (usually 10–30 µg/ml of BrdU; keep cells under lightproof conditions), fix cells in 70% ethanol, on ice.
2. Centrifuge approximately $2 \times 10^6$ cells from the ethanol, suspend the cell pellet in 1 ml of HBSS containing 0.5% Tween 20 and 100 units of RNase. Incubate for 30 min at 37 °C. Centrifuge.
3. Resuspend the cell pellet in 1 ml of 2 M HCl containing 0.5% Tween. After 20 min at room temperature add 5 ml of HBSS, centrifuge, and drain thoroughly. Resuspend the cells again in 5 ml of HBSS (or even better in high molarity buffer, e.g. 0.2 M phosphate buffer at pH 7.6, to neutralize the acid, centrifuge, and drain thoroughly[a].
4. Suspend cell pellet in 100 µl of HBSS containing 0.1% Tween 20, 0.5% bovine serum albumin (diluting buffer) and the anti-BrdU antibody (follow the instructions provided by the supplier regarding the dilution, time, and temperature of the incubation). After incubation (generally 30–60 min at room temperature), add 5 ml of diluting buffer and centrifuge.
5. Suspend cell pellet in 100 µl of the diluting buffer containing properly diluted fluoresceinated secondary antibody (goat anti-mouse IgG). Incubate 30 min, add 5 ml of the diluting buffer, centrifuge[b].
6. Suspend the pellet in 1 ml of HBSS containing 10 µg/ml of propidium iodide. Analyse cells in the flow cytometer, using blue light excitation (488 nm) and measuring green (fluorescein, 530 + 15 nm) and red (PI, above 620 nm) fluorescence.

[a] Because differences in chromatin structure between various cell types affect the DNA denaturability, the DNA denaturation step may be modified for optimal results. Thus, either a higher HCl concentration (3 or 4 M) is required, or heat-induced DNA denaturation is sometimes preferred. In the latter case, in step 3, instead of treatment with acid, the cells are suspended in 1 ml of low ionic strength buffer (e.g. 0.1 mM cacodylate, pH 6.0) and heated for 10 min at 80 or 95 °C (35).
[b] Step five may be omitted when the primary antibody labelled with fluorescein is used.

*Figure 8* illustrates typical results when exponentially labelled cells are pulse labelled with BrdU. Kinetic parameters can be estimated by subsequently measuring the time of appearance of the cohort of BrdU-labelled cells in $G_2$, S, and $G_1$ (8, 36, 37).

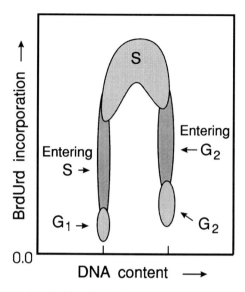

**Figure 8.** Populations of cells identified after pulse labelling with BrdU and bivariate analysis of BrdU immunofluorescence and DNA content. S-phase cells exposed to this precursor during the entire duration of the pulse are maximally labelled. The cells entering and leaving S phase during the pulse show variable labelling. Detection of BrdU labelled cells at different times after the pulse, simultaneous with their DNA content estimate, enables one to measure numerous kinetic parameters (37).

i. *Advantages*
(a) This is a very straightforward, direct estimate of cell kinetics, applicable both *in vivo* and *in vitro*.
(b) The assay is very sensitive: a 5 min pulse of BrdU is adequate to label the cells.

ii. *Limitations*
(a) The sensitivity of DNA *in situ* to denaturation varies between cell types, as a result of cell treatment with various drugs, etc. Thus, the step of DNA denaturation, critical for the detection of BrdU, cannot be fully standardized, and the BrdU-associated fluorescence is not stoichiometrically related to the extent of BrdU incorporation.
(b) BrdU, especially when cells are inadvertently exposed to light, may perturb their cycle.

### 4.2.2 Detection of BrdU by Hoechst 33258 fluorescence quenching

Several methods which utilize BrdU to analyse the cell-cycle progression are based on the detection of this precursor via quenching of HO 258 fluorescence (34) and on counterstaining of DNA with another dye, whose

*3: Mammalian cell-cycle analysis*

fluorescence is unaffected by BrdU (38–40). *Figure 9* schematically illustrates the character of the data generated by this approach.

---

**Protocol 10.** Detection of BrdU by Hoechst 33258 fluorescence quenching

1. Incubate cells continuously with BrdU (10–50 μM), for up to three cell-doubling times, in the dark[a]. After incubation the cells may be transferred to a solution containing culture medium, 10% serum, and 10% dimethyl sulfoxide, and stored at −20 to −40 °C, indefinitely.
2. Suspend $4-8 \times 10^5$ cells in 1 ml of a solution containing:
   - 100 mM Tris
   - 154 mM NaCl
   - 1 mM $CaCl_2$
   - 0.5 mM $MgCl_2$
   - 0.1% Nonidet P-40
   - 0.2% BSA
   - 1.2 μg/ml of HO 258, at pH 7.4, on ice
3. After 15 min, from a 100 × concentrated stock solution, add ethidium bromide (EB), to a final concentration of 1.5–2.0 μg/ml. The optimal EB concentration may vary depending on the cell type (38). Addition of RNase A (100 units per 1 ml) improves specificity of DNA staining with EB.
4. Measure the blue (HO 258) and red (EB) cell fluorescence 15 min after addition of EB, under excitation with UV light. The best reported results have been obtained with instruments using a mercury lamp as an excitation source (UG 1 filter).

[a] By comparison with cell proliferation (growth curves) in parallel, BrdU-untreated control cultures, check that this precursor does not have any undesirable effect on cell proliferation.

---

i. *Advantages*

(a) HO 258 fluorescence quenching is stoichiometrically related to BrdU incorporation. Thus, this method, can measure the degree of thymidine substitution by BrdU.

(b) The staining procedure is rapid. Since no centrifugations are required, few cells are lost.

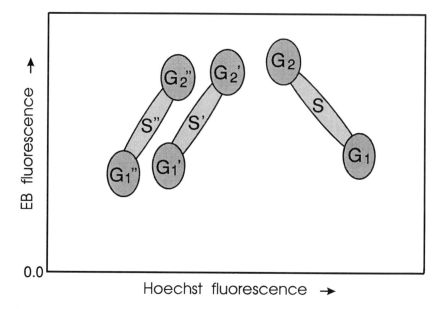

**Figure 9.** Analysis of cell-cycle progression by continuous cell labelling with BrdU and bivariate analysis of their stainability with Hoechst 33258 and ethidium bromide (EB). When cells are allowed to enter S phase (e.g. after mitotic stimulation) and progress through three generations in the presence of BrdU, each consecutive cell cycle (C, C', C") can be distinguished based on the progressively increasing degree of quenching of Hoechst fluorescence by the incorporated BrdU. Because fluorescence of EB is minimally affected by BrdU, it is possible to discriminate cells in $G_1$, S, and $G_2$ of the respective cycles.

ii. *Limitations*

(a) Despite appearing deceptively simple, this protocol requires very stringent conditions (cell number, HO 258 and EB concentration, excitation and fluorescence detection wavelengths).

(b) The method is suitable for the analysis of cells synchronized in $G_0$ or $G_1$ (e.g. lymphocytes or fibroblasts during mitogenic stimulation) and to measure their kinetic parameters for an extended period of time, equivalent to one or two cell-cycle times. It is of more limited value for studying cell kinetics in exponentially growing asynchronous cultures.

# Acknowledgements

Supported by NIH Grant RO1 CA 28704. I thank Dr Frank Traganos for his comments and suggestions, and Ms Irene Logsdon for her assistance in the preparation of the manuscript.

# References

1. Howard, A. and Pelc, S. R. (1953). *Heredity* (Suppl), **6**, 261–73.
2. Cleaver, J. E. (1967). *Thymidine metabolism and cell kinetics*. North Holland, Amsterdam.
3. Quastler, H. and Sherman, F. G. (1959). *Exp. Cell Res.*, **17**, 420–38.
4. Tannock, I. F. (1970). *Cancer Res.*, **30**, 2470–6.
5. Steel, G. G. (1977). *Growth kinetics of tumours: cell population kinetics in relation to growth and treatment of cancer*. Clarendon Press, Oxford.
6. Melamed, M. R., Lindmo, T., and Mendelsohn, M. L. (ed.) (1990). *Flow cytometry and sorting*. Wiley–Liss, New York.
7. Darzynkiewicz, Z. and Crissman, H. A. (ed.) (1990). *Flow cytometry*. Academic Press, San Diego.
8. Dean, P. N., Dolbeare, F., Gratzner, H., Rice, G., and Gray, J. W. (1983). *Cell Tissue Kinet.*, **17**, 427–36.
9. Rogers, A. W. (1973). *Techniques of autoradiography*. Elsevier, North Holland, Amsterdam.
10. Maurer, H. R. (1981). *Cell Tissue Kinet.*, **4**, 111–20.
11. Simpson-Herren, L. (1987). In *Techniques in cell cycle analysis*, (ed. W. Gray and Z. Darzynkiewicz), pp. 1–30. Humana Press, Clifton, New Jersey.
12. Shackney, S. E. and Ritch, P. S. (1987). In *Techniques in cell cycle analysis* (ed. J. W. Gray and Z. Darzynkiewicz), pp. 31–46. Humana Press, Clifton, New Jersey.
13. Mendelsohn, M. L. (1962). *J. Natl Cancer Inst.*, **28**, 1015–29.
14. Braunschweiger, P. G. (1987). In *Techniques in cell cycle analysis* (ed. J. W. Gray and Z. Darzynkiewicz), pp. 47–72. Humana Press, Clifton, New Jersey.
15. Schiffer, L. M., Markoe, A. M., and Nelson, J. S. R. (1976). *Cancer Res.*, **36**, 2415–18.
16. Okada, S. (1967). *J. Cell Biol.*, **34**, 915–16.
17. Hedley, D. W. (1990). In *Flow cytometry* (ed. Z. Darzynkiewicz and H. A. Crissman), pp. 139–48. Academic Press, San Diego.
18. Darzynkiewicz, Z., Sharpless, T., Staiano-Coico, L., and Melamed, M. R. (1980). *Proc. Natl Acad. Sci. USA*, **77**, 6670–96.
19. Darzynkiewicz, Z., Traganos, F., and Melamed, M. R. (1980). *Cytometry*, **1**, 98–108.
20. Darzynkiewicz, Z., Traganos, F., Sharpless, T., and Melamed, M. R. (1977). *Cancer Res.*, **37**, 4635–40.
21. Clevenger, C. V., Epstein, A. L., and Bauer, K. D. (1987). *J. Cell Physiol.*, **130**, 335–43.
22. Bauer, K. D. (1990). In *Flow cytometry* (ed. Z. Darzynkiewicz and H. A. Crissman), pp. 235–47. Academic Press, San Diego, Calif.
23. Bruno, S., Gorczyca, W., and Darzynkiewicz, Z. (1992). *Cytometry*, **13**, 496–501.
24. Gerdes, J., Lemke, H., Baisch, H., Wacker, H.-H., Schwab, U., and Stein, H. (1984). *J. Immunol.*, **133**, 1710–15.
25. Kurki, P., Ogata, K., and Tan, E. M. (1988). *J. Immunol. Meth.*, **109**, 49–59.
26. Celis, J. E., Fey, S. J., Larsen, P. M., and Celis, A. (1984). *Proc. Natl Acad. Sci. USA*, **81**, 3128–32.

27. Freeman, J. W., McGrath, P., Bonada, V., Selliah, N., Ownby, H., Maloney, T., Busch, K., and Busch, H. (1991). *Cancer Res.*, **51**, 1973–78.
28. Lachman, H. M. and Skoultchi, A. I., (1984). *Nature* (London), **310**, 592–4.
29. Negri, C., Chiesa, R., Cerino, A., Bestagno, M., Sala, C., Zini, N., Maraldi, N. M., and Astaldi Ricotti, G. C. B. (1992). *Exp. Cell Res.*, **200**, 452–9.
30. Wang, E. (1985). *J. Cell Biol.*, **101**, 1695–701.
31. Puck, T. T. and Steffen, J. (1963). *Biophys. J.*, **3**, 379–97.
32. Wright, N. A. and Appleton, D. R. (1980). *Cell Tissue Kinet.*, **13**, 643–63.
33. Darzynkiewicz, Z., Traganos, F., and Kimmel, M. (1987). In *Methods in cell cycle analysis* (ed. J. W. Gray and Z. Darzynkiewicz), pp. 291–336. Humana Press, Clifton, New Jersey.
34. Latt, S. A. (1973). *Proc. Natl Acad. Sci. USA*, **70**, 3395–9.
35. Moran, R., Darzynkiewicz, Z., Staiano-Coico, L., and Melamed, M. R. (1985). *J. Histochem. Cytochem.*, **33**, 821–7.
36. Dolbeare, F., Kuo, W.-L., Besker, W., Vanderlaan, M., and Gray, J. W. (1990). In *Flow cytometry* (ed. Z. Darzynkiewicz and H. A. Crissman), pp. 207–16. Academic Press, San Diego.
37. Gray, J. W., Dolbeare, F., and Pallavicini, M. G. (1990). In *Flow cytometry and sorting* (ed. M. R. Melamed, T. Lindmo, and M. L. Mendelsohn), pp. 445–68. Wiley–Liss, New York.
38. Bohmer, R. M. and Ellwart, J. (1981). *Cell Tissue Kinet.*, **14**, 653–60.
39. Poot, M., Hoehn, H., Kubbies, M., Grossman, A., Chen, Y., and Rabinovitch, P. S. (1990). In *Flow cytometry* (ed. Z. Darzynkiewicz and H. A. Crissman), pp. 185–98. Academic Press, San Diego.
40. Rabinovitch, P. S., Kubbies, M., Chen, Y. C., Schindler, D., and Hoehn, H. (1988). *Exp. Cell Res.*, **174**, 309–18.

# 4

# Analysis of the cell cycle in *Saccharomyces cerevisiae*

BRUCE FUTCHER

## 1. Introduction

The basic mechanisms of cell-cycle control seem to be similar in all eukaryotes. The yeasts *S. cerevisiae* and *Schizosaccharomyces pombe* are two of the simplest and most powerful model systems for studying eukaryotic cell-cycle control. Progress has been rapid, and the lessons learned have often been directly relevant to mammals and other eukaroyotes. Some methods useful for studying the yeast cell cycle are described here. However, these can be applied only in the context of the usual techniques of yeast molecular genetics and cell biology (1, 2). Similar methods for *S. pombe* are described in Chapter 5, while methods for preparing synchronous cultures of both yeasts are described in Chapter 2.

This chapter deals with two kinds of investigations. Section 3, 'Monitoring the cell cycle', is directed at situations in which one has a yeast mutant, a drug, or some other condition that kills or arrests yeast cells. The problem then is to find out (a) whether the arrest is due to a cell-cycle specific defect; (b) if so, where in the cycle the cells have arrested; and (c) which cell-cycle specific process has been affected. Advice is given on how to arrest cells in preparation for cell-cycle analysis (Section 3.2). The arrest is considered as cell-cycle specific arrest if cell enlargement continues (Section 3.3), and if there is a uniform terminal morphology as assayed by budding index and cell morphology (Section 3.4), DNA content (Section 3.5), microtubule morphology (Section 3.6), and nuclear morphology (Section 3.7). These assays also give a preliminary indication of the point of arrest. With this information in hand, more detailed information can be obtained using reciprocal shift experiments (Section 3.8) and protein kinase assays (Section 3.9). This narrows the range of possible affected processes.

Section 4, 'Analysis of heterologous genes in yeast', is directed at investigations in which yeast are used as hosts to analyse heterologous cell-cycle genes. This is a relatively new area, and so it is more difficult to say which approaches will or will not work. The section describes the cloning of new heterologous cell-cycle genes by complementation of yeast cell-cycle mutations (Section 4.1), the testing and classification of heterologous genes by complementation of yeast mutations (Section 4.2), and very briefly, methods for identifying new yeast genes homologous to known mammalian genes (Section 4.3).

## 2. The cell cycle of *S. cerevisiae*

A normal *S. cerevisiae* cell cycle and its landmark events is shown in *Figure 1* (adapted from ref. 3). 'Start', in $G_1$ phase, is an event which commits cells to a round of division. After 'Start' has occurred, environmental factors such as mating pheromone and poor nutrients are no longer able to prevent division. 'Start' is also the co-ordination point for several semi-independent pathways of cell-cycle events (see *Figure 2*). Budding is an obvious visible sign that a cell has passed 'Start', and, in general, the larger the bud, the closer the cell is to division. The short mitotic spindle forms shortly after 'Start'. As in other fungi, the nuclear envelope never breaks down. Because individual chromosomes are not visible by light microscopy, there is no assay for metaphase in *S. cerevisiae*. Anaphase can be seen by staining the DNA with the fluorescent dye DAPI (see Section 3.3.7 below).

The rate-limiting process for progress through the cycle is growth and protein synthesis (4). Cells can pass through 'Start' only when they have achieved a certain critical size, which is about 35 $\mu m^3$ for wild-type haploid

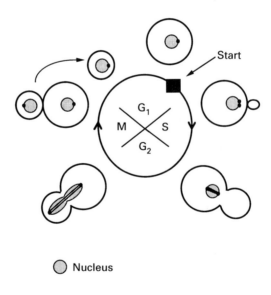

○ Nucleus

• Spindle pole body

**Figure 1.** Normal *S. cerevisiae* cell cycle and its landmark events.

## 4: Analysis of cell cycle in S. cerevisiae

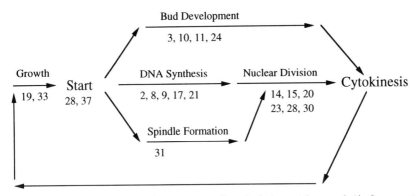

**Figure 2.** A dependency map of the yeast cell cycle (adapted from ref. 3). Consecutive arrows represent dependent events (e.g. DNA synthesis is dependent on 'Start'; nuclear division is dependent on DNA synthesis). Parallel arrows represent independent pathways (e.g., bud development is independent of DNA synthesis). Numbers below the arrows indicate examples of *cdc* mutants defective in a particular process (e.g. *cdc*2, *cdc*8, *cdc*9, etc. are defective in DNA synthesis).

cells in rich medium. Slowly growing cells have a relatively long generation time, and the extra cycle time is spent mostly in $G_1$ phase, before 'Start'. Daughter cells (cells which have never produced a bud) are smaller than mother cells, and so have longer $G_1$ phases. This mother–daughter asymmetry increases at lower growth rates (5, 6).

In the early 1970s, Hartwell and co-workers isolated a large number of cell division cycle (*cdc*) mutants (reviewed in ref. 3). These are temperature-sensitive lethal mutants which are unable to complete some cell-cycle-specific event, and so arrest with a terminal phenotype that is at least 80% homogeneous. For instance, *cdc*15 causes a terminal arrest at a late stage of mitosis: almost all the cells are large, budded, have the 2N (replicated) DNA content, and have long mitotic spindles. There are now mutants in more than 60 different *cdc* genes affecting most cell-cycle specific processes.

Analysis of these mutants has led to a dependency map of the cycle (*Figure 2*). After 'Start', the cell cycle consists of several semi-independent pathways, such as DNA replication, bud formation, and spindle formation. Some events depend on certain earlier events because of 'checkpoints' which monitor the completion of the earlier events (7). For instance, the dependence of nuclear division on DNA replication is maintained by a checkpoint system.

## 3. Monitoring the cell cycle

### 3.1 Introduction

The initial analysis of non-dividing or aberrantly dividing cells revolves around two questions: Is there a defect in some process specific to the cell cycle? If so, which process is defective? The main indication of a cell-cycle specific defect is a uniform arrest morphology, and the techniques described below are intended for these situations. Several different cell-cycle parameters must be assayed to accurately describe the state of the cells.

Whether a defect is 'specific' to the cell cycle sometimes becomes a question of definition and semantics. For instance, since 'Start' requires growth to a critical size, many mutations that slow down growth and protein synthesis prevent 'Start', and give a homogeneous Cdc⁻ $G_1$-arrest phenotype. These mutants include *cdc19* (which encodes pyruvate kinase), *cdc35* (adenylate cyclase), *cdc33* (mRNA cap-binding protein, eIF-4E), and *cdc60* (leucyl tRNA synthetase). Although these certainly qualify as *cdc* mutants, processes such as glycolysis and protein synthesis are generally not considered 'specific' to the cell cycle.

Similarly, difficulties are often encountered in the analysis of proteins with multiple roles. Phosphatases, for example, may be required for one or more essential processes specific to the cycle, and also for essential non-specific processes. Cells defective for such multi-role proteins may give a heterogeneous arrest: that is, different cells in the culture may arrest for different reasons, and so in different states. It is extremely difficult to disentangle any cell-cycle-specific defect from the non-specific defect in such cases. In fact, attempts to do so can waste a tremendous amount of time, since if there is a heterogenous arrest, there is usually no good reason to think there is a specific defect in the cell cycle.

## 3.2 Preparation of cells arrested at cell-cycle blocks

### 3.2.1 Growth of cells

Typically, cells are cultured in a liquid medium containing glucose as a carbon source to give rapid growth. At lower growth rates, the small daughter cells have longer generation times than their larger mothers, and so appropriate arrest times become long and difficult to predict. Cells in colonies on plates are heterogeneous in physiological state, and so are unsuitable. For each type of strain and medium, cultures will grow to a characteristic final density: for instance, a final density of $2 \times 10^8$ cells/ml is typical at stationary phase in YEPD medium (1% yeast extract, 2% peptone, 2% glucose). Cells still grow rapidly ('exponential' or 'log' phase) until they achieve 10% or more of their final density (e.g. $2 \times 10^7$ cells/ml in YEPD). Arrested cells are prepared by shifting exponentially growing cells to the restrictive condition and incubating for 1.5–2 generation times to ensure that all the cells in the culture reach the arrest point. Analysis of the arrest phenotype should be performed at this time. After long times at the restrictive temperature, some cell lysis often occurs, and phenotypes may become bizarre and uninterpretable. Drug-arrested cells are often able to overcome the arrest, but it may take many hours.

### 3.2.2 Temperature shifts with *cdc* mutants

*cdc* mutants can be arrested by shifting the temperature from 23 °C (permissive temperature) to 37 °C (restrictive temperature). The use of an air incubator rather than a waterbath slows the rate of temperature change and helps to avoid heat-shock effects.

### 3.2.3 Drug arrests

Arresting cells at particular cell-cycle stages with drugs is a useful way to synchronize cells. It is also necessary for reciprocal shift experiments (see Section 3.8 below).

*i. α-factor arrest*

The mating pheromone α-factor is produced by *MAT*α cells, and arrests *MAT*a cells at 'Start' by inhibiting some aspect of Cln-Cdc28 function (see Section 3.9). The pheromone has no effect on *MAT*α cells because they lack the appropriate receptor. α-factor is a simple 13 amino acid peptide (TRP–HIS–TRP–LEU–GLN–LEU–LYS–PRO–GLY–GLN–PRO–MET–TYR, form. wt. 1684), and is soluble in ethanol. The peptide can be purchased (Sigma T6901), but if quantities greater than 15 mg are to be used, it is much more economical to have the peptide synthesized. Purification (e.g. two HPLC C18 columns) is necessary to remove antagonistic by-products of the synthesis.

Cells have multiple mechanisms for recovering from α-factor arrests. One of the most important is the Bar1 protease, which specifically destroys α-factor. Because of this protease, the concentration of α-factor required to arrest cells for a given time depends on the cell concentration (9). At about $1 \times 10^7$ cells/ml, 10 μg/ml (6 μM) α-factor will arrest the cells for about 3 h. The Bar1 protease may be less active at low pH, so α-factor arrest experiments are often performed with medium buffered with sodium succinate to about pH 5 (or less). About 100-fold less α-factor is required to arrest cultures of *bar1* mutant cells.

Cells are released from the arrest by harvesting them by centrifugation (4000 *g*, 5 min, room temperature), washing once in fresh medium, and resuspending them in fresh medium. The first buds appear about 40 min later.

*ii. Hydroxyurea*

Hydroxyurea is an inhibitor of ribonucleotide reductase, an essential enzyme of DNA precursor metabolism, and so inhibits DNA synthesis, arresting cells in S phase. It is soluble in water, and is used at a final concentration of 0.1–0.2 M. Cells are released by washing with fresh medium, and resuspending in fresh medium.

*iii. Nocodazole*

Nocodazole is an inhibitor of microtubule polymerization, and so prevents mitosis. It is soluble in DMSO (dimethyl sulfoxide), and is used at a final concentration of 15 μg/ml. Jacobs et al. (10) used medium containing 1% DMSO to keep the nocodazole soluble. These workers reported that either too little or too much nocodozale failed to arrest cells, and also that there was batch-to-batch variability in the amount needed. Cells are released by washing with fresh medium, and resuspending in fresh medium.

## 3.3 Measurement of cell size

Since most *cdc* mutations that block division do not stop growth or protein synthesis, *cdc* mutant cells generally increase in mass and volume during cycle arrest. Cells held at a cell-cycle block for 2–4 generation times appear much larger than unblocked controls. If blocked cells do not become enlarged, the genetic defect may be in general metabolism (e.g. glycolysis, protein synthesis) rather than in a cell-cycle-specific process. These mutants arrest as relatively small unbudded cells that increase in size only very slowly, if at all.

To measure cell size, take microphotographs, and measure in arbitrary units the length ($L$) and width ($W$) (taken as the maximum width perpendicular to the length) of the cells with the smallest buds (to derive the 'critical size' for budding). The volume of each cell is then:

$$V = (\pi/6)LW^2, \tag{1}$$

assuming the cell to be a prolate ellipsoid. An absolute volume can be obtained if objects of known size (e.g. beads from Coulter Electronics) are also photographed. Cell size distributions (in arbitrary units) can be visualized by forward light scattering using a flow cytometer. The best instrument for cell-size measurements is a Coulter Channelyzer, which measures the volume of electrolyte displaced by cells and other particles, and gives cell-size distributions in absolute units.

Size measurements are useful for mutants that alter the critical size for 'Start'; those that reduce critical size produce small cells, and are called *whi* ('wee') mutants (8). These generally have a decreased ratio of $G_1$ to $G_2$ cells, but have normal generation times (since the cell's ability to synthesize protein is unaltered).

## 3.4 Budding

### 3.4.1 Assaying budding

The cell-cycle state of a culture is most easily characterized by the percentage of budded cells. A bud signals that a cell is past 'Start', the $G_1$-phase commitment point for division. Sonicate cells, then examine them by phase-contrast or differential-interference-contrast (Nomarski) microscopy. Large buds are obvious; but the first trace of a bud is a very tiny protrusion that can

4: *Analysis of cell cycle in S. cerevisiae*

be hidden on the far side of a cell. Rapidly cycling populations have about 50% budded and 50% unbudded cells. If blocked cells have a uniform morphology (i.e. greater than 80% budded, or greater than 80% unbudded) then the arrest is probably at a specific point in the cell cycle.

### 3.4.2 Clumpiness

Most strains are 'clumpy' to some extent: daughter cells stay attached to their mothers for some time after cell division. Before assaying budding, separate mothers from recently abcised daughters by brief (2–5 sec) sonication. Some strains are flocculant, meaning that cells (not necessarily mothers and daughters) stick together. Flocculation depends on calcium and perhaps other divalent ions, and so can be stopped by washing the cells once or twice in 10 mM EDTA or EGTA, and sonicating.

### 3.4.3 Quantitation

For quantitative measurements it is best if the person counting the proportion of budded cells does not know which sample is which. This can make a shockingly large difference to the result.

### 3.4.4 Sick cells

Cultures of sick cells often have a slightly abnormal distribution of budded vs. unbudded cells, usually for some complicated and indirect reason. One should be wary of concluding that a mutation causing a slightly abnormal percentage of budded cells is necessarily a cell-cycle mutation.

## 3.5. DNA content and FACS analysis

### 3.5.1 Measuring proportions of 1N and 2N cells by flow cytometry

Flow cytometry measures DNA content per cell, and so distinguishes between $G_1$ and $G_2/M$ cells. In a typical procedure, the cells are permeabilized and fixed with 70% ethanol, cellular RNA is destroyed by treatment with RNase A at an elevated temperature, the DNA is stained with a fluorescent dye, and then the fluorescence per cell is measured with a flow cytometer (FACS, fluorescence activated cell sorter). The general nucleic acid dye propidium iodide is normally used. Since more then 95% of the nucleic acid in a yeast cell is RNA, it is essential that this RNA be destroyed with RNase A before staining with this dye. This is aided by high temperatures partly because the RNase is more active, and partly because yeast cells often contain a considerable amount of double-stranded RNA, which is not readily digested by RNase A at low temperatures. *Protocol 1* gives a simple procedure; more complex procedures giving better resolution of 1N and 2N peaks have been described (11). A similar procedure for fission yeast cells is detailed in Chapter 5.

**Protocol 1.** FACS analysis

*Materials*

- 95% ethanol
- 50 mM sodium citrate, pH 7
- RNase A: 10 mg/ml in 10 mM Tris, 1 mM EDTA, pH 8.0
- Propidium iodide: 16 µg/ml, in 50 mM sodium citrate, pH 7

*Method*

1. Grow yeast to exponential phase (e.g. $2 \times 10^7$ cells ml), or block at the desired point, then place samples on ice.
2. Harvest $1 \times 10^7$ cells by centrifugation (e.g. 4000 $g$, 5 min, 4 °C). Resuspend in 3 ml water. Sonicate briefly (2–5 sec).
3. Vortex the cell suspension. While vortexing, slowly add 7 ml of 95% ethanol.
4. Incubate overnight at 4 °C (or a shorter time at room temperature).
5. Harvest cells by centrifugation (e.g. 4000 $g$, 5 min.), and resuspend in 5 ml of 50 mM sodium citrate, pH 7. Centrifuge again, and resuspend final cell pellet in 1 ml 50 mM sodium citrate, pH 7. Sonicate briefly.
6. Add RNase to a final concentration of 0.25 mg/ml. Incubate at 50 °C for 1 h.
7. Add 50 µl of 20 mg/ml proteinase K. Incubate at 50 °C for 1h. This step gives improved resolution of 1N and 2N peaks and is recommended although not essential.
8. At room temperature, add 1 ml of 50 mM sodium citrate, pH 7, containing 16 µg/ml propidium iodide.
9. Analyse with a fluorescence activated cell sorter.

### 3.5.2 Interpretation of flow cytometry and potential difficulties

(a) A culture of rapidly cycling wild-type cells should give 1N and 2N peaks of approximately equal area. Such a culture should be used as a control for any experiment.

(b) Since the DNA content of a yeast cell is very low, a well-maintained instrument and a knowledgeable operator are important.

(c) There is always some fluorescence from the cytoplasm and/or cell wall. Larger cells have more of this background fluorescence, and so their peaks are shifted to the right. In extreme cases, the 1N peak of very large arrested haploid cells may be shifted to the position of the 2N peak of wild-type cells, thus giving a misleading result. The position of the peak cannot be considered an absolute measure of the amount of DNA in the cell, and should be checked against appropriate controls and tested after arrests of different times.

(d) Some cells are resistant to permeablization and fixation (e.g. stationary-phase cells, spores). Stained cells may be a selected subfraction of the population, and so may give a misleading result. This can be checked by fluorescence microscopy: all the cells in the sample should fluoresce.

(e) A flow cytometer can measure cell size (by forward light scattering) as well as DNA content, and these often complement. For instance, if a cycling population has more 2N cells than normal, and the cells are large, it indicates a slow step in $G_2$ or M. On the other hand, if the proportion of 2N cells is larger than normal, but the cells are small, it indicates a fast step at 'Start' (i.e. a reduced critical size).

(f) Mitochondrial DNA provides part of the signal and broadens the peaks. Since the mitochondrial DNA usually continues to replicate, the peaks shift to the right (to higher fluorescence) as incubation at the block increases. It is sometimes desirable to create strains that lack mitochondrial DNA by growing the yeast in sublethal concentrations of ethidium bromide, and picking petite mutants (*Protocol 2*).

---

**Protocol 2.** Generating $\varrho^0$ mutants

*Materials*
- 10 mg/ml ethidium bromide in water
- YEPD plates: 1% yeast extract, 2% peptone, 2% glucose, 2% agar
- YEPGE plates: 1% yeast extract, 2% peptone, 2% glycerol, 2% ethanol, 2% agar

*Method*
1. Spread about 500 cells on a YEPD plate.
2. Scrape out a depression of about 50 μl volume at the centre of the plate.
3. Put about 50 μl of 10 mg/ml ethidium bromide in the depression.
4. Incubate the plate at 30 °C for about 2 days.
5. Choose several of the small white colonies near the well of ethidium bromide. Restreak these on (a) a YEPD plate; and (b) a YEPGE plate. Authentic $\varrho^0$ mutants should grow on the YEPD plate, but be bone-

**Protocol 2.** *Continued*

white rather than creamy in colour, and should fail to grow on the YEPGE plate[a].

[a] Step 5 establishes that a strain is $\rho^-$ (i.e., lacks mitochondrial function), but does not establish that it is $\rho^0$ (i.e. totally lacks mitochondrial DNA. Nevertheless, this is almost always true of mutants made in this way.

## 3.6 Microtubule morphology

Immunofluorescence staining for tubulin is an excellent way to discern the state of a cell (12, 13). $G_1$-phase cells have a few bundles of microtubules emanating from the cytoplasmic face of the spindle pole body (which is embedded in the nuclear envelope). Later in the cycle, approximately at the time of S phase, there is a short spindle spanning the spherical nucleus. This short spindle has a distinctive bar-like appearance. After the beginning of anaphase, the nucleus is elongated, and is spanned by a thinner spindle.

Detailed protocols for tubulin staining have been published (1, 14). An anti-yeast α-tubulin monoclonal antibody created by Kilmartin, YOL1/34, is commercially available from Accurate Chemical and Scientific Corp. (cat. #MAS078) and gives excellent results.

## 3.7 Staining DNA with DAPI

The fluorescent dye 4′,6-diamidino-2-phenylindole (DAPI) is fairly specific for DNA. It can be used to see whether cells have entered anaphase. It is particularly effective when used in combination with tubulin staining, so that both the spindle morphology and the nuclear morphology of a single cell can be seen. Live cells can be stained simply by adding DAPI to a final concentration of 1–10 µg/ml to an exponential-phase culture in YEPD, and incubating at 30 °C for 1 h. For stronger staining, fix and permeabilize the cells, and then stain as in *Protocol 3*. For cells that have already been fixed and stained with an antibody (e.g. anti-tubulin), stain for 5 min, with 1 µg/ml DAPI in phosphate-buffered saline (PBS) containing 10 mg/ml bovine serum albumin, and then wash once with the PBS/BSA.

**Protocol 3.** DAPI staining

*Materials*

- 95% ethanol
- Formaldehyde solution
- 1 mg/ml aqueous solution of 4′,6-diamidino-2-phenylindole (DAPI)

*4: Analysis of cell cycle in S. cerevisiae*

*Method*
1. Sonicate cells. Harvest about $2 \times 10^7$ cells by centrifugation (e.g. 4000 *g*, 5 min, 4 °C)
2. Follow either (a) or (b).
   (a) *Fix with ethanol*: resuspend in 0.3 ml water, and add 0.7 ml of 95% ethanol. Incubate for about 45 min at room temperature. (Shorter times will also work, but may produce weaker staining.)
   
   **or**
   
   (b) *Fix with formaldehyde*: resuspend in 1 ml 3% formaldehyde, 50 mM sodium phosphate, pH 7; incubate for about 45 min at room temperature.
3. Pellet cells by centrifugation (4000 *g*, 5 min, room temperature). Resuspend cells in 1 ml distilled water. Pellet cells again.
4. Resuspend the cells in 100 μl of 1–10 μg/ml DAPI (in water). Stain for about 10 min.
5. Pellet cells and resuspend in 1 ml distilled water. Sonicate briefly if necessary.
6. Observe by UV fluorescence. The optimum excitation wave-lengths for DAPI are 340–365 nm, and the strongest emission wavelengths are 450–488 nm.

## 3.8 Reciprocal shift experiments

An unknown cell-cycle arrest point can be ordered with respect to a known arrest point by reciprocal shift experiments. Two different reversible blocks induced by different conditions are required. For example, the primary defect of a new *cdc* mutant, *cdc*x, might be ordered with respect to DNA synthesis as follows (*Figure 3*): In the first reciprocal shift experiment, cells are arrested at the *cdc*x block point by holding at the restrictive temperature for at least one generation time. After arrest, hydroxyurea (which prevents DNA synthesis) is added, and the culture is shifted down to the permissive temperature. After one generation time of incubation, the cell number is counted—it would either have doubled from the number at the first arrest, or have stayed the same. In the reciprocal experiment, cells are arrested in S phase using hydroxyurea, shifted to the restrictive temperature for *cdc*x, and the hydroxyurea washed out, still at the restrictive temperature. After one generation time of incubation, the cell number will either double or stay the same. If the cell number doubled in the first experiment but not the second (as in Figure 3), then the *cdc*x arrest point is after DNA synthesis but before cell division. If the cell number stayed the same in the first experiment but doubled in the second, then *cdcx* arrest point is after division, but before S phase. If the cell number stayed the same in both experiments, then *cdcx* arrests cells in S phase.

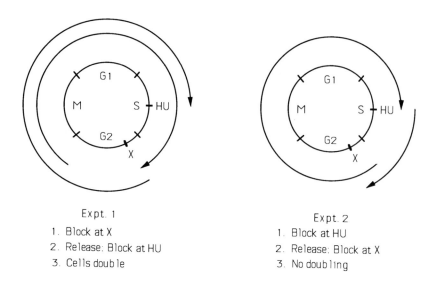

Figure 3. Reciprocal shift experiments. (For details see text.)

## 3.9 Histone H1 kinase assays

'Start', spindle formation, spindle elongation, and other cell-cycle events require the activity of the Cdc28 protein kinase (known as Cdc2 in *S. pombe* and other organisms). The Cdc28 polypeptide itself is probably not active as a protein kinase. Rather, the activity requires a complex of Cdc28 with a regulatory protein called a cyclin. One set of cyclins ($G_1$ cyclins) regulates Cdc28 activity at 'Start', and a separate set (mitotic cyclins) regulates Cdc28 activity through $G_2$ and mitosis. Histone H1 is the substrate used to assay the Cdc28 activity. The level of histone H1 kinase activity at 'Start' is much lower than at mitosis. There are several different ways to assay Cdc28 histone H1 protein kinase activity.

(a) A crude cell extract can be assayed (15, 16). Much of the histone H1 kinase activity in a yeast cell is Cdc28-dependent, so even a crude extract often gives a fair reflection of the amount of Cdc28 activity. This method is unable to distinguish between the various forms of Cdc28 activity; this may be advantageous in some experiments (for instance, when it is not known which form is relevant), and disadvantageous in others. Since background levels of histone H1 kinase activity are high in this assay (because of other protein kinases in the cell) it is relatively insensitive, and only useful as a measure of mitotic levels of activity.

(b) Cdc28 kinase complexes can be specifically precipitated with 'p13$^{suc1}$' beads. These are agarose beads with the Suc1 protein covalently crosslinked to them (17, 18). The 13 kDa *S. pombe* Suc1 protein (and its *S. cerevisiae* homologue, Cks1) binds to Cdc2 or Cdc28 complexes. It is widely believed that Suc1 binds directly to the Cdc28 polypeptide, but it is also possible that it binds instead to the cyclin or to some other component of the complex. Cyanogen bromide activated beads with crosslinked Suc1 are commercially available (Oncogene Sciences, NY). Since cyanogen bromide activation of agarose can lead to some non-specific protein binding, other crosslinking agents may give an even more specific reagent. Suc1 beads are used simply by incubating them in a cell extract at 4 °C for about an hour, much as an immunoprecipitation would be done. If necessary, the activity can be eluted from the beads using excess soluble Suc1. A Suc1 bead assay is much more sensitive and specific than an assay based on a crude extract. A possible disadvantage is that there may be forms of the kinase complex that fail to bind to Suc1 beads, though all complexes tested so far do bind, with the possible exception of the Cln3–Cdc28 complex. A second possible disadvantage is that there may be other proteins that bind.

Suc1 beads are the method of choice for obtaining quantities of semi-purified mitotic Cdc28 kinase activity. For maximum activity, cells should be arrested in mitosis with nocodazole. Cdc28 complexes can then be obtained from a cell extract.

(c) Kinase complexes can be precipitated with an antibody against the Cdc28 or against one of the cyclin subunits. Even an anti-Cdc28 antibody usually does not precipitate all kinds of complexes, as the epitope will not necessarily be exposed. Antibodies against different proteins and epitopes often inhibit the kinase activity to different extents.

*Protocol 4* below was developed to assay the Cdc28 kinase activity immunoprecipitated with the Cln1, Cln2, and Cln3 $G_1$ cyclins. These activities are extremely weak, and so this rather complex protocol is designed to maximize activity and minimize background. The antibody used is a monoclonal antibody produced by the 12CA5 cell line (commercially available from the Berkeley Antibody Company (BAbCo). The antibody reacts with a nine amino acid epitope (YPYDVPDYA) which is tandemly triplicated and used to tag the Clns at their C-termini (19). With minor alterations, the protocol can be adapted to other antibodies, or to p13$^{suc1}$ beads.

**Protocol 4.** Assay for Cln-associated histone H1 kinase activity

*Materials*
- *S. cerevisiae* strain bearing an epitope-tagged cyclin or *CDC*28 gene
- 12CA5 monoclonal antibody (ascites fluid)
- Competitor peptide (YPYDVPDYA) for negative control
- Lysis buffer:
    50 mM Tris–HCl, pH 7.5
    250 mM NaCl
    50 mM NaF
    5 mM EDTA
    0.1% NP-40
    1 mM PMSF
    0.6 mM dimethylaminopurine
    1 µg/ml leupeptin
    1 µg/ml pepstatin
    10 µg/ml TPCK
    10 µg/ml soybean trypsin inhibitor
- Kinase buffer:
    50 mM Tris–HCl
    10 mM $MgCl_2$
    1 mM DTT
- Unlabelled ATP
- [$\gamma$-$^{32}$P]ATP, 3000 Ci/mmol
- Acid-washed glass beads (425–600 $\mu m^3$, Sigma)
- Protein A–Sepharose beads
- 2 × Laemmli SDS–PAGE sample dye (120 mM Tris–HCl, pH 6.8, 4% SDS, 20% glycerol, 10% $\beta$-mercaptoethanol, 0.04% bromophenol blue)
- SDS–polyacrylamide gel
- Coomassie blue
- Autoradiography equipment

*Method*
*NB* Perform steps 2–16 at 0–4 °C unless otherwise indicated.
1. Use about 2 × $10^9$ growing cells for each extract; fewer cells give too low a protein concentration.
2. Harvest cells by centrifugation (e.g. 4000 *g*, 5 min, 0–4 °C) and wash with ice-cold water. Wipe off excess water.

## 4: Analysis of cell cycle in S. cerevisiae

3. Resuspend the pellet in a volume of lysis buffer similar to the volume of the cell pellet (about 200 μl per $2 \times 10^9$ cells). Excessive dilution impairs activity[a].

4. Transfer the cell suspension to a cold tube. The size and shape of the tube is important; $13 \times 100$ mm glass test tubes, 2 ml polypropylene microcentrifuge tubes, and $12 \times 35$ mm flat-bottom glass vials all work well. Ordinary 1.5 ml polypropylene microcentrifuge tubes work poorly, apparently because of the narrow, conical bottom. Add acid-washed glass beads (425–600 $\mu m^3$, Sigma; smaller beads do not work) to the cell suspension until the beads come to about 1 mm below the meniscus. The ratio of bead volume to liquid volume is important.

5. Vortex at top speed four times for 20–30 sec each, interspersed with cooling in an ice–water bath. Breakage may be checked by phase-contrast microscopy (broken cells lose refractility and may appear black). About 50% breakage should be achieved after 40 sec, and 75–90% after 80 sec.

6. Transfer the liquid phase to a 1.5 ml microcentrifuge tube on ice using a 1 ml micropipette tip (smaller tips clog with beads). Rinse the remaining beads with 100 μl of lysis buffer, and pool.

7. Centrifuge at top speed in a microcentrifuge (15 000 $g$) for 2 min.

8. Transfer liquid to a new 1.5 ml centrifuge tube on ice. Centrifuge for 10 min at top speed (15 000 $g$).

9. Transfer the liquid to another 1.5 ml centrifuge tube. Avoid the thick layer near the bottom of the tube. The lipid pellicule at the top of the tube is not particularly harmful[a].

10. Assay the protein concentration of the extracts by the Bradford Coomassie Blue assay. The concentration should be about 20 mg/ml. Lower concentrations may yield little or no Cln-associated kinase activity.

11. For each assay, use a set amount of protein (calculated from Step 10) (at least 2 mg). Add 0.1–0.25 μl of 12CA5 ascites fluid (depending on the concentration of antibody) to each sample of cell extract[a,b]. Incubate on ice for about 1 h.

12. Wash protein A–beads twice with lysis buffer to remove any unbound protein A. Add 15 μl of protein A–bead slurry (1:1 beads to buffer) to each sample. Rock tubes at 4 °C for 1 h.

13. Wash the immunoprecipitate four times with 1 ml of lysis buffer without the protease inhibitors. Pellet the beads using minimum $g$ forces. Otherwise, small, non-specific precipitates formed spontaneously in yeast extracts will also be collected in the pellet.

**Protocol 4.** *Continued*

With a variable speed microcentrifuge, set the speed to 3000 r.p.m. (about 700 *g*) and turn on for 3–5 sec only. For a fixed speed microcentrifuge, pulse the centrifuge on for a fraction of a second, and let coast to a stop[a].

14. After the 4th wash, resuspend the beads in 1 ml of 1 × kinase buffer, and transfer to a fresh 1.5 ml tube; this cuts down background.
15. Wash once more with 1 × kinase buffer. Remove all liquid by aspirating from the bottom of the tube with a 27-gauge needle attached to a vacuum line.
16. Pre-incubate the pellet at 30 °C for 10 min.
17. Start the reaction by adding a 5 μl mix containing 10 μCi [γ-$^{32}$P]ATP, 1 μg histone H1 (Boehringer–Mannheim), and the desired concentration of unlabelled ATP (1 μm to enhance phosphorylation of co-precipitated substrates; 5–100 μm depending on the amount of activity, to favour linearity and histone H1 phosphorylation; and 100 μm if p13$^{suc1}$ beads are used), all in 1 × kinase buffer. Incubate at 30 °C for 10 min.
18. Stop the reaction by adding 10 μl of 2 × Laemmli SDS–PAGE sample dye. Either keep tubes on ice until ready to run gel, or freeze at −70 °C.
19. Boil the samples for 1–2 min, and load one-half of each reaction mix (7.5 μl) very carefully into an SDS–polyacrylamide gel. Even a small amount of sample spilled into the upper buffer chamber can make the whole gel radioactive.
20. Electrophorese the samples. Run the dye front off the gel. The unincorporated $^{32}$P will run into the lower buffer reservoir (dispose of accordingly).
21. Fix the gel, stain with Coomassie blue, and dry. Expose to X-ray film for an autoradiogram. For quantitation, the histone H1 bands (which will be visible) can be cut out of the dried gel and assayed using a liquid scintillation counter.

[a] The cell suspension can be frozen at −70 °C after step 3. In fact, the procedure can be interrupted by freezing at several points without losing activity, i.e. cell pellet (Step 2), crude lysate (Step 9), washed immunoprecipitate (Step 13).
[b] If p13$^{suc1}$ beads are used, then about 10 μl of beads (typically containing 5 μg/μl Suc1) should be used with the extract containing several hundred micrograms of cellular protein in place of steps 11, 12 and see footnote *c*.
[c] A useful negative control is to perform Step 11 using an antibody that has been mixed with 20–50 μg of the epitope peptide. Since the antibody has a higher avidity for the triple epitope tag than for the single epitope peptide, the incubation for the negative control should not be too long (no more than 1 h).

4: *Analysis of cell cycle in S. cerevisiae*

# 4. Analysis of heterologous genes in yeast

Because of the amazing generality of many basic eukaryotic cell-cycle control mechanisms, many mammalian cell-cycle genes can function and be studied in yeast. In fact, the human *cdc2* gene was first discovered by functional complementation in yeast (20), and since that time a number of mammalian and other cyclins have been cloned in the same way. The dihybrid screen for interacting proteins (21) is another example of a powerful method for discovering and studying heterologous genes using the yeast system.

## 4.1 Cloning heterologous homologues of yeast genes by functional complementation

This approach requires a conditionally lethal yeast strain, such as a *cdc* mutant. The strain must have at least one auxotrophic marker, such as *leu2* or *ura3*. The approach also requires a cDNA library in an appropriate yeast expression vector carrying a complementing prototrophic marker. The cells are grown at the permissive temperature, and transformed with DNA from the cDNA library. They are then spread on selective plates. In most cases, a dual selection is applied simultaneously for the prototrophic marker on the vector (e.g. *LEU2*) and also for complementation of the *cdc* mutation. Sometimes it is not possible to apply both selections simultaneously, and so clones transformed to prototrophy are selected, and then a second round of selection is applied later for complementation of the *cdc* mutation.

### 4.1.1 General considerations

This is a risky approach, because there may be no homologue that can function in yeast, and even if there is, it may not be in the available libraries. It is also a difficult approach, because yeast transformation is relatively inefficient. However, it can sometimes find more distantly related homologues than the polymerase chain reaction. There are three limiting factors in transformation: plasmid DNA, yeast cells, and medium. First, there may be $10^6$ transformants per microgram of plasmid DNA at best, but more typically $10^4$–$10^5$. Second, there is rarely more than 1 transformant per 1000 yeast cells. Third, if too many yeast cells are spread on a plate, crowding reduces the effective transformation frequency; optimally, there should be only about $10^7$ cells spread per millilitre of medium. Interesting mammalian genes have been identified by screening as few as $5 \times 10^4$ yeast transformants, but, when warranted by the complexity of the library, $10^6$–$10^7$ transformants is a better goal.

### 4.1.2 Libraries

Several mammalian cDNA libraries in yeast expression vectors have been described (22–25). *S. pombe* (26) and Drosophila (27) cDNA libraries in yeast expression vectors have also been described. The vectors include a yeast origin of replication, a selectable marker (either *LEU2* or *URA3*), a strong constitutive or inducible promotor, a transcriptional terminator, and one or more cloning sites.

### 4.1.3 Transformation

There are three transformation methods that might be used. With any of them, the number of plasmid-transformed cells should be assayed in each experiment by selecting only for the marker on the vector.

(a) Electroporation (28) is easy and gives good efficiency on a small scale. However, it is difficult to scale-up, and is not yet a practical option for screening libraries.

(b) The lithium acetate procedure (29) as modified by the addition of single-stranded carrier DNA (30–32), is simple, gives good transformation efficiencies, can be carried out on a large scale, and has been used to obtain complementing mammalian genes. It is the best choice for workers inexperienced with yeast, and also the best choice if a relatively small number of clones (up to $10^6$) are to be screened. Because crowding of cells reduces transformation efficiency, one should be satisfied with about $3 \times 10^4$ transformants per 14 cm plate, though it may be possible to get more.

(c) The spheroplast method (*Protocol 5*) is much more complex and difficult than the lithium acetate method, and requires practice for perfection, but gives very high efficiencies, can be performed on any scale, and has been used to obtain complementing mammalian genes. The transformed colonies obtained are embedded in agarose, so they cannot be replica-plated directly should that be required. However, the layer of agar containing the transformants can be homogenized using a blender, and the resulting cell suspension can be spread on ordinary plates.

---

**Protocol 5.** *S. cerevisiae* spheroplast transformation—$10^7$ transformants

*Materials*

- 1.1 M sorbitol (ideally, buffered at pH 5.5 with 20 mM sodium phosphate)
- 44% PEG 4000 (w/v in water; filter sterilized)
- STC: 1.1 M sorbitol, 10 mM $CaCl_2$, 10 mM Tris, pH 7.6
- Lipofectin (BRL)
- Plates: about 100 14 cm selective SD (1) plates (e.g., -ura plates)

## 4: Analysis of cell cycle in S. cerevisiae

containing 1 M sorbitol; a plate of this size contains 45–50 ml of medium
- SD regeneration agar: about 4 l of selective SD medium containing 3% agar and 1 M sorbitol
- Sample recipe: 6.7 g yeast nitrogen base with ammonium sulfate, 30 g agar, 182 g sorbitol, 850 ml water, and required amino acids, pH 5.5; autoclave; add 100 ml of autoclaved 20% glucose
- Lyticase (Sigma): dissolve 50 000 units in 50% glycerol buffered at pH 7 with 20 mM sodium phosphate. Store at $-20\ °C$
- All glassware, solutions, etc. must be absolutely free of detergent.

*Method*

NB Carry out steps 2–13 at room temperature unless otherwise indicated.

1. Grow cells to about $3 \times 10^7$ cells per ml in rich medium such as YEPD (Section 3.2.1). If the cell density becomes too high, the spheroplasting will not work. To obtain $10^7$ transformants, use 2 litres ($6 \times 10^{10}$) cells.

2. Harvest by centrifugation (e.g. 4000 $g$, 7 min) in sterile bottles. Wash once in 1.1 M sorbitol. Resuspend in 1.1 M sorbitol at a concentration of $3 \times 10^8$ cells per ml. Add $\beta$-mercaptoethanol to 0.02% (v/v).

3. Add 2500 units of Lyticase for every $10^{10}$ cells. Other enzymes (Glusulase, Zymolyase) are either less effective at removing the cell wall, or produce excessive lysis and killing. Incubate at 30 °C for 1–2 h.

4. The most critical aspect in the transformation is to remove as much of the cell wall as possible without causing too much cell lysis. Assay the degree of spheroplasting every half-hour by diluting 100 µl of the cell suspension into 900 µl of (a) 1.1 M sorbitol; and (b) water (no detergent). The sorbitol suspension should be cloudy, while the water suspension should go clear. Also, examine the two suspensions by phase-contrast microscopy. There should be 95–100% cell lysis in the water dilution. There should be about 5% lysis in the sorbitol dilution. If there is no lysis in the sorbitol dilution, let the incubation go a bit longer.

5. Harvest the spheroplasts by low-speed centrifugation (e.g. 4000 $g$, 5 min). Wash the spheroplasts twice by resuspending in 1.1 M sorbitol and centrifuging.

6. Resuspend the spheroplasts in STC and centrifuge for 5 min at 4000 $g$.

7. Resuspend the spheroplasts in STC at a concentration of $3 \times 10^9$ cells per ml.

**Protocol 5.** *Continued*

8. Add DNA to the spheroplasts[a,b]. For the most efficient use of the transforming plasmid DNA, use 5–20 µg per $10^{10}$ cells. However, the final total concentration of DNA must be about 40 µg per ml. Thus, for 6 × $10^{10}$ spheroplasts (in 20 ml STC) use about 120 µg of plasmid DNA, and 800 µg of sterile carrier double-stranded calf thymus DNA. If plasmid DNA is not limiting, then more should be used, and will give a larger number of transformants. Mix DNA with the spheroplast suspension. Incubate for 10 min at room temperature.

9. To increase transformation efficiency about 10-fold, add 0.2–1 volume of STC containing Lipofectin (BRL) to the transformation mixture, such that the final concentration of Lipofectin is 20 µg/ml (e.g. for a 20 ml transformation, add 5 ml of STC containing 100 µg per ml Lipofectin.) Incubate for 10 min.

10. Add 5–10 volumes of 44% PEG 4000. Mix. Incubate for 15 min at room temperature.

11. Centrifuge to harvest spheroplasts (4000 $g$, 5 min). Resuspend the spheroplasts in 1.1 M sorbitol at 3 × $10^9$ per ml.

12. For each 200–300 µl of cells, use 45 ml of molten selective regeneration agar at a temperature of 47.5 °C. Much hotter agar will kill the cells; cooler will solidify. Put the 200–300 µl of cells into a tube containing 45 ml of SD regeneration agar, mix thoroughly but very quickly, and pour on to a large (14 cm diameter) sorbitol plate.

13. Carry out a control for spheroplast survival and regeneration. Dilute the transformed spheroplasts in 1.1 M sorbitol, and plate about 3 × $10^3$ spheroplasts in non-selective regeneration agar (just add the appropriate amino acid to the tube). Survival should be 10–20%; if it is less than this, too many cells were killed.

[a] A control should be performed for transformation frequency. Dilute the transformed spheroplasts in 1.1 M sorbitol, and plate about 3 × $10^5$, 3 × $10^6$, and 3 × $10^7$ spheroplasts in SD regeneration agar that selects for the transforming vector, but not for any additional complementing activity.

[b] A DNA-free control should also be carried out to ensure there is not a significant number of pre-existing prototrophs.

### 4.1.4 The 'co-segregation' or 'plasmid loss' test

Yeast clones growing on selective plates in the restrictive condition do not necessarily contain an interesting plasmid. Instead, they may be Cdc⁺ revertants that are prototrophic because of an otherwise irrelevant library plasmid. Test whether the Cdc⁺ phenotype is due to the library plasmid, by growing each yeast clone in the double-permissive condition (e.g. + leucine

4: Analysis of cell cycle in S. cerevisiae

medium, permissive temperature), spreading a few hundred colonies on double-permissive plates, and replica-plating to plates lacking leucine at the permissive temperature. This allows one to isolate subclones of the original transformant that have now lost the library plasmid. One now asks whether the plasmid⁻ subclones can grow at the restrictive temperature—i.e. whether the $Cdc^+$ phenotype co-segregates with the $Leu^+$ phenotype. If the $Leu^-$ strain cannot grow at the restrictive temperature, it is good evidence that a complementing activity is present on the plasmid. If it can grow, then the yeast clone is probably a $Cdc^+$ revertant, and the plasmid is of no interest.

If the plasmid marker is URA3, then a simplified version of the co-segregation test can be done. The drug 5-fluoro-orotic acid (5-FOA) is toxic to URA3 strains, so plates containing both 5-FOA and uracil select for ura3 cells (1). A plate containing many different yeast clones to be tested can simply be replica plated to a 5-FOA plate incubated at the restrictive temperature. Clones that grow ($Ura^-$ but $Cdc^+$) are probably revertants (i.e., they lost the URA3 plasmid while maintaining the $Cdc^+$ phenotype), while clones that cannot grow ($Ura^-$ and $Cdc^-$) are the ones desired. 5-FOA can be obtained from Sigma or Pharmacia, or much less expensively from SCM Specialty Chemicals. The Genetics Society of America purchases 5-FOA in bulk in intervals, and members of the GSA can obtain it even more economically: contact the Society for details.

### 4.1.5 Isolation and re-testing of complementing plasmids

DNA is extracted from the yeast and *E. coli* is transformed to ampicillin resistance by electroporation (yeast cells contain an inhibitor that reduces the efficiency of most other methods of transforming *E. coli*.). The primary yeast transformants often contain more than one plasmid. Therefore, it is best to examine at least six *E. coli* transformants per yeast clone. Finally, the plasmid DNA from a single *E. coli* transformant is used to re-transform the parental yeast strain. All of the prototrophic transformants should also be $Cdc^+$.

## 4.2 Testing known heterologous genes for their ability to complement mutations in yeast

When a plant or animal homologue of a yeast cell-cycle gene has been cloned, it is natural to ask whether the heterologous gene can complement a mutation in the corresponding yeast gene. For instance, many larger eukaryotes contain at least one Cdc28/Cdc2 homologue that can complement lethal *cdc28* mutations of *S. cerevisiae* and *cdc2* mutations of *S. pombe*; such complementation is popular as a test for 'real' homologues (33). Although the conclusions that can be drawn from complementation, or lack thereof, are quite limited, the complementation test is useful as a means of classifying genes. Also, when complementation succeeds, it means functional experiments can be carried out on the heterologous protein in yeast.

Complementation tests are performed by cloning the gene or cDNA to be tested into a yeast expression vector. These vectors usually have a strong constitutive or inducible promoter, followed by one or more unique restriction sites, followed by a transcription terminator. Suitable vectors have been extensively reviewed (34–37). Mutant yeast cells are transformed and grown under permissive conditions with the plasmid carrying the test gene, using selection for the prototrophic marker on the plasmid (e.g. *LEU2* or *URA*3). Once transformants have been isolated, they are tested under restrictive conditions for the mutation to see if the test gene complements. If it does, the co-segregation test described above is usually performed to show that when the plasmid is lost, the ability to grown under restrictive conditions is also lost. When complementation fails, it is important to check by Northern analysis that an mRNA of the appropriate size is present. If possible, the presence of the heterologous protein should be assayed.

## 4.3 Analysis of heterologous genes not having known homologues

Many interesting mammalian genes such as *myc*, p53, retinoblastoma, and SV40 T-antigen do not have known *S. cerevisiae* homologues. Two approaches have been used to find out more about the function of these genes using the yeast system. The first is to over-express the foreign gene in yeast, and ask whether this has any phenotypic effect. In the cases of p53 (38, 39) and SV40 T-antigen, over-expression is somewhat toxic to yeast cells. However, this toxicity may or may not be related to the normal role of the protein.

The other approach is to look for mutations that make viability of the mutant strain dependent on expression of the foreign gene. Such mutations may identify yeast genes with the same role as the heterologous gene. Three types of screens for such mutations are described below. Although these and other screens have been very useful for finding yeast genes redundant with one another (briefly reviewed in ref. 40), there have been no clear successes in finding yeast genes redundant with foreign genes.

*i. Dependence on expression from a conditional promoter*
A strain is constructed where the test gene is expressed from a conditional promotor such as the *GAL*1. The strain is mutagenized (in galactose medium) and spread on galactose plates. Colonies are replica plated to glucose plates, where the promoter is off. Colonies that grow on the galactose but not the glucose plates are candidate mutants.

*ii. Inability to lose a URA3-marked plasmid carrying the foreign gene*
The test gene is cloned into an autonomously replicating, *URA3* vector. A strain carrying the plasmid is mutagenized and spread on plates. Colonies are replica plated to 5-FOA plates. Colonies that cannot grow on the 5-FOA

plates (i.e. mutants unable to lose the *URA3* plasmid and still survive) are candidate mutants.

*iii. Inability to lose an ADE3-marked plasmid carrying the foreign gene*
This screen is similar to the one described above, but colonies are screened by colour instead of by 5-FOA sensitivity.

## Acknowledgements
This work was supported by NIH grants GM45410 and GM39978.

## References
1. Rose, M. D., Winston, F., and Hieter, P. (1990). *Methods in yeast genetics: a laboratory course manual.* Cold Spring Harbor Laboratory Press, Cold Spring Harbor, NY.
2. Guthrie, C. and Fink, G. R. (ed.) (1991). *Methods in enzymology: guide to yeast genetics and molecular biology*, Vol. 184, Academic Press, NY.
3. Pringle, J. R. and Hartwell, L. H. (1981). In *The molecular biology of the yeast Saccharomyces: life cycle and inheritance* (ed. J. N. Strathern, E. W. Jones, and J. R. Broach), pp. 97–142. Cold Spring Harbor Press, Cold Spring Harbor, NY.
4. Johnston, G. C., Pringle, J. R., and Hartwell, L. H. (1977). *Exp. Cell Res.*, **105**, 79.
5. Hartwell, L. H. and Unger, M. W. (1977). *J. Cell Biol.*, **75**, 422.
6. Carter, B. L. A. and Jagadish, M. N. (1978). *Exp. Cell Res.*, **112**, 15.
7. Hartwell, L. H. and Weinert, T. A. (1989). *Science*, **246**, 629.
8. Sudbery, P. E., Goodey, A. R., and Carer, B. L. A. (1980). *Nature*, **288**, 401.
9. Thorner, J. (1981). In *The molecular biology of the yeast Saccharomyces: life cycle and inheritance* (ed. J. N. Strathern, E. W. Jones, and J. R. Broach), pp. 143–180. Cold Spring Harbor Laboratory Press, Cold Spring Harbor, NY.
10. Jacobs, C. W., Adams, A. E. M., Szaniszlo, P. J., and Pringle, J. R. (1988). *J. Cell Biol.*, **107**, 1409.
11. Dien, B. S. and Srienc, F. (1991). *Biotechnology Progress*, **7**, 291.
12. Kilmartin, J. V. and Adams, A. E. M. (1984). *J. Cell Biol.*, **98**, 922.
13. Huffaker, T. C., Thomas, J. H., and Botstein, D. (1988). *J. Cell Biol.*, **106**, 1997.
14. Pringle, J. R., Adams, A. E. M., Drubin, D. G., and Haarer, B. K. (1991). In *Methods in enzymology: guide to yeast genetics and molecular biology* (ed. C. Guthrie and G. R. Fink), Vol. 194, pp. 565–602. Academic Press, NY.
15. Langan, T. A., Gautier, J., Lohka, M., Hollingsworth, R., Moreno, S., Nurse, P., Maller, J., and Sclafani, R. A. (1989). *Mol. Cell. Biol.*, **9**, 3860.
16. Surana, U., Robitsch, H., Price, C., Schuster, T., Fitch, I., Futcher, A. B., and Nasmyth, K. (1991). *Cell*, **65**, 145–61.
17. Arion, D., Meijer, L., Brizuela, L., and Beach, D. (1988). *Cell*, **55**, 371–8.
18. Brizuela, L., Draetta, G., and Beach, D. (1987). *EMBO J.*, **6**, 3507.

19. Tyers, M., Tokiwa, G., Nash, R., and Futcher, A. B. (1992). *EMBO J.*, **11**, 1773.
20. Lee, M. G. and Nurse, PO. (1987). *Nature*, **327**, 31.
21. Fields, S. and Song, O. (1989). *Nature*, **340**, 245.
22. Becker, D. M., Fikes, J. D., and Guarente, L. (1991). *Proc. Natl Acad. Sci. USA*, **88**, 1968.
23. Colicelli, J., Nicolette, C., Birchmeier, C., Rodgers, L., Riggs, M., and Wigler, M. (1991). *Proc. Natl Acad. Sci. USA*, **88**, 2913.
24. Schild, D., Brake, A. J., Kiefer, M. C., Young, D., and Barr, P. J. (1990). *Proc. Natl Acad. Sci. USA*, **87**, 2916.
25. Elledge, S. J. and Spottswood, M. R. (1991). *EMBO J.*, **10**, 2653.
26. Fikes, J. D., Becker, D. M., Winston, F., and Guarente, L. (1990). *Nature*, **346**, 291.
27. Leopold, P. and O'Farrell, P. H. (1991). *Cell*, **66**, 1207.
28. Becker, D. M. and Guarente, L. (1991). In *Methods in enzymology: guide to yeast genetics and molecular biology* (ed. C. Guthrie and G. R. Fink), Vol. 194, pp. 182–187. Academic Press, NY.
29. Ito, H., Fukuda, Y., Murata, K., and Kimura, A. (1983). *J. Bacteriol.*, **153**, 263.
30. Schiestl, R. H. and Gietz, R. D. (1989). *Curr. Genet.*, **16**, 339.
31. Gietz, D., St. Jean, A., Woods, R. A., and Schiestl, R. H. (1992). *Nucleic Acids Res.*, **20**, 1425.
32. Elble, R. (1992). *Biotechniques*, **13**, 18.
33. Meyerson, M., Enders, G. H., Wu, C-L., Su, L-K., Gorka, C., Nelson, C., Harlow, E., and Tsai, L-H. (1992). *EMBO J.*, **11**, 2909.
34. Rose, M. D. and Broach, J. R. (1991). In *Methods in enzymology: guide to yeast genetics and molecular biology* (ed. C. Guthrie and G. R. Fink), Vol. 194, pp. 195–230. Academic Press, NY.
35. Romanos, M. A., Scorer, C. A., and Clare, J. J. (1992). *Yeast*, **8**, 423.
36. Schneider, J. C. and Guarente, L. (1991). In *Methods in enzymology: guide to yeast genetics and molecular biology* (ed. C. Guthrie and G. R. Fink), Vol. 194, pp. 373–388. Academic Press, NY.
37. Schena, M., Picard, D., and Yamamoto, K. (1991). In *Methods in enzymology: guide to yeast genetics and molecular biology* (ed. C. Guthrie and G. R. Fink), Vol. 194, pp. 389–398. Academic Press, NY.
38. Bischoff, J. R., Casso, D., and Beach, D. (1992). *Mol. Cell. Biol.*, **12**, 1405.
39. Nigro, J. M., Sikorski, R., Reed, S. I., and Vogelstein, B. (1992). *Mol. Cell. Biol.*, **12**, 1357.
40. Bender, A. and Pringle, J. (1991). *Mol. Cell. Biol.*, **11**, 1295.

# 5

# Methods for analysis of the fission yeast cell cycle

S. A. MACNEILL and P. A. FANTES

## 1. Introduction

The fission yeast *Schizosaccharomyces pombe* is a unicellular ascomycete fungus. Physiological analysis of the cell cycle of *S. pombe* was initiated by Mitchison (1) who appreciated the advantages of working with cells that were large enough to permit detailed microscopic analysis, but which were as easy to culture as bacteria, which were themselves popular experimental organisms at the time. In particular, Mitchison realized that the growth of *S. pombe* cells by length extension made the organism ideal for investigating the relationship between cell growth and cell division.

Genetic analysis of the cell cycle began with the isolation almost 20 years later, by Nurse and co-workers, of the first fission yeast cell-cycle mutants (2–5). These mutants fell into two classes. The first class were conditional lethal mutations that arrested cell-cycle progress at a particular point when shifted to their restrictive temperature (2, 3), while the second class affected the regulation of cell-cycle progress (4, 5). At present over 50 genes have been identified as being involved in a variety of cell-cycle events (reviewed in ref. 6).

The purpose of this chapter is to bring together commonly-used methods for studying the mitotic cell cycle of fission yeast. General molecular biology and genetical methods for fission yeast are described in ref. 7 and 8, respectively. Details of the agreed system of genetic nomenclature, *which differs from that used in budding yeast*, can be found in ref. 9.

## 2. Life cycle and genetics

The life cycle of the fission yeast is diagrammed in *Figure 1*. When provided with an adequate supply of nutrients, haploid cells grow and enter the mitotic cell cycle (*Figure 1*; top left), growing by apical extension (initially at one end, then later at both) and dividing by medial fission. The cells are rod-shaped,

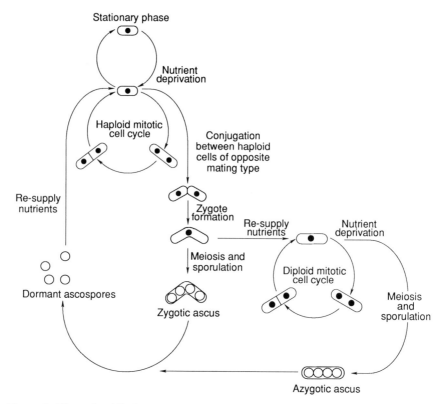

**Figure 1.** Life cycle of fission yeast. Top left, haploid mitotic cell cycle. Centre and lower left, haploid cells mating to form a diploid zygote, followed by meiosis and sporulation leading to zygotic ascus formation. Lower right, re-entry of diploid zygotes into mitotic cycle. See text for details.

with haploids being around 3–4 μm in diameter and 14 μm in length at the time of cell division (so that new-born cells are around 7 μm long). Diploid cells are larger. The events of the cell cycle are typically eukaryotic (6); there are discrete $G_1$, S, $G_2$, and M-phases occupying 0.1, 0.1, 0.7, and 0.1 of the cycle, respectively (where the total length of the cycle is represented as 1.0). In rapidly growing cells, $G_1$ and S phase are completed soon after mitosis but before cell separation is complete. As a result the nuclei of new-born cells at cell division are already in $G_2$. Mitosis is marked by formation of an intranuclear spindle and chromosome condensation (visible under the light microscope: ref. 10). The nuclear membrane does not break down during mitosis.

On exhaustion of nutrients, several alternative developmental fates are possible (reviewed in ref. 11). If cells of only a single mating type are present then these cells exit the mitotic cell cycle from either $G_1$ or $G_2$ and accumulate in stationary phase. In contrast, if cells of both mating types are present then those arrested in $G_1$ are able to conjugate to form a diploid zygote which then undergoes meiosis and sporulation to generate an ascus containing four haploid spores, which lie dormant till the nutritional climate improves (*Figure 1*; centre and lower left). If diploid zygotes are transferred to fresh medium prior to the initiation of meiosis then they are able to re-enter the mitotic cell cycle and grow and divide normally (*Figure 1*; lower right). Subsequently if such diploids are again deprived of nutrients they are able to undergo meiosis and sporulation without intervening conjugation.

These features of its life cycle make *S. pombe* particularly amenable to genetic analysis: genetic mapping can be carried out by tetrad dissection or random spore analysis of meiotic products, and complementation analysis performed using zygotically-derived diploids (8). Over 500 *S. pombe* genes have been identified, many of which have been placed on the genetic map of the organism (12) and approximately 150 sequenced. A gene database for *S. pombe* correlating physical and genetic information about known genes has recently been established in preparation for the genome sequencing project (13).

## 3. Monitoring cell-cycle progress

### 3.1 Growing fission yeast cells

Fission yeast cells can be grown in either complex or defined media (7). Approximate generation times for wild-type haploid cells growing in liquid media at a range of temperatures are given in *Table 1*. For most purposes 32 °C is a convenient growth temperature. Below 25 °C cells grow only slowly and above 37 °C growth is not possible. Larger cultures (100 ml and above) are generally inoculated from a smaller preculture (10 ml of medium) set up the previous day, although 2–3-day-old precultures can be used. After incubation overnight the preculture should be in early stationary phase (that is, greater than $1 \times 10^7$ cells/ml). The size of inoculum required (to ensure that the main culture is in mid-exponential growth the following day) will be dependent upon growth medium and incubation temperature, and also on the properties of the cells under study, but will generally fall between 1/200 and

**Table 1.** Generation times (min) for cells growing in liquid media

| Temperature | Minimal medium | Complex medium |
|---|---|---|
| 25 °C | 260 | 210 |
| 29 °C | 180 | 150 |
| 32 °C | 150 | 130 |
| 35 °C | 140 | 120 |

1/50 of the volume of the main culture. Gentle shaking is required for larger cultures but is unnecessary for precultures. For physiological experiments defined medium is best, despite being relatively time-consuming to prepare. Recipes for both complex and defined (minimal) media can be found in ref. 7.

Cells are best viewed under a phase-contrast microscope with at least 200 × magnification (that is, a 20 × objective with 10 × eyepiece magnification) although for more detailed analysis of cell ultrastructure a 100 × oil-immersion objective is required (giving a total magnification of 1000 ×). Dark-field optics are best for visualizing the septa in dividing cells (see below).

## 3.2 Determining cell number

For physiological experiments it is important that the cells are in the mid-exponential phase of growth. This corresponds to a cell number of between $1 \times 10^6$–$1 \times 10^7$ cells per millilitre of culture. Ideally, at the start of an experiment the cell number should be closer to the lower end of this range (between 1–2 × $10^6$ cells/ml), as this allows for the cells to undergo at least two cell number doublings while growing exponentially. Cell number can be determined by haemocytometer counts or by using an electronic particle counter (such as a Coulter counter). For haemocytometer counting cells with visible septa are counted as one cell, while two adherent daughter cells are counted as two (14). To distinguish between these two classes the appearance of the notch that marks the beginnings of septum cleavage is traditionally taken as the point at which one cell becomes two (*Figure 2*). For counting cells in an electronic particle counter a chamber with a 70–100 µm orifice is used. Prior to counting it is necessary to sonicate the sample gently to break up clumps of cells and to separate adherent daughters (14). The degree of clumping varies greatly depending on the growth conditions and the genetic properties of the strain in question: some cell-cycle mutant strains, for example, are particularly prone to clumping. To count using an electronic particle counter, transfer 100 µl of culture to 10 ml of Isoton II (Coulter Electronics) or similar using a micropipette. Be sure that the culture is mixed well prior to taking the sample as settling out of cells can markedly affect the reading obtained. Next, sonicate the cells to break up clumps. The pulse conditions required to achieve maximal declumping (that is, the time and amplitude of the pulse) will depend on the properties of the sonicator being used (its power and the size of the probe). Once optimal sonication conditions have been determined it is still worthwhile to occasionally check the extent of declumping.

An alternative means of determining culture growth is to measure the optical density ($OD$) of the culture at 595 nm using a spectrophotometer (14). The $OD$ of the culture relates to its biomass and by this method an $OD_{595}$ reading of 0.1 corresponds to approximately 2 × $10^6$ wild-type cells per

*5: Fission yeast cell cycle*

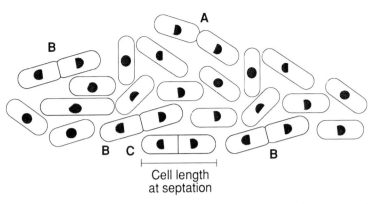

**Figure 2.** Growing fission yeast cells. A number of cells are shown in schematic form to illustrate the distinction between adherent daughter cells (labelled A) in the final stages of cell separation, cells in which septum cleavage has begun but which has yet to be completed (B), and cells in which the septum has formed but which have not yet begun cleavage (C). Cell length at division is determined by measuring only cells of type C (for wild-type cells this is generally around 14 μm). For cell number counts by particle counter, sonication separates adherent daughters, which are therefore counted as two cells. For cell number counts using a haemocytometer, cells with visible septa are counted as one cell while adherent daughters are counted as two.

millilitre of culture. However, the relationship between cell number and $OD_{595}$ varies between different strains and growth conditions and also between spectrophotometers, making it is essential to establish a relationship between cell number and $OD_{595}$ before beginning work. Note in particular that in cell-cycle arrest, biomass continues to increase (and, therefore, so does $OD_{595}$) without a corresponding increase in cell number.

### 3.3 Cell length and septation (cell plate) index

As fission yeast cells grow solely by apical extension the length of a cell is related to its age. New-born wild-type cells are generally around 7–8 μm in length, while cells commence cell division at around 14 μm. Cell length can be measured using an eyepiece micrometer. To obtain an accurate measure of cell length at cell division it is necessary to measure only cells that are in the process of division, that is, cells in which the septum is formed but which have not, as yet, commenced septum cleavage (*Figure 2*). The same applies to determining the septation or cell-plate index: for a wild-type population of cells in mid-exponential growth this is generally between 5–10%.

### 3.4 Measuring DNA content

In a variety of situations it is necessary to know whether individual cells, or populations of cells, have an unreplicated (1C, for haploids) or replicated (2C) DNA content, corresponding to the $G_1$/early S phases and the late

S/$G_2$-phases of the cell cycle, respectively. This is important, for instance, in the characterization of cell-cycle mutants, where the DNA content per cell after arrest is informative about the primary defect in the mutant cells, and also for following changes in DNA content with time during the course of an experimental treatment. Measurement of DNA content of individual cells can be achieved using flow cytometry (see *Protocol 4*). However, a frequently-used alternative is to measure the DNA content of a population of cells chemically using the diphenylamine method (15, 16) as described by Bostock (17) and Gendimenico (18), and then to calculate from this the mean DNA content per cell, as described below (*Protocol 1*).

---

**Protocol 1.** Determination of bulk DNA content

*Materials*

- 0.5 M perchloric acid
- diphenylamine reagent: 1.5 g diphenylamine (Sigma D2385), 100 ml glacial acetic acid, 1.5 ml concentrated sulphuric acid, 0.5 ml 1.6% acetaldehyde
- deoxyadenosine

*Method*

1. Grow cells to mid-exponential phase. Determine cell number and sample $2.5 \times 10^8$ cells (approximately 10 µg of DNA for a wild-type haploid).

2. Pellet the cells by centrifugation (1000 $g$, 5 min, 4 °C) and discard the supernatant. Resuspend the pellet in 1 ml of ice-cold 0.5 M perchloric acid (PCA) at 0 °C to remove cold acid-soluble material. Incubate at 0 °C for 30 minutes.

3. Collect cold acid-insoluble material by centrifugation in a microcentrifuge at 4 °C for 5 minutes and wash twice with 1 ml of ice-cold 0.5 M PCA.

4. Resuspend the pellet in 250 µl of 0.5 M PCA and hydrolyse nucleic acids by heating to 70 °C for 20 min. Mix the tubes every few minutes during this incubation period.

5. Centrifuge (microcentrifuge, 4 °C, 10 min) and remove and retain the supernatant.

6. Repeat Steps 4 and 5, finally pooling the supernatant fractions (total volume 500 µl).

7. To the pooled supernatants add 1.0 ml of freshly-prepared diphenylamine reagent. At the same time prepare a series of standards using deoxyadenosine as follows. Make up solutions of 0, 2, 4, 6, 8, 10, and

12 μg/ml deoxyadenosine in 0.5 M PCA. To 500 μl of each add 1.0 ml of diphenylalanine reagent.
8. Incubate in the dark at 50 °C for 4 h. Determine absorbance at 600 nm. For the standards, 1 μg deoxyadenosine/ml is equivalent to 2.605 μg DNA/ml.

Using this method haploid cells in $G_1$ (which have a 1C DNA content) are estimated to have between 15–20 fg of DNA per cell. Cells in $G_2$ (2C DNA content) have between 25–35 fg per cell (17). Since almost all the cells in an exponentially-growing population are in $G_2$, the mean DNA content of cells in a rapidly-growing culture is 25–35 fg. The variability of the values obtained is a reflection of the fact that the assay is not entirely reproducible (discussed in ref. 19). In addition, it has been noted that the 4C value obtained for diploid cells in mid-exponential growth is almost always less than twice that of the 2C value determined for haploids in similar conditions, for unknown reasons.

## 3.5 Monitoring DNA replication

When the behaviour of cells during a particular experimental treatment is under investigation, the rate of DNA synthesis is often a more useful parameter than the total DNA content. Like other fungi, fission yeast cells lack thymidine kinase and so their DNA cannot be exclusively labelled using radioactive precursors. Two methods for monitoring DNA replication in fission yeast using the incorporation of labelled precursors have been described however (20). The first of these (*Protocol 2*) measures the incorporation of [6-$^3$H]uridine into DNA following pulse labelling of cell cultures, while the second (*Protocol 3*) provides a means of detecting DNA replication in single cells by autoradiography. Both methods incorporate treatments to degrade and remove labelled cellular RNA.

**Protocol 2.** Monitoring DNA replication via [6-$^3$H]uracil uptake

*Materials*
- [6-$^3$H]uridine (20–30 Ci/mmol, DuPont Ltd., NEN 156)
- Unlabelled uridine (Sigma U3750)
- Alkali resuspension buffer: 0.5 M NaOH, 10 mm EDTA, 100 μg/ml calf thymus DNA (Sigma D1501)
- 60% perchloric acid (PCA)
- 0.1 M NaCl in 70% (v/v) ethanol
- Optiphase scintillant (Pharmacia)

**Protocol 2.** Continued

*Method*

1. Grow cells to mid-exponential phase in minimal medium: label 0.5 ml of cell culture for 10–20 min by addition of 10 μCi [6-$^3$H]uridine. Stop by adding 0.5 ml of 1 mg/ml unlabelled uridine. Further details concerning the kinetics of uridine uptake can be found in ref. 20.
2. Pellet the cells (microcentrifuge, 4 °C, 5 min) and wash the pellet twice with 0.5 ml unlabelled 1 mg/ml uridine. Freeze at −20 °C till required.
3. Thaw samples and pellet cells (microcentrifuge, 4 °C, 5 min). Discard supernatant and resuspend pellets in 0.3 ml alkali resuspension buffer and incubate at 40 °C for 90 min with occasional mixing. The purpose of the alkali treatment is to hydrolyse labelled cellular RNA: by the end of the process (that is, after Step 8 below) greater than 90% of the radioactivity remaining is in the form of cytosine and thymine, and virtually none remains as uracil.
4. Cool to 0 °C and acidify by adding 120 μl 60% perchloric acid (PCA), precipitate on ice for at least 1 h.
5. Pellet the sample (microcentrifuge, 4 °C, 10 min) and discard the supernatant.
6. Wash the pellet once with ice-cold 0.5 M PCA containing 100 μg/ml uridine, then twice with ice-cold 0.1 M NaCl in 70% (v/v) ethanol.
7. Dry the pellets under vacuum and hydrolyse in 0.5 ml 0.5 M PCA for 30 min at 70 °C with frequent mixing.
8. Centrifuge (microcentrifuge, 4 °C, 10 min), remove 400 μl of the supernatant and add to 10 ml of Optiphase scintillant, or equivalent, and count.

---

**Protocol 3.** Autoradiography of single labelled cells

*Materials*

- [8-$^3$H]guanosine (5–15 Ci/mmol, ICN 24030)
- 0.5 M Perchloric acid (PCA)
- RNase A (Sigma R5503)
- Unlabelled guanosine (Sigma G6752)
- 0.4% formaldehyde in 0.1 M sodium phosphate buffer, pH 7.0
- Poly-L-lysine (Sigma P8920)
- Photographic emulsion (Ilford type K2)
- Kodak No. 1 safe-light filter (IBI Ltd.)

*Method*

1. Grow cells to mid-exponential phase in minimal medium: pulse label cells for 10–40 min by adding 50 µCi/ml of [8-$^3$H]guanosine. For reasons not understood it is not possible to use [6-$^3$H]uridine in place of guanosine in this protocol. Stop by adding an equal volume of ice-cold 0.5 M PCA and leave on ice for at least 5 min. (Cells can be left at this stage for several hours if necessary.)

2. Collect the cells by centrifugation (microcentrifuge, room temperature, 5 min), wash twice with water and resuspend in 0.3 M NaCl, 0.03 M trisodium citrate (pH 7.0) containing 200 µg/ml preboiled RNase A. Incubate at 37 °C for 60 min with occasional mixing.

3. Collect the cells by centrifugation, wash twice with 100 µg/ml guanosine and resuspend in 0.4% formaldehyde buffer. Incubate at room temperature for 30 min.

4. Collect the cells by centrifugation, wash once with water, and resuspend in 0.2 M NaOH. Incubate at 25 °C for 100 min.

5. Collect the cells by centrifugation and wash three times with water. Apply the cells to a microscope slide coated with 0.1% (w/v) aqueous solution of poly-L-lysine and dip into photographic emulsion (pre-heated at 45 °C for 20–30 min) in the dark (a Kodak No. 1 safe-light filter is appropriate). To coat microscope slides with poly-L-lysine, briefly cover the surface of the slide with a drop of poly-L-lysine before aspirating off. Allow the slide to dry at room temperature.

6. Expose the slides at 4 °C. Develop (generally 3–4 week exposure times are required) and view under the microscope. In asynchronously growing populations of cells, grains of radioactivity are seen clustered over the nuclei of septated cells.

A quantitative measure of the cell-cycle specificity of the labelling can be obtained by labelling synchronous cultures, treating the cells as if for radiography, and then counting the residual radioactivity by scintillation counting instead of autoradiography (20).

## 3.6 Flow cytometry

Flow cytometry provides a rapid and accurate way of determining the relative DNA content of populations of fission yeast cells (21, 22). Cells are fixed in ethanol, treated with ribonuclease and finally stained with propidium iodide (*Protocol 4*). Analysis is carried out using an excitation wavelength of 488 nm and a detection wavelength of 510–550 nm.

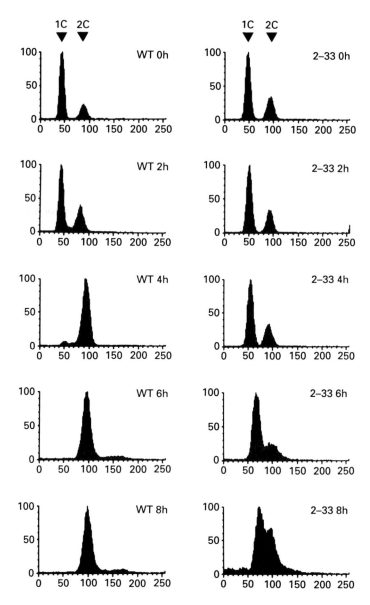

**Figure 3.** Flow cytometric analysis of propidium iodide stained fission yeast cells. In this experiment (S. M. and P. Nurse, unpublished results) two strains (one wild-type, the other carrying the temperature-sensitive *cdc2-33* allele) were first starved of nitrogen for 18 h at 25 °C, then refed by adding $NH_4Cl$ (at 0 hours, see below) and transferred directly to the restrictive temperature of 35 °C. Nitrogen starvation results in most of the cells accumulating in $G_1$ (with a 1C DNA content): this is clearly seen in the 0 hour samples. At 2–4 h following refeeding the wild-type cells (data in left-hand column) replicate their DNA and go on to divide. By 6 hours after refeeding the cell number in the wild-type culture has

5: *Fission yeast cell cycle*

---

**Protocol 4.** Preparation of samples for flow cytometry

*Materials*
- 50 mM sodium citrate buffer, pH 7.0
- DNase-free RNase A (Sigma R5503)
- 50 mM trisodium citrate, pH 7.0
- Propidium iodide (Sigma P4170)

*Method*
1. Grow cells in minimal medium to mid-exponential phase ($1 \times 10^6$–$1 \times 10^7$ cells per ml), determine cell number.
2. Sample $\sim 5 \times 10^7$ cells; pellet cells by centrifugation at 1000 $g$ for 5 min at 4 °C and resuspend at $1 \times 10^7$ cells/ml in 70% (v/v) ethanol. Incubate at 4 °C for at least 1 h.
3. Take 1 ml of suspension and wash once with 1 ml of 50 mM sodium citrate buffer pH 7.0 before resuspending in 0.5 ml of the same buffer containing 100 µg/ml preboiled DNase-free RNase A (Sigma R5503). Incubate at 37 °C for 2 h, or overnight.
4. Add 0.5 ml of 50 mM trisodium citrate (pH 7.0) containing 4 µg/ml propidium iodide (Sigma P4170). Incubate for 1 h in the dark at 4 °C. (Cells can be stored in the dark for at least one week at this stage if required).
5. Prior to flow cytometric analysis, lightly sonicate the cells.

---

It is particularly important for the interpretation of flow cytometric data obtained in experimental situations that the positions of the 1C and 2C peaks can be unambiguously identified. In the absence of the appropriate controls this is by no means straightforward, particularly since longer cells fluoresce more than shorter cells (22). Sample data from cells prepared as described in *Protocol 4* is shown in *Figure 3*. In this instance DNA content was monitored over a period of 8 hours following release of cells from a $G_1$ arrest brought about by nitrogen starvation (see figure legend for details). Notice how the positions of the peaks on the *x*-axes in *Figure 3* move gradually to the right

---

almost doubled. The single 2C peak is indicative of a rapidly-growing culture: since DNA replication occurs simultaneously with cell separation no 1C peak is visible. In contrast, cells carrying the *cdc2–33* allele (data in right-hand column) are unable to undergo replication or enter mitosis and divide. The cell number in the culture remains constant over at least the first 6 h of the experiment, although by this time the cells have begun to leak through the $G_1$–S block. By 8 hours after refeeding the number of cells in S/$G_2$ has increased markedly. (Data gathered using a Becton-Dickinson FACScan and analysed using Consort-30 software.)

during the course of the experiment, indicating that the DNA content of the cells is increasing. In order to locate the 1C and 2C peaks cdc mutant strains can be used: *cdc*10–129 cells arrest after 4 hours at 35 °C with 1C DNA content, while *cdc*2–33 cells mostly arrest in $G_2$ with 2C DNA content, although some of the cells arrest in $G_1$ with 1C DNA. Beyond these times cells may be begin to leak through the imposed blocks.

## 3.7 Measuring p34$^{cdc2}$ H1 kinase activity

Mitotic p34$^{cdc2}$ histone H1 kinase activity can be conveniently assayed in crude protein extracts using the method of Moreno, Hayles, and Nurse (23); see also Chapter 13. Protein extracts are prepared by breaking the cells using acid-washed glass beads in a buffer containing a variety of phosphatase inhibitors. Histone H1 and [γ-$^{32}$P]ATP are added to the extract which is incubated for 10 minutes prior to SDS–PAGE. Histone H1 migrates with $M_r$ of 20–25 kDa. Phosphorylation of the protein is temperature-sensitive in extracts prepared from strains carrying certain temperature-sensitive p34$^{cdc2}$ proteins (23).

---

**Protocol 5.** Measuring p34$^{cdc2}$ activity in crude extracts

*Materials*

- STOP buffer: 150 mM NaCl, 1 mM NaN$_3$, 10 mM EDTA (pH 8.0), 50 mM NaF
- Acid-washed glass beads (Sigma G9268)
- HB15 buffer: 60 mM β-glycerophosphate, 15 mM *p*-nitrophenyl-phosphate, 25 mM MOPS (pH 7.2), 15 mM EGTA, 15 mM MgCl$_2$, 1 mM DTT, 0.1 mM sodium vanadate (pH 8.0), 1% (v/v) Triton X-100, adjusted to pH 7.2 with NaOH and with protease inhibitors as follows: leupeptin (final concentration 20 μg/ml), aprotinin (20 μg/ml), pepstatin A (2 μg/ml), and PMSF (1 mM added fresh each time from the stock).
- Leupeptin (10 mg/ml stock in water, Sigma L2884)
- Aprotonin (10 mg/ml stock in water, Sigma A1153)
- Pepstatin A (10 mg/ml stock in water, Sigma P 4265)
- PMSF (100 mM stock in ethanol, Sigma P7626)
- HB5 buffer (as HB15 buffer, but EGTA at 5 mM)
- KIN buffer: HB5 buffer containing 2 mg/ml histone H1 (Boehringer–Mannheim 223459) and 200 μm ATP including 40 μCi/ml [γ-$^{32}$P]ATP (10 μCi/μl stock, Amersham PB10168)

- SDS–PAGE sample buffer
- 12% SDS–PAGE gel and equipment

*Method*

1. Grow up cells in minimal media to mid-exponential phase; determine cell number per millilitre of culture.

2. Harvest $2.5 \times 10^8$ cells by centrifugation at 1000 $g$ for 5 min at 4° C: discard supernatant and wash the pellet once with ice-cold STOP buffer. Store pellet on ice till required. For assaying $p34^{cdc2}$ activity in cell-cycle mutant strains held at the restrictive temperature it is particularly important that these steps are completed quickly before the cells have the opportunity to escape from the imposed cell-cycle block.

3. Resuspend cells in 20 µl of HB15 buffer by vortexing, add 1 ml of ice-cold acid-washed glass beads and vortex vigorously for 90 sec.

4. Add 1 ml ice-cold HB15 buffer and mix well. Remove 400 µl of liquid to a prechilled microcentrifuge tube and centrifuge for 30 sec at 4 °C.

5. Transfer supernatant to fresh tube and spin in a microcentrifuge at 4 °C. Transfer supernatant to a fresh prechilled tube and determine protein concentration by the Bradford, or similar, method.

6. Adjust the concentration of the extract to 0.5 µg/µl with ice-cold HB15: take 4 µl to a fresh screw-cap microfuge tube and add 6 µl of HB5 buffer. Mix gently.

7. Incubate for 15 min at the appropriate temperature before adding 10 µl of KIN buffer, mix gently, and incubate for a further 10 min.

8. Stop the reaction by adding 20 µl of SDS–PAGE sample buffer and freezing at −20 °C. Prior to SDS–PAGE, boil the sample for 3 min and briefly centrifuge to collect the sample at the bottom of the tube. Electrophorese the samples on a 12% SDS–PAGE gel. Fix, stain, and destain the gel to visualize the histone bands, then dry the gel and expose to film (see Chapter 13 for details).

Not all temperature-sensitive $p34^{cdc2}$ mutant proteins exhibit thermolabile histone H1 kinase activity under these conditions: the mutant $p34^{cdc2\text{-}M26}$ and $p34^{cdc2\text{-}L7}$ proteins are relatively thermolabile, whereas the $p34^{cdc2\text{-}33}$ protein is not (23). Enoch *et al.* (24) have substituted purified chicken lamin B protein for histone to demonstrate that fission yeast $p34^{cdc2}$ present in crude extracts will also phosphorylate this protein in a temperature-dependent manner. An alternative method to that presented in *Protocol 5* for assaying $p34^{cdc2}$ histone kinase activity is to purify p34$^{cdc2}$ either by immunoprecipitation using anti-p34$^{cdc2}$ antibodies (25) or by binding to p13$^{suc1}$ beads (26).

## 4. Cell staining

### 4.1 Nuclear staining

A number of compounds can be used to stain the nuclear DNA of fission yeast cells. The most widely used is DAPI (4′,6′-diamidino-2-phenylindole dihydrochloride; ref. 27). A method for DAPI staining of formaldehyde-fixed cells is presented below. An example of DAPI staining of cells is shown in *Figure 4A*.

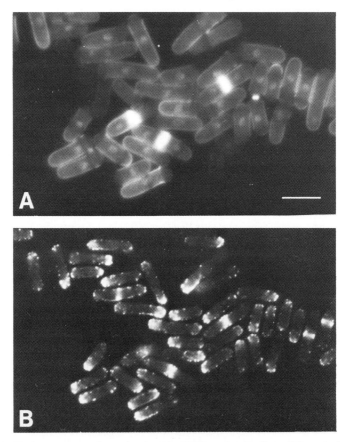

**Figure 4.** Fission yeast cells stained with DAPI and calcofluor (A) and rhodamine–phalloidin (B). Scale bar = 10 μm. See text and refs 30 and 31 for details. Photomicrograph kindly provided by J. S. Hyams, University of London, UK.

**Protocol 6.** DAPI staining of nuclear DNA of formaldehyde-fixed cells

*Materials*
- 30% formaldehyde
- PBS: 10 mM sodium phosphate (pH 7.2), 150 mM NaCl, 1 mM $NaN_3$
- Triton X-100
- Poly-L-lysine (Sigma P8920, see *Protocol 3*)
- DAPI mounting solution: 1 µg/ml DAPI (Sigma D1388), 1 mg/ml *p*-phenylenediamine dihydrochloride (Sigma P1519), 50% glycerol (v/v)

*Method*
1. Grow cells to mid-exponential phase of growth ($1 \times 10^6$–$1 \times 10^7$ cells/ml) in minimal medium at desired temperature.
2. Sample 900 µl of cells and fix with 100 µl of 30% formaldehyde for 30 min. For cells growing at restrictive temperatures it is important to incubate in formaldehyde at the same temperature to prevent escape from the imposed block.
3. Pellet the cells by centrifugation (microcentrifuge, room temperature, 5 min) and wash once with 1 ml of PBS, then once with 1 ml of PBS containing 1% Triton X-100 before resuspending in 100 µl PBS. At this stage the cells can be stored at 4 °C.
4. Apply cells to a poly-L-lysine coated coverslip to form a monolayer and leave to dry at room temperature.
5. Lower the coverslip (with the cell monolayer facing downwards) on to a small drop of DAPI mounting solution on a slide, seal round the edges, and examine under the fluorescence microscope. Slides prepared in this way can be stored in the dark at 4 °C for several weeks. The DAPI mounting solution can be stored at −20 °C in the dark for several months.

It is also possible to stain live cells with DAPI. To do this simply wash the cells three times in water and apply to the slide as described above. Other stains used for nuclear DNA include propidium iodide (used in FACS analysis, see *Protocol 4*) which stains nuclear and mitochondrial DNA as well as RNA (22), and ethidium bromide which has been used preferentially to stain the nucleolus, allowing it to be clearly distinguished from the rest of the nuclear DNA (28).

## 4.2 Staining the cell wall and septum

Calcofluor is used to stain the cell walls and septa of growing cells. New-born cells stain brightly with calcofluor at their old end: the new end is not stained until later in the cell cycle, at a point in $G_2$ known as NETO (for new end take off: see ref. 29), and is seen as a dark hemisphere. At NETO the deposition of new cell-wall material at the new end results in this end also staining with calcofluor: the dark hemisphere is displaced from the cell tip and appears as a dark band (known as the birth scar) around the cylindrical cell body. The appearance of this band thus indicates that a cell has passed the point of NETO. To stain cells with calcofluor (29), grow cells to mid-exponential phase, take a 1 ml sample, centrifuge and wash once with 1 ml of ice-cold 1% saline, before finally resuspending in 100 µl of ice-cold saline containing 1 mg/ml calcofluor (Sigma F6259). Leave the cells on ice for 15 min before examining under the fluorescence microscope. An example of cells stained with calcofluor (and DAPI) is shown in *Figure 4A*.

## 4.3 Staining actin structures

The behaviour of actin structures in fission yeast has been analysed by a number of methods and provides a useful cell-cycle stage marker. Rhodamine-conjugated phalloidin (30) as well as indirect immunofluorescence using anti-actin antibodies (31) have both been used to analyse actin structures. The results obtained from each method are essentially identical, but the former is more rapid and convenient, and is set out in *Protocol 7*. Phalloidin staining of growing cells sees staining either at one end of the cell (always the old end), at both ends after NETO (although staining at the new end is less intense), or at the cell equator. G-actin is not stained by this method. By comparing phalloidin staining patterns with those obtained with DAPI and calcofluor (see above) the distribution of actin through the cell cycle has been carefully charted (30). An example of rhodamine–phalloidin staining of cells is shown in *Figure 4B*.

---

**Protocol 7.** Rhodamine–phalloidin staining of F-actin

*Materials*
- PM buffer: 35 mM $KH_2PO_4$ (pH 6.8), 0.5 mM $MgSO_4$
- Triton X-100
- Rhodamine-conjugated phalloidin
- Poly-L-lysine coated coverslip (*Protocol 3*)

*Method*

1. Grow 50 ml cells to mid-exponential phase of growth ($1 \times 10^6$–$1 \times 10^7$ cells/ml) in minimal medium at the desired temperature.
2. Fix cells by adding 5 ml of freshly-prepared 30% formaldehyde in PM buffer to the culture. Continue shaking for 30 min.
3. Pellet the cells by centrifugation at 1000 *g* for 5 min at room temperature and wash the pellet three times, each time with 10 ml of PM buffer.
4. Permeabilize the cells by resuspending in 10 ml of PM containing 1% Triton X-100 for 30 sec.
5. Repeat Step 3, finally resuspending the pellet, or part thereof, in 50–100 μl of 20 μg/ml rhodamine-conjugated phalloidin. Incubate 1 h in the dark on a rotary inverter.
6. Apply the cells on to a poly-L-lysine coated coverslip and allow to dry, before inverting on to a small drop of PM buffer (containing 1 μg/ml DAPI and 1 mg/ml *p*-phenylenediamine if required, see *Protocol 6*). Seal round the edges and examine under the fluorescence microscope.

To visualize both areas of cell-wall growth and F-actin simultaneously, resuspend the cells at Step 5 in 50 μl of rhodamine–phalloidin plus 50 μl 1 mg/ml calcofluor (see Section 4.2). Note that rhodamine–phalloidin binds perhaps ten times more efficiently than fluorescein–phalloidin.

## 4.4 Immunofluorescence microscopy

### 4.4.1 Visualizing microtubule structures by indirect immunofluorescence

The ability to visualize microtubule structures in fission yeast using anti-tubulin antibodies by indirect immunofluorescence has proved enormously valuable in the analysis of the cell cycle and cell-cycle mutants, in particular in allowing one to distinguish between the $G_2$ phase and mitosis, and also for studying the precise terminal phenotypes of a variety of mitotic mutants. The most commonly used antibody in these studies is the mouse monoclonal YOL1/34 described by Kilmartin *et al.* (32), which is now available commercially (from Accurate Chemical and Scientific Corp.). Using this reagent, the behaviour of microtubules through the cell cycle of wild-type cells, as well as in a number of cell-cycle mutants, has been carefully mapped (33). Detailed protocols describing the preparation of cells for indirect immunofluorescence can be found in ref. 7 and 33.

### 4.4.2 General considerations

In recent years a number of groups have used indirect immunofluorescence microscopy as a means of determining the subcellular localization of specific proteins using polyclonal antisera raised, mostly in rabbits, against bacterially-expressed proteins or synthetic peptides (24, 34–36). Generally affinity-purified rather than crude antisera have been used for these studies, with similarly treated pre-immune sera providing a negative control. For a wide-ranging discussion of issues to be considered when attempting indirect immunofluorescence in yeast see ref. 37.

An alternative approach to the question of subcellular localization, which is useful in conjunction with indirect immunofluorescence, is to ask whether the protein under investigation is located predominantly in the nucleus or cytoplasm. This can be done by Western blotting of fractionated cell extracts. A number of cell-cycle proteins have been analysed in this way. Fractionation is accomplished by centrifugation through a 10–30% linear Percoll gradient (38, 39).

### 4.5 Electron microscopy

Electron microscopy (EM) has been used to study a number of aspects of fission yeast cell ultrastructure through the cell cycle. Protocols for the preparation of cells for electron microscopy together with examples of structures visualized in this way can be found in ref. 40–42.

## 5. Synchronizing cells

Although a number of different methods have been used over the years to obtain a population of synchronously-dividing fission yeast cells, centrifugal elutriation is now the method of choice. Methods for obtaining synchronously-dividing populations of cells are dealt with in Chapter 2.

## 6. Cell-cycle inhibitors

### 6.1 Hydroxyurea

Hydroxyurea is a potent inhibitor of DNA precursor metabolism that acts by binding to the catalytic subunit of ribonucleotide reductase (43). Fission yeast cells treated with hydroxyurea (Sigma H8627) undergo cell-cycle arrest and become highly elongated (44). Hydroxyurea is added to either defined or complex (minimal) medium to a final concentration of 12 mM. Cells treated with hydroxyurea at this concentration will arrest for up to 4–6 h, but beyond this time the cells may begin to leak through the cell-cycle block. At higher concentrations hydroxyurea has toxic effects.

*5: Fission yeast cell cycle*

## 6.2 Benomyl and related compounds

Benomyl (methyl-1-butylcarbamoyl-benzimidazol-2-yl carbamate; DuPont), MBC (methylbenzimadazol-2-yl carbamate; DuPont), and TBZ (thiabendazole; Sigma T8904) are microtubule inhibitors that block mitosis in fission yeast. Benomyl is used at a final concentration of 20 µg/ml (from a stock solution of 10 mg/ml in dimethylsulfoxide): following addition of benomyl, the medium must be autoclaved (to generate active MBC, see below) prior to use. Transfer of cells from minimal medium to medium containing 20 µg/ml benomyl at 25 °C results in the fraction of binucleate cells dropping from ~ 5% to zero within 30 min (45) and remaining low for 3.5 h thereafter. At later times, however, (3.5–4 h after transfer to benomyl-containing medium) a gradual increase in cell number is seen, although this varies from culture to culture. This effect cannot be prevented by increasing the initial benomyl concentration which can lead to cells becoming abnormally swollen or branched. The sensitivity of cells to benomyl decreases at temperatures above 25 °C. MBC may be used in place of benomyl at a final concentration of 20 µg/ml, in which case autoclaving is not required. The action of both benomyl and MBC is strongly temperature-dependent. TBZ is used at a final concentration of 100 µg/ml from a stock solution at 10 mg/ml prepared in dimethylsulfoxide (46). Note that nocodazole, a widely used microtubule inhibitor in both mammalian cells and budding yeast, has not been used to block mitosis in fission yeast successfully.

## 6.3 Cycloheximide

Although not a specific inhibitor of cell-cycle progress, cycloheximide has been used to test the protein synthesis requirement for completion of certain cell-cycle events (45). From a stock solution (5 mg/ml in water) cycloheximide (Sigma C6255) is added to give a final concentration of 100 µg/ml. This concentration of cycloheximide has been reported to be sufficient to inhibit the rate of uptake of radiolabelled leucine by more than 90% within 2 min of addition (47).

# 7. Cell-cycle genetics

## 7.1 Cell-cycle mutants

Cell-cycle mutants fall into a number of classes, the details of which mostly lie outside the scope of this chapter (reviewed in ref. 8). The first cell-cycle mutants to be isolated defined genes whose functions were required for cell-cycle progress but which played no part in cell growth (2, 3). As a result the mutant cells underwent cell elongation when shifted to the restrictive temperature but did not divide. The isolation of regulatory mutants altered at the $G_2$–M transition followed soon after (4, 5), and at present over 50 genes have been defined as being involved in cell-cycle events (8). The following sections briefly summarize several important aspects to be considered when analysing or using a cell-cycle mutant for the first time as, for example, when preparing to embark on experiments intended to clone the gene, or a higher eukaryotic homologue, by complementation.

## 7.2 Characterization of new mutants

### 7.2.1 General points concerning analysis of mutants

Several properties of any mutant strain need to be considered before setting out on any physiological, cytological, or genetic analysis. In the case of mutants with conditional lethal defects, such as the cdc mutants described above, establishing optimal permissive growth conditions as well as restrictive conditions is vitally important. Almost all the temperature-sensitive (that is, heat-sensitive) mutants isolated to date grow well at 25 °C, and most at 28–29 °C. Beyond these temperatures, however, the ability to grow and divide varies enormously from strain to strain: some mutants will arrest at 32 °C, while some are still able to grow and divide slowly at temperatures as high as 35 °C (though typically not without a marked loss of viability). These properties are, of course, determined by the effect of the mutation on the wild-type protein function, which cannot be generalized. Cold-sensitive mutations have also been isolated (48, 49). These generally grow well at 32–35 °C, but are defective at 20 °C. When applying a cell-cycle block by shifting cells to 20 °C, it is important to bear in mind that the time taken for all the cells in a population to arrest will be substantially longer than would be required to block cells carrying a temperature-sensitive mutation, as the growth rate at 20 °C is much reduced even when compared to 25 °C (generation time of approximately 8 h).

The type of growth medium used (whether complex or defined) can also exert effects upon the mutant phenotype that need to be characterized. These effects can be subtle (such as altering the restrictive temperature by a degree or so) or quite marked (completely masking the mutant phenotype: see, for example, ref. 50, 51). A number of temperature-sensitive mutant strains are also sensitive to osmotic conditions, being suppressed by the presence of sorbitol in media used in some transformation methods (see below). The frequency of reversion of a lethal mutation is also of some importance since reversion at high frequency ($> 10^{-5}$ events per generation) can make the interpretation of results difficult as well as making the strain difficult to handle (see for example ref. 52).

### 7.2.2 Transition point determination

For conditional mutants that arrest cell-cycle progress, it is possible to determine the time of action of the gene product by determining its transition point, defined as the last point of action of a gene-controlled step (53). Transition points, in conjunction with other types of analysis, can provide valuable information about the likely defect of a new mutant, and can be determined by shifting either an asynchronous culture, or samples from a synchronous culture, to the restrictive temperature and determining at which cell-cycle stage cells can complete the ongoing cell cycle. Ideally transition points should be determined for more than one allele of any gene if these exist (53).

### 7.2.3 Ordering gene functions by reciprocal shifts

This method has proved useful in determining the order of gene functions and their relation to cell-cycle stages defined by specific inhibitors (53). For example, to investigate the latter point, cells carring a temperature-sensitive cell-cycle mutation may be arrested at their permissive temperature using the DNA synthesis inhibitor hydroxyurea and then, when all the cells have become blocked, transferred to their restrictive temperature, at which point the hydroxyurea is washed out. If the cells are capable of cell division under these circumstances then the cell-cycle arrest point lies before the hydroxyurea block, whereas if no division is possible the function lies after the hydroxyurea block. To generate the maximum amount of information from such experiments it is necessary to do both reciprocal shifts. In addition, it is vital to establish that the condition used to arrest cell-cycle progress is readily reversible. This has been shown to be the case for both hydroxyurea (44) and benomyl (45), but the degree of reversibility shown by different cell-cycle mutants varies greatly (45). The limitations of such experiments are considered at length in ref. 53.

## 7.3 Cloning of cell-cycle genes

### 7.3.1 Cloning by complementation

Complementation of a conditionally lethal strain using a plasmid library provides the most straightforward means of gene cloning in fission yeast and has been the method of choice for cloning almost all the cell-cycle genes isolated to date. Originally transformation was achieved using a protoplasting method (54), adapted from that developed for budding yeast, but in recent years a variety of more effective methods have been developed (see *Table 2*). Transformation efficiencies range from about $10^3$–$10^4$ transformants/μg of plasmid DNA for lithium chloride (56, 57), $10^4$–$10^5$ for traditional protoplasting methods (54) to up to $10^6$/μg for electroporation (59, 60), lithium acetate (58), or lipofectin-addition methods (55).

A number of vectors suitable for the construction of genomic libraries for cloning by complementation have been described in recent years (reviewed in ref. 7), but plasmids pDB262 and pWH5 are of particular merit (61, 62). pDB262 and pWH5 both carry the bacteriophage λ cI repressor gene

**Table 2.** Transformation methods for fission yeast

| Method | Efficiency (per μg) | Reference |
|---|---|---|
| Protoplasting method | $10^4$–$10^5$ | 54 |
| Protoplasting/lipofectin | $10^5$–$10^6$ | 55 |
| Lithium chloride | $10^3$–$10^4$ | 56, 57 |
| Lithium acetate | $10^5$–$10^6$ | 58 |
| Electroporation | $10^5$–$10^6$ | 59, 60 |

together with a tetracycline-resistance gene fused to the λ Pr promoter, in addition to a yeast selectable marker and autonomously replicating sequence. Insertion of DNA into restriction sites within the λ cI gene (*Hind*III or *Bcl*I in the case of pDB262; *Hind*III, *Bcl*I, or *Sma*I in pWH5) inactivates the repressor thus relieving the repression of the tetracycline gene to provide a positive selection for recombinant plasmids during gene bank construction. pWH5 is also ampicillin-resistant. Following transformation, the cells are plated on selective media and incubated either at the permissive temperature until colonies form, at which point these can be replica plated to restrictive conditions, or can be placed directly at the restrictive temperature. An alternative to placing the cells directly at the restrictive temperature is to incubate them for a short period (typically overnight) at the permissive temperature before transferring to the restrictive temperature, thereby allowing the cells time to recover from the transformation procedure before imposing selection. The choice of protocol to follow at this stage will depend upon the results of pilot experiments using the mutant strain of interest. In cases where a selection other than temperature is desired (resistance to an inhibitor for example) similar considerations will apply.

Once clones able to grow at the restrictive temperature have been identified it is essential to show that this phenotype is conferred by the plasmid and not by a chromosomal suppressor mutation. This is done by checking for co-segregation of the rescue phenotype and the nutritional marker carried by the plasmid (*Protocol 8*).

---

**Protocol 8.** Checking for co-segregation of plasmid and phenotype

1. Streak a single transformant colony on selective medium to purify, then re-streak a single colony on to non-selective media (complex media is ideal, see ref. 7) and incubate at the permissive temperature until colonies form.

2. Replica plate on to selective media. Incubate 1–2 days at both the permissive and restrictive temperatures.
3. Score the plates for growth:
   - If the rescue phenotype is associated with the plasmid then all the colonies capable of growth at the restrictive temperature will be prototrophic. The opposite should also be true, that is, that all the prototrophic colonies ought to be capable of growth at the restrictive temperature. However, this result can be complicated by the ability of individual cells to pick up more than one plasmid by transformation. Relaxing selection for the autotrophic marker while growing the cells at the restrictive temperature prior to testing for co-segregation can allow unnecessary plasmids to be lost.
   - In cases where growth at restrictive conditions is due to back mutation or unlinked chromosomal suppressing mutations all the colonies will show the rescue phenotype while only some will be prototrophic.

In a small minority of cases reported to date it has proved impossible to isolate the desired gene by this approach. The simplest explanation for this is that the wild-type gene sequence is not intact in any of the plasmid libraries used, which serves to emphasize the wisdom of screening several libraries. It may also be the case that overexpression (from a multicopy plasmid) of the wild-type gene has a deleterious effect on cell growth. Several ways of overcoming this problem, none of which rely upon prior knowledge of a gene's function, have been suggested. For example, Hiraoka *et al.* (63), faced with difficulties in cloning the wild type $nda3^+$ gene (which encodes β-tubulin), first cloned, by complementation, the mutant allele $nda3$-K311 and then isolated the wild-type gene by hybridization, subsequently demonstrating that its overexpression was lethal to the cell. An alternative strategy for isolating the gene of interest, in cases where genetic mapping data is available, is to clone a nearby gene first by complementation and then to chromosome walk to the desired locus (1 cM = approximately 10 kb: see ref. 64). An example of this approach is described in ref. 65.

### 7.3.2 Recovering plasmids from yeast transformants

Once transformants have been identified that are able to grow under the relevant restrictive conditions the next step is to recover plasmids from the cells and to purify these through bacteria. In the past plasmid recovery has been rather variable, probably because many plasmids are prone to rearrangement in yeast. Such rearrangements can result in the disruption of functions required for plasmid propagation in bacteria. The method given below (*Protocol 9*) has proved successful in the vast majority of cases.

**Protocol 9.** Recovery of plasmids from fission yeast (66)

*Materials*
- 20 mM citrate phosphate buffer, pH 5.6
- Zymolase 20T (ICN 32–092–1)
- SDS
- 5 × TE buffer: 50 mM Tris–HCl (pH 7.6), 5 mM EDTA (pH 8.0)
- 5 M potassium acetate (pH 5.5)
- GeneClean kit (Bio 101–3105; Stratech Scientific)
- *E. coli* (e.g. JA226)

*Method*
1. Grow up a single transformant colony at permissive temperature in 10 ml of selective medium for 2 days.
2. Harvest the cells by centrifugation (1000 $g$, 5 min, room temperature) and resuspend the pellet in 1.5 ml of 20 mM citrate phosphate buffer, pH 5.6, containing 2.5 mg/ml zymolyase 20T.
3. Incubate at 37 °C for 2 h. Test cells for sensitivity to detergent by mixing 5 μl of suspension with 5 μl of 20% SDS on a microscope slide. The cells should become phase-dark under these conditions.
4. Pellet the cells for 2 min in a microcentrifuge. Discard the supernatant and resuspend the pellet in 300 μl of 5 × TE containing 1% SDS. Heat to 65 °C for 10 min, then add 100 μl of 5 M potassium acetate pH 5.5.
5. Incubate on ice for 30 min, then centrifuge at 16 000 $g$ for 15 minutes at 4 °C. Carefully remove the supernatant (400 μl) and discard the pellet. At this stage the supernatant is treated according to the protocol of Vogelstein and Gillespie (67). The GeneClean kit (Bio 101–3105; Stratech Scientific) is a convenient commercial source of the required reagents.
6. Transform the purified DNA into *E. coli*. Traditionally *recBC E. coli* strains such as *E. coli* JA226 have been used for this purpose. Non-*recBC* strains can be used however, although this may lead to difficulties in recovering certain plasmids.

It is essential to recheck the ability of plasmids recovered in this way to function in yeast. Fission yeast cells can take up more than one plasmid by transformation and plasmid rearrangements are common so not all the recovered plasmids will necessarily confer the phenotype on retesting. On retesting, all the transformants obtained with a plasmid should show the rescued phenotype.

5: *Fission yeast cell cycle*

### 7.3.3 Integrative mapping

Once a plasmid has been isolated with rescuing activity it is essential to confirm that the cloned sequence corresponds to the gene of interest. In cases where several non-overlapping plasmids are obtained this assumes even greater importance, as in many cases the isolation of the desired gene is accompanied by one or more extragenic suppressors (see for example ref. 68–70). To demonstrate that the correct gene has been cloned the recovered plasmids are retransformed into the mutant strain and mitotically-stable integrants isolated (see below). Integrant strains are then analysed genetically.

---

**Protocol 10.** Integrative mapping

1. Retransform the rescuing plasmid(s) into the original mutant strain. In order to facilitate homologous recombination at a higher frequency than normal for a circular plasmid, the plasmid can be linearized by restriction enzyme cleavage (within the genomic DNA) prior to transformation. Alternatively, the rescuing DNA can be cloned into an integrative vector (see ref. 62).

2. Identify putative integrative transformants. Although these will generally grow faster than transformants in which the plasmid is being maintained as an episome, and so can be readily identified, it is essential to confirm their stability by testing for stable inheritance of the nutritional marker (see *Protocol 8*). It is advisable also to check that the plasmid has integrated at the correct site by Southern blotting.

3. Cross the integrant strain (by standard methods: see ref. 8) to (a) an otherwise wild-type auxotroph of opposite mating type, and (b) the mutant strain used to isolate the plasmid (again the opposite mating type is required). After 2–3 days at 25 °C, prepare the spores by helicase treatment (8) and plate on complex media (7). Incubate till colonies form.

4. Score products by replica plating:
   - For cross (a), if the plasmid has integrated at the mutant locus all the products of the cross should be able to grow under restrictive conditions. If the plasmid has integrated elsewhere in the genome (because the plasmid carried an unlinked suppressor gene) then a high proportion (50%) of the meiotic products will be unable to grow under restrictive conditions.
   - For cross (b), all those clones capable of growth at the restrictive temperature (50% of the total) should be prototrophic.

Note that it is not always necessary to isolate the plasmid before integrative mapping: spontaneous integrants derived from the original mitotically-unstable transformants, identified by their faster growth rate, can be used at this stage. The use of integrants isolated by retransformation is recommended, however.

### 7.3.4 Delimiting the functional region

Fission yeast genes can be particularly close together in the chromosome, often within 1 kb, making it particularly important to unambiguously identify the open reading frame of the gene of interest. This is most commonly achieved simply by testing subclones (often generated for the purpose of DNA sequencing) for activity, although transposon mutagenesis has also been used (71, 72). Note that in a number of cases truncated protein products have been shown to have some degree of rescuing activity (see for example ref. 73, 74) so care is required in the interpretation of results obtained from such experiments. Subclones can also be used to map the location of the lesions within mutant chromosomal alleles (by marker rescue: see ref. 52, 75). This is particularly useful for narrowing down the region to be sequenced in order to identify the nature of the mutation (see Section 7.4.3).

### 7.3.5 Cloning fission yeast genes by complementation in budding yeast

To date few fission yeast cell-cycle genes have been cloned by complementation of budding yeast cell-cycle mutants, largely because of the difficulties imposed by the existence of introns within *S. pombe* genes that are unlikely to be spliced in *S. cerevisiae*. This problem has been successfully circumvented (76) by the use of cDNA rather than genomic DNA plasmid libraries, such as that described in ref. 77.

## 7.4 Manipulation of cloned genes

### 7.4.1 Overexpression

Overexpression of a cloned gene can be a powerful method for investigating gene function, and a number of promoter elements can be used in fission yeast to achieve a high level of expression. These can be divided into two categories: constitutive and regulatable. A number of constitutive promoters have been used to direct high-level expression of cell-cycle genes: these include the adh promoter, derived from the endogenous alcohol dehydrogenase gene $adh1^+$ (78), as well as the SP6 bacteriophage (79), and SV40 early viral promoters (80).

In recent years a number of regulatable fission yeast promoters have been isolated. The first of these to be described, the nmt promoter, is still the best characterized and probably the most useful of the group (81, 82), particularly since modified versions of the promoter, giving three different levels of expression, are now available (83). The nmt promoter is derived from the $nmt1^+$ gene; $nmt1^+$ appears to have a role in thiamine biosynthesis as cells deleted for the gene are thiamine auxotrophs (81). Transcription from the nmt promoter is repressed in the presence of thiamine and induced in its absence, although induction is preceded by a lag period of 9–12 h at 32 °C during which time the cells' internal thiamine pools are depleted.

---

**Protocol 11.** Inducing overexpression from the nmt promoter

1. Grow cells up in minimal media supplemented with 5 µM thiamine (Sigma T4625) till they reach mid-exponential phase (1 × $10^6$–1 × $10^7$ cells per ml).
2. Determine cell number and sample the required number of cells (for inoculation of the main culture; see below) by centrifugation (1000 $g$, 5 min, 4 °C): as the culture will continue to grow during induction (undergoing at least 4 cell-number doublings during the period of induction at 32 °C) the initial inoculum will be small.
3. Discard the supernatant and wash the pellet three times with minimal medium. At each wash vortex the cells thoroughly.
4. Finally resuspend the cells in minimal medium and incubate overnight at 25–32 °C. Induction of the promoter occurs 9–12 h after transfer to thiamine-free medium at 32 °C.

---

### 7.4.2 Deletion or disruption

The deletion or disruption of genes carried on the chromosome is a particularly powerful means of investigating gene function in yeast and has been widely used to create null mutants of many genes. In fission yeast, although the efficiency of gene replacement versus non-homologous integration of the introduced DNA is not as high as that observed with budding yeast, gene deletions or disruptions have been described for most, if not all, of the well-studied cell-cycle genes. General considerations regarding strains and constructs appropriate for gene deletion/disruption experiments are described in refs. 7 and 62.

### 7.4.3 Cloning mutant alleles

While direct sequencing of polymerase chain reaction (PCR) products provides the most rapid means of determining the DNA sequence of mutant alleles, two other methods have been used in fission yeast to clone mutant alleles from the chromosome for sequencing by making use of recombination between the genomic locus and plasmids introduced by transformation. These methods are gapped plasmid repair (52, 85, 86) and targetted plasmid integration and recovery (52). The advantage that these methods hold over PCR is that the mutant gene can subsequently be re-introduced into yeast to test function without the need to identify a clean copy of the allele that lacks the extra mutations that arise from error-prone PCR amplification.

## 7.5 Isolation of higher eukaryotic homologues

### 7.5.1 Cloning by complementation

Although the methods that can be used to clone higher eukaryotic homologues of fission yeast cell-cycle genes by complementation are mostly identical to those described above for cloning their fission yeast prototypes, several additional points must be considered.

- It is essential to screen cDNA rather than genomic DNA libraries: although one higher eukaryotic intron has been shown to be spliced in fission yeast (87), this now appears to have been an isolated case.
- A greater number of transformants must be screened to take into account the more complex genomes of higher eukaryotes.
- cDNA libraries in non-yeast vectors may be prone to rearrangement in yeast (see below).
- The ability of a cDNA to rescue may depend upon the extent to which the higher eukaryotic protein has diverged in structure from its fission yeast homologue: efficient rescue may only be possible at semi-permissive temperatures, or at temperatures close to those at which the higher eukaryotic protein normally functions.
- The efficiency of expression of the cDNA will also play a part in influencing its ability to rescue: not all promoters found in cDNA expression vectors will function in yeast, although some do (such as the SV40 early viral promoter; ref. 80).

As with cloning fission yeast genes, once transformants growing under restrictive conditions have been obtained, it is desirable (if possible) to check for co-segregation of the rescue phenotype and the nutritional marker, and necessary to demonstrate that all the transformants obtained upon retesting a recovered plasmid show the rescued phenotype (see Section 7.3.2). If possible, the rescuing ability should be checked against a true null allele (a non-functional gene deletion) as well as a conditional-lethal mutation.

The feasibility of cloning higher eukaryotic functional homologues of cell-cycle genes by complementation was first demonstrated by Lee and Nurse (88) who cloned a cDNA encoding human p34$^{cdc2}$ by a procedure that involved co-transformation of a pre-existing human cDNA library along with a fission yeast vector (pDB262) carrying a selectable marker. In this instance expression of the cDNAs was by the SV40 viral early promoter. The co-

## 5: Fission yeast cell cycle

transformation method has the advantage that it obviates the need to construct a cDNA library in a fission yeast vector specifically for use in the complementation experiments. One disadvantage is that one cannot test for co-segregation of rescue phenotype and nutritional marker as the two are carried on separate plasmids. However, it is possible to demonstrate mitotic instability of the rescue phenotype by streaking the transformed cells to single colonies on fully-supplemented complex media at the permissive temperature and then replica plating to the restrictive temperature. Only a small fraction of the colonies will be able to grow at the higher temperature, the remainder having lost the plasmid. With many cell-cycle mutants microscopic analysis of transformant colonies will also reveal instability. Co-transformation has recently been used successfully by Okayama and co-workers to clone cDNAs encoding homologues of the $p80^{cdc25}$ and $p107^{wee1}$ proteins (89, 90). A second problem that can arise with using pre-existing higher eukaryotic cDNA libraries in co-transformation experiments is that the parent plasmids may be prone to rearrangement by recombination in fission yeast. (This can also be a problem with some fission yeast vectors.) This problem was overcome by Jimenez and colleagues (79, 91) who, when faced with difficulties in recovering rearranged plasmids in bacteria, modified a *Drosophila* cDNA library by the insertion of additional sequences into a unique restriction site in the plasmid used to construct the library. The additional sequences consisted of an adaptor molecule comprising an *S. pombe* selectable marker together with an autonomously replicating sequence. The disadvantage of this approach is that the sequence complexity of the library diminishes with each cloning and amplification step. Nevertheless the adapted library was successfully used to clone cDNAs encoding *Drosophila* functional homologues of both the $p34^{cdc2}$ and $p80^{cdc25}$ proteins (79, 91). In this case transcription of the cDNAs was promoted by the bacteriophage SP6 promoter. The ideal solution to such problems is, of course, to construct the required cDNA library in a fission yeast vector containing a suitably-placed promoter element. This approach has recently been taken by B. Edgar, C. Norbury, and P. Nurse at Oxford (personal communication) who have constructed several higher eukaryotic cDNA libraries in an nmt-promoter replicating vector.

### 7.5.2 Cloning by physical methods: PCR, hybridization, and antibody screens

Each of these approaches has been taken to isolate structural homologues of fission yeast cell-cycle genes. Successful PCR strategies are dependent upon the design of suitable degenerate oligonucleotide primers (84), something that is most easily done when several protein sequences have already been determined. One approach that has proved successful is to seek to initially isolate a budding yeast functional homologue by complementation (92) and then design PCR primers corresponding to sequences conserved between the two yeasts using the protein sequence information thus obtained (93).

## 8. Summary

We have presented here a summary of the methods commonly used for physiological, cytological, and genetical analysis of the fission yeast cell cycle. Although the range of methods is not comparable, as yet, to that available for analysis of the cell cycle of the budding yeast, it is only likely to be a matter of time before additional methods are refined for use in fission yeast. The advantage of using *S. pombe* over *S. cerevisiae* is that its cell cycle is a more typically eukaryotic one, with discrete $G_1$, S, and $G_2$ phases leading to a mitosis marked by formation of an intranuclear mitotic spindle. In budding yeast, defining the time of mitotic initiation is particularly difficult as a short spindle persists for much of the cell cycle (discussed in ref. 94). In addition, the chromosomes of *S. pombe* undergo visible condensation upon entry into mitosis, whereas those of *S. cerevisiae* do not. These conserved aspects of its cell cycle make the organism a particularly attractive model system in its own right, and also for identifying and investigating the behaviour of heterologous genes from higher eukaryotic cells.

## Acknowledgements

We would like to thank our colleagues for their advice and encouragement during the preparation of this article. In particular we are grateful to the following for reading and commenting on the manuscript: Dr J. Creanor (University of Edinburgh), and Drs J. Hayles and C. J. Norbury (University of Oxford). Research in our laboratory is funded by the Medical Research Council (MRC) and Cancer Research Campaign. S.A.M. holds an MRC Post-Doctoral Training Fellowship.

## References

1. Mitchison, J. M. (1957). *Exp. Cell Res.*, **13**, 244.
2. Nurse, P., Thuriaux, P., and Nasmyth, K. A. (1976). *Mol. Gen. Genet.*, **146**, 167.
3. Nasmyth, K. A. and Nurse, P. (1981). *Mol. Gen. Genet.*, **182**, 119.
4. Nurse, P. (1975). *Nature*, **256**, 547.
5. Nurse, P. and Thuriaux, P. (1980). *Genetics*, **96**, 627.
6. Fantes, P. A. (1989). In *Molecular biology of the fission yeast* (ed. A. Nasim, P. Young, and B. F. Johnson), pp. 128–204. Academic Press, London.

7. Moreno, S., Klar, A., and Nurse, P. (1991). In *Methods in enzymology*, Vol. 194 (ed. C. Guthrie and G. R. Fink), pp. 795. Academic Press, London.
8. Gutz, H., Heslot, H., Leupold, U., and Loprieno, N. (1974). In *Handbook of genetics* (ed. R. C. King), Vol. I, pp. 395–446. Plenum, New York.
9. Kohli, J. (1987). *Curr. Genet.*, **11**, 575.
10. Robinow, C. F. (1977). *Genetics*, **87**, 491.
11. Egel, R. (1989). In *Molecular biology of the fission yeast* (ed. A. Nasim, P. Young, and B. F. Johnson), pp. 31–74. Academic Press, London.
12. Munz, P., Wolf, K., Kohli, J., and Leupold, U. (1989). In *Molecular biology of the fission yeast* (ed. A. Nasim, P. Young, and B. F. Johnson), pp. 1–30. Academic Press, London.
13. Lennon, G. G. and Lehrach, H. (1992). *Curr. Genet.*, **21**, 1.
14. Mitchison, J. M. (1970). In *Methods in cell physiology*, Vol. 4 (ed. D. M. Prescott), pp. 131. Academic Press, London.
15. Burton, K. (1956). *Biochem J.*, **62**, 315.
16. Burton, K. (1968). In *Methods in enzymology*, Vol. 12 (ed. L. Grossman and K. Holdave), pp. 45. Academic Press, London.
17. Bostock, C. R. (1970). *Exp. Cell Res.*, **60**, 16.
18. Gendimenico, G. J., Bouquin, P., and Tramposch, K. M. (1988). *Anal. Biochem.*, **173**, 45.
19. Creanor, J. and Mitchison, J. M. (1990). *J. Cell Sci.*, **96**, 435.
20. Nasmyth, K. A., Nurse, P., and Fraser, R. S. S. (1979). *J. Cell Sci.*, **39**, 215.
21. Costello, G., Rodgers, L., and Beach, D. (1986). *Curr. Genet.*, **11**, 119.
22. Sazer, S. and Sherwood, S. W. (1990). *J. Cell Sci.*, **97**, 507.
23. Moreno, S., Hayles, J., and Nurse, P. (1989). *Cell*, **60**, 321.
24. Enoch, T., Peter, M., Nurse, P., and Nigg, E. (1991). *J. Cell. Biol.*, **112**, 797.
25. Simanis, V. and Nurse, P. (1986). *Cell*, **45**, 261.
26. Brizuela, L., Draetta, G., and Beach, D. (1987). *EMBO J.*, **6**, 3507.
27. Williamson, D. H. and Fennell, D. J. (1975). In *Methods in cell biology*, Vol. 12 (ed. D. M. Prescott), pp. 335. Academic Press, London.
28. Umesono, K., Hiraoka, Y., Toda, T., and Yanagida, M. (1983). *Curr. Genet.*, **7**, 123.
29. Mitchison, J. M. and Nurse, P. (1985). *J. Cell Sci.*, **75**, 357.
30. Marks, J. and Hyams, J. S. (1985). *Eur. J. Cell Biol.*, **39**, 27.
31. Marks, J., Hagan, I. M., and Hyams, J. S. (1986). *J. Cell Sci. Suppl.*, **5**, 229.
32. Kilmartin, J. V., Wright, B., and Milstein, C. (1982). *J. Cell Biol.*, **93**, 576.
33. Hagan, I. M. and Hyams, J. S. (1988). *J. Cell Sci.*, **89**, 343.
34. Alfa, C. E., Ducommun, B., Beach, D., and Hyams, J. S. (1990). *Nature*, **347**, 680.
35. Booher, R. N., Alfa, C. E., Hyams, J. S., and Beach, D. H. (1989). *Cell*, **58**, 485.
36. Adams, A. E. M. and Pringle, J. R. (1984). *J. Cell Biol.*, **98**, 934.
37. Pringle, J. R., Adams, A. E. M., Drubin, D., and Haarer, B. (1991). In *Methods in enzymology*, Vol. 194 (ed. C. Guthrie and G. R. Fink), pp. 565. Academic Press, London.
38. Hirano, T., Hiraoka, Y., and Yanagida, M. (1988). *J. Cell Biol.*, **106**, 1171.
39. Adachi, Y. and Yanagida, M. (1989). *J. Cell Biol.*, **108**, 1195.
40. Tanaka, K. and Kanbe, T. (1986). *J. Cell Sci.*, **80**, 253.

41. Kanbe, T., Kobayashi, I., and Tanaka, K. (1989). *J. Cell Sci.*, **94**, 647.
42. Kanbe, T., Hiraoka, Y., Tanaka, K., and Yanagida, M. (1990). *J. Cell Sci.*, **96**, 275.
43. Elledge, S. J., Zhou, Z., and Allen, J. B. (1992). *Trends Biochem. Sci.*, **17**, 119.
44. Mitchison, J. M. and Creanor, J. (1971). *Exp. Cell Res.*, **67**, 368.
45. Fantes, P. A. (1982). *J. Cell Sci.*, **55**, 383.
46. Toda, T., Yamamoto, M., and Yanagida, M. (1981). *J. Cell Sci.*, **52**, 271.
47. Polanshek, M. M. (1977). *J. Cell Sci.*, **23**, 1.
48. Toda, T., Umesono, K., and Hiraoka, A. (1983). *J. Mol. Biol.*, **168**, 251.
49. Ohkura, H., Adachi, Y., Kinoshita, N., Niwa, O., Toda, T., and Yanagida, M. (1988). *EMBO J.*, **7**, 1465.
50. Ogden, J. E. and Fantes, P. A. (1986). *Curr. Genet.*, **10**, 509.
51. Krek, W., Marks, J., Schmitz, N., Nigg, E., and Simanis, V. (1992). *J. Cell Sci.*, **102**, 43.
52. Carr, A. M., MacNeill, S. A., Hayles, J., and Nurse, P. (1989). *Mol. Gen. Genet.*, **218**, 41.
53. Pringle, J. R. (1978). *J. Cell Physiol.*, **95**, 393.
54. Beach, D. and Nurse, P. (1981). *Nature*, **290**, 140.
55. Allshire, R. C. (1990). *Proc. Natl Acad. Sci. USA*, **87**, 4043.
56. Itoh, H., Fukada, Y., and Murata, K. (1983). *J. Bact.*, **153**, 163.
57. Broker, M. (1987). *BioTech.*, **5**, 6.
58. Okayama, K., Okayama, N., Kume, K., Jinno, S., Tanaka, K., and Okayama, H. (1990). *Nucl. Acids Res.*, **18**, 6485.
59. Hood, M. and Stachow, C. (1990). *Nucl. Acids Res.*, **18**, 688.
60. Prentice, H. (1992). *Nucl. Acids Res.*, **20**, 621.
61. Wright, A., Maundrell, K., Heyer, W.-D., Beach, D., and Nurse, P. (1986). *Plasmid*, **15**, 156.
62. Russell, P. (1989). In *Molecular biology of the fission yeast* (ed. A. Nasim, P. Young and B. F. Johnson), pp. 244–272. Academic Press, London.
63. Hiraoka, Y., Toda, T., and Yanagida, M. (1985). *Cell*, **39**, 349.
64. Gygax, A. and Thuriaux, P. (1984). *Curr. Genet.*, **8**, 85.
65. Uzawa, S., Samajima, I., Hirano, T., Tanaka, K., and Yangida, M. (1990). *Cell*, **62**, 913.
66. Hagan, I. M., Hayles, J. and Nurse, P. (1989). *J. Cell Sci.*, **91**, 587.
67. Vogelstein, B. and Gillespie, D. (1979). *Proc. Natl Acad. Aci. USA*, **76**, 615.
68. Hayles, J., Beach, D., Durkacz, D., and Nurse, P. (1986). *Mol. Gen. Genet.*, **202**, 291.
69. Ohkura, H., Kinoshita, N., Miyatani, S., Toda, T., and Yanagida, M. (1989). *Cell*, **57**, 997.
70. Fernandez-Sarabia, M.-J., McInerny, C. J., Harris, P., Gordon, C., and Fantes, P. A. (1993). *Mol. Gen. Genet.*, **238**, 241.
71. Warbrick, E. and Fantes, P. (1991). *EMBO J.*, **10**, 4291.
72. Warbrick, E. and Fantes, P. (1992). *Mol. Gen. Genet.*, **232**, 440.
73. Russell, P. and Nurse, P. (1986). *Cell*, **45**, 145.
74. Aves, S. J., Durkacz, B., Carr, A. M., and Nurse, P. (1985). *EMBO J.*, **4**, 457.
75. Hughes, D. A., MacNeill, S. A., and Nurse, P. (1992). *Mol. Gen. Genet.*, **231**, 344.
76. Forsburg, S. L. and Nurse, P. (1991). *Nature*, **351**, 245.

77. Fikes, J. D., Becker, D. M., Winston, F., and Guarente, L. (1990). *Nature*, **346**, 291.
78. Russell, P. and Hall, B. D. (1983). *J. Biol. Chem.*, **258**, 143.
79. Jimenez, J., Alphey, L., Nurse, P., and Glover, D. (1989). *EMBO J.*, **9**, 3565.
80. Jones, R. H., Moreno, S., Nurse, P., and Jones, N. C. (1988). *Cell*, **53**, 659.
81. Maundrell, K. (1990). *J. Biol. Chem.*, **265**, 10857.
82. Tommasino, M. and Maundrell, K. (1991). *Curr. Genet.*, **20**, 63.
83. Basi, G., Schmid, E., and Maundrell, K. (1993). *Gene* **123**, 131.
84. Innis, M. A., Gelfand, D. A., Sninsky, J. J., and White, T. J., (ed.) (1990). *PCR protocols: a guide to methods and applications*. Academic Press, London.
85. Russell, P. and Nurse, P. (1987). *Cell*, **49**, 559.
86. Szankasi, P., Heyer, W.-D., Schuchert, P., and Kohli, J. (1988). *J. Mol. Biol.*, **204**, 917.
87. Kaufer, N. F., Simanis, V., and Nurse, P. (1985). *Nature*, **318**, 78.
88. Lee, M. G. and Nurse, P. (1987). *Nature*, **327**, 31.
89. Nagata, A., Igarashi, M., Jinno, S., Suto, K., and Okayama, H. (1991). *New Biologist*, **3**, 959.
90. Igarashi, M., Nagata, A., Jinno, S., Suto, K., and Okayama, H. (1991). *Nature*, **353**, 80.
91. Alphey, L., White-Cooper, H., Dawson, I., Nurse, P., and Glover, D. M. (1992). *Cell*, **69**, 977.
92. Russell, P., Moreno, S., and Reed, S. I. (1989). *Cell*, **57**, 295.
93. Sadhu, K., Reed, S. I., Richardson, H., and Russell, P. (1989). *Proc. Natl Acad. Sci. USA*, **87**, 5139.
94. Nurse, P. (1985). *Trends Genet.*, **2**, 5.

# 6

# Cell-cycle analysis using the filamentous fungus *Aspergillus nidulans*

B. R. OAKLEY and S. A. OSMANI

## 1. Introduction

The filamentous fungus *Aspergillus nidulans* has proved to be an excellent experimental system for the identification, isolation, and analysis of genes that are essential to cell-cycle progression and regulation. Studies in *A. nidulans* have led to the discovery of several key elements of cell-cycle progression and regulation that were not first identified in other experimental systems (1). It is a homothallic ascomycetous fungus that has both a sexual and an asexual life cycle (*Figure 1*). The classical genetics and molecular genetics of *A. nidulans* have been previously described (2) and are, in many respects, as advanced as the yeast systems. It is the purpose of this chapter to describe some of the techniques that can be utilized to study the cell cycle in *A. nidulans*.

Typically, although dependent on growth media and temperature, *A. nidulans* has a $G_1$ phase of 15 min, an S phase of 40 min, a clearly defined $G_2$ of 40 min, and a mitotic phase (M phase) of 5 min, giving a nuclear division generation time of 100 min (3). During early growth from asexual spores, called conidia, there is no cytokinesis after mitosis in *A. nidulans*; the daughter nuclei remain in a common cytoplasm (*Figure 1*). By counting the number of nuclei in a cell it is thus possible to determine the mitotic history of that cell. One nucleus demonstrates no mitosis has occurred, two nuclei indicate one mitosis completed, four nuclei two mitoses, and so on. This mode of growth has been useful in isolating and studying mutations that are specific to the nuclear division cycle (ref. 4 and *Figure 1*). A collection of cell-cycle specific mutations has been generated (4), many of which can be obtained from the Fungal Genetics Stock Center, as can strains that contain mutations that affect the tubulin genes of this species.

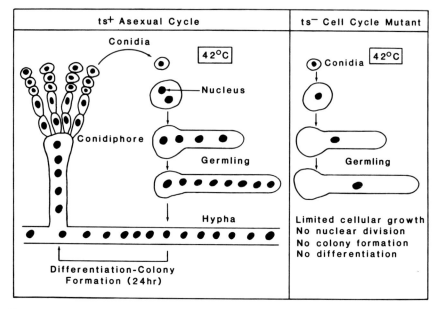

**Figure 1.** Life cycle of *Aspergillus nidulans*. Depicted on the left diagrammatically is the asexual life cycle of *A. nidulans* when conidia are inoculated on to solid medium. In liquid culture the differentiation programme is not initiated and cells continue growth as hyphae. The effect of a temperature-sensitive mutation on nuclear division and cellular growth is shown to the right. Mutations that prevent nuclear division, but not short-term growth, are defined as cell-cycle specific mutations. Those mutations that arrest in interphase are called *nim* mutants (for *n*ever *in m*itosis) and those that cause arrest in mitosis are called *bim* mutants (for *b*locked *in m*itosis).

## 2. Microscopic examination of the cell cycle and scoring of mitosis

The mitotic portion of the cell cycle is more easily defined in *Aspergillus nidulans* than in some other lower eukaryotes used for cell-cycle research (see Chapters 4 and 5). In *A. nidulans*, as in higher organisms, chromosomes condense and the mitotic spindle forms at the end of $G_2$. Mitosis lasts approximately 5% of the cell cycle (approximately 5 min at 37 °C), and the chromosomes decondense and the spindles break down at the onset of $G_1$. The mitotic index can be defined by examining chromosomal condensation (chromosome mitotic index), mitotic spindle formation (spindle mitotic index), or both. *A. nidulans* is coenocytic (*Figure 1*) and the nuclei within each hyphal segment tend to enter mitosis in synchrony. It is thus not appropriate to define the mitotic index as the fraction of nuclei in mitosis, but, rather, as the fraction of hyphal segments in mitosis.

6: Cell-cycle analysis using A. nidulans

## 2.1 Determination of chromosome mitotic index

The chromosomes of *A. nidulans* are small and are more difficult to stain and visualize than those of higher organisms. High-resolution optics and care in aligning the microscope are necessary. Among the most reliable methods of staining chromosomes are procedures that have long been in use, such as Feulgen staining and aceto–orcein staining. These procedures are laborious, but have the advantage that condensed, mitotic chromosomes stain more intensely than interphase nuclei allowing one to quickly determine whether a nucleus is in mitosis or interphase. These procedures have been described in detail by Robinow and Caten (5) and by Morris *et al.* (6) and we will not discuss them here. More recently the DNA-binding dye DAPI (4′, 6-diamidino-2-phenylindole) has been used to stain chromosomes for visualization by fluorescence microscopy. Chromosomes can be stained satisfactorily with DAPI by a number of protocols. *Protocol 1* from Oakley and Rinehart (7) for DAPI staining of germlings or hyphae growing in liquid culture (6)

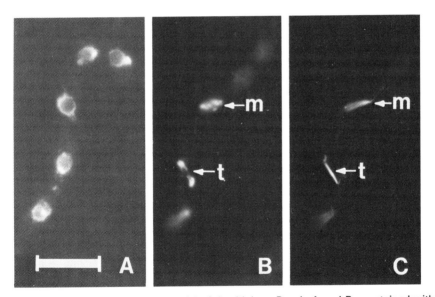

**Figure 2.** Mitotic and interphase nuclei of *A. nidulans*. Panels A and B are stained with DAPI. The nuclei in panel A are in interphase. Chromatin is not condensed and the large nucleolus displaces chromatin, resulting in a large unstained region in each nucleus. The nuclei in B are in mitosis. The chromosomes have condensed. One nucleus (m) is in medial nuclear division and the chromosomes are arranged along the mitotic spindle. In the other nucleus (t), the chromosomes have moved to the poles of the mitotic spindle. Panel C shows the same field as B but it is an immunofluorescence micrograph of material stained with an α-tubulin antibody. The spindles are rod-shaped and faint cytoplasmic microtubules extend from their ends. Magnifications for all panels are the same. Scale in A = 10 μm. Reproduced by permission from ref. 20.

gives good staining of nuclear chromatin as well as mitochrondrial genomes. DAPI staining is much more rapid than older techniques, but identifying mitotic nuclei is somewhat more difficult. Fortunately, since nuclei in a single hyphal segment are generally in mitosis at the same time, if it is difficult to determine whether a nucleus is in mitosis, an adjacent nucleus may be easier to score. Examples of DAPI-stained interphase and mitotic nuclei are shown in *Figure 2*.

---

**Protocol 1.** DAPI staining of chromosomes[a]

1. Prepare a stock solution of 1.5 μg/ml DAPI in distilled water. This solution is stable for several weeks at 4 °C in the dark.

2. Add 0.1 ml of 10% glutaraldehyde (electron microscopy grade) to a 0.9 ml sample of hyphae in a microcentrifuge tube. Allow the sample to fix for 10 min. (The glutaraldehyde should be at the same temperature as the sample and the sample should be fixed as quickly as possible after it is taken from the culture. Since mitosis only lasts approximately 5 min, and temperature shocks may inhibit the entry into mitosis, delay or significant temperature changes may dramatically reduce the mitotic index.) Subsequent steps may be carried out at room temperature.

3. Wash the hyphae by centrifugation (> 1000 g, 1 min) decant the supernatant and resuspend the pellet in distilled water. Leave the pellet in water for 10 min. Repeat this washing step.

4. Wash the hyphae twice for 15 min in acetone to lower background staining.

5. Wash the hyphae twice for 10 min in distilled water.

6. Centrifuge (> 1000 g, 1 min) and then resuspend the hyphae in DAPI solution (0.015 μg/ml in distilled water, made by diluting stock solution). Since DAPI may bleach, DAPI staining should be carried out in the dark if possible and DAPI-stained material should be stored in the dark.

7. Inhibit fading of the DAPI staining with antifading agents such as Citifluor AF1 (City University, London or Maravac Ltd, cat # CM512–1).

[a] From Oakley and Rihehart (7)

---

## 2.2 Determination of spindle mitotic index

The spindle mitotic index is best determined by staining germlings or hyphae for immunofluorescence microscopy with an antibody against α- or β-tubulin. Many of the monoclonal antibodies that have been made against brain tubulins do not bind to *A. nidulans* tubulins. Two monoclonal antibodies that stain *A. nidulans* microtubules and are available commercially are YOL1/34 (Accurate Cat. # MAS 0786 or Sera Labs) which was made against yeast α-tubulin (8) and DM1A (Sigma) which was made against mammalian α-tubulin (9). Ascites fluids often contain antibodies that recognize *A. nidulans* proteins other than α-tubulin and thus may give high backgrounds or false signals in fluorescence microscopy and multiple bands on Western blots. Supernatants of cultures producing monoclonal antibodies usually give fewer problems.

A number of problems must be overcome for staining to be successful. First, the mitotic spindles are small and good optics are necessary. Second, secondary antibody preparations, even those that have been affinity purified, usually have antibodies that recognize *A. nidulans* proteins. This causes a problem with background staining, but can be overcome by preparing an acetone powder of *A. nidulans* proteins and using the powder to adsorb the antibodies that bind to *A. nidulans* proteins (*Protocol 2*).

Another significant problem is that the hyphal wall of *A. nidulans* is impermeable to antibodies. For successful staining one must digest the wall to make it permeable; generally the enzyme Novozym 234 is used for digestion. Novozym 234 is produced by a fungus, *Trichoderma*, and the commercially available preparations contain large amounts of proteases. Extended digestion with Novozym 234 thus leads to loss of microtubules by proteolysis, while underdigestion leaves the wall impermeable to antibodies. Since the cell-wall digesting ability and protease content of Novozyme 234 varies greatly from batch to batch, digestion times that give satisfactory staining must be determined empirically for each batch. The protease activity of Novozym 234 is so high that inhibiting the activity with commercial protease inhibitors is difficult and prohibitively expensive. We inhibit the protease activity with egg white which is inexpensive and contains potent protease inhibitors. We also include Driselase (Sigma) in the wall-digesting solution because it is relatively low in proteases and the combination of Novozym 234 and Driselase digests the wall of *A. nidulans* more rapidly than does Novozyme 234 alone. The tips of growing hyphae and germlings are digested more easily than older hyphal segments. Staining of microtubules at hyphal tips, but not at older segments of hyphae, is an indication of underdigestion. Staining of microtubules away from tips along with fragmented or absent microtubules at hyphal tips is an indication of overdigestion. Finally, a procedure for reducing the protease activity of Novozym 234 has been recently published (10), and this reduced-protease Novozym 234 should be useful for immunofluorescence microscopy although we have not yet tested it.

**Protocol 2.** Immunofluorescence staining of *A. nidulans* mitotic spindles

*Materials*

- Medium
- Conidia
- 16% paraformaldehyde solution (prepared in a fume hood within one week of use):
  — Add 16 g paraformaldehyde[a] to 90 ml double-distilled water.
  — Warm to 60 °C.
  — Add saturated NaOH, dropwise, until the paraformaldehyde dissolves (about 10 drops; the dissolution occurs quickly, but a small amount of precipitate will remain); add double-distilled water to a final volume of 100 ml; remove precipitate by filtration (e.g.through a Whatman No. 1 filter).
- Fixative solution (prepare immediately before use):

  | | |
  |---|---|
  | — Double-distilled water | 5.9 ml |
  | — 1 M Pipes (adjust pH to 6.7 with KOH) | 1.0 ml |
  | — 0.25 M EGTA (adjust pH to 7.0 with KOH) | 2.0 ml |
  | — DMSO | 1.0 ml |
  | — 1 M MgSO$_4$ | 0.1 ml |
  | — 16% paraformaldehyde solution | 10.0 ml |

- Novozym 234 solution (prepare in advance and store at −70 °C:
  — Add 500 mg of Driselase[b] (Sigma) to 5 ml of double-distilled water.
  — Mix and leave on ice for 15 min.
  — Centrifuge at 2000 *g* for 5 min at 4° C and recover the supernatant.
  — Add 100 mg of Novozym 234 (Interspex Products cat. no. 0412–1).
  — Add 0.1 ml of 0.25 M EGTA and mix thoroughly.
  — Aliquot in microcentrifuge tubes (0.5 ml/tube), freeze at −70 °C.
- Digestion solution (prepare less than 5 min before use):
  — Thaw microcentrifuge tube of Novozym 234 solution, keep on ice.
  — Mix with an equal volume of egg white (normally stored frozen); the solution is initially viscous, but becomes less so as the egg white dissolves.
- Extraction solution (100 ml):

  | | |
  |---|---|
  | — 1 M Pipes, pH 6.7 | 10.0 ml |
  | — 0.25 M EGTA, pH 7.0 | 10.0 ml |

|  |  |
|---|---|
| — Nonidet P-40 (or Triton X-100) | 0.1 ml |
| — Double-distilled water | 79.9 ml |

- PEM (100 ml):

|  |  |
|---|---|
| — Double-distilled water | 84.5 ml |
| — 1 M Pipes, pH 6.7 | 5.0 ml |
| — 0.25 M EGTA, pH 7.0 | 10.0 ml |
| — 1 M MgSO$_4$ | 0.5 ml |

- Coverslips (prepare in advance):
  - Soak coverslips in 95% ethanol for 1 h and rinse thoroughly in double-distilled water.
  - Soak in 0.1 M HCl for 1 h and rinse thoroughly in double-distilled water.
  - Autoclave the coverslips on filter paper in glass Petri dishes.
- Small (coverslip-sized) Coplin jars (e.g. VWR Scientific, cat. no. 25452–002)
- Parafilm
- Primary and secondary antibodies
- Citifluor AF1 (Citifluor Ltd or Maravac)
- DAPI

*Method*

A. *Growing conidia on coverslips*[c]

1. Inoculate liquid medium with conidia ($5 \times 10^5$/ml).
2. Place the sterile coverslips in a sterile Petri dish and cover with the inoculated medium (about 10 ml/100 mm Petri dish).
3. Incubate at 37 °C without shaking. Spores will adhere to the coverslips and germinate. Useful germlings are present after approximately 6 h at 37 °C. Long hyphae generally do not adhere as well to coverslips or stain as well as germlings.

B. *Fixation and staining*

*NB* Unless otherwise stated, all procedures are carried out at room temperature. Fixation, extraction, and washes are carried out in small (coverslip-sized) Coplin jars.

1. Warm fixative solution in a covered Coplin jar to the incubation temperature of the culture.
2. Rapidly transfer the coverslips to the fixative solution.

**Protocol 2.** *Continued*

3. Remove the Coplin jar to room temperature and allow fixation to proceed for 45 min.
4. Wash the coverslips three times, 5 min each time, in PEM.
5. Place each coverslip face down on a 200 μl drop of digestion solution on Parafilm in a Petri dish, making sure no bubbles are present.
6. Digest at 25 °C. Digestion times are critical and need to be determined for each batch of Novozym 234. Determine the optimal digestion time, by varying the digestion times from 15 to 90 min, in 15 min increments.
7. Wash the coverslips in PEM, once for 5 min, then twice for 10 min.
8. Place the coverslips in the extraction solution for 5 min.
9. Wash the coverslips in PEM once for 5 min, then once for 15 min.
10. Place each coverslip on a 200 μl drop of primary antibody diluted in PEM to the supplier's recommended concentration. Incubate for 1 h.
11. Wash the coverslips three times, each for 10 min, in PEM. (Use different Coplin jars for each coverslip when using different primary antibodies.)
12. Place each coverslip on a 200 μl drop of secondary antibody diluted in PEM[d]. Incubate for 1 h in the dark.
13. Wash the coverslips three times, for 10 min each, in PEM, keeping them in the dark.
14. Dip each coverslip in double-distilled water, to remove excess salts, and mount in an anti-fading agent such as Citifluor AF1.
15. If desired stain with DAPI to visualize spindles and chromatin. Place the coverslip on a 200 μl drop of DAPI solution (0.02 μg/l in double-distilled water or PEM), for 10 min, in the dark. Mount the coverslip in anti-fading medium.

[a] Paraformaldehyde is a solid which dissolves, in aqueous solution, producing formaldehyde. Formaldehyde oxidizes to give formic acid. Paraformaldehyde may be purchased in sealed ampoules (e.g. E. M. Sciences cat. no. 15700) and should be opened immediately before use).
[b] Driselase is sold adsorbed to starch granules. This procedure dissolves the Driselase and removes starch granules which may increase background staining.
[c] Details of conidia preparation are given in ref. 6.
[d] To eliminate secondary antibody reactions with *A. nidulans* proteins, pre-adsorb dilute secondary antibody solutions with an *A. nidulans* acetone powder, prepared as described in ref. 11. Adsorb the secondary antibody with one-fifth vol. of acetone powder overnight at 4 °C.

# 3. Methods based on transformation of *A. nidulans*

## 3.1 Transformation of *A. nidulans*

It is possible to transform *A. nidulans* with linear DNA, plasmid DNA, and cosmid DNA. Numerous cloned nutritional markers can be used to select for transformation events (12), the majority of which are normally integrative. In our laboratories we typically use the *A. nidulans pyr*G (orotidine-5'-phosphate decarboxylase) gene or the equivalent *Neurospora crassa* gene, *pyr*4, to complement the *pyr*G89 mutation of *A. nidulans* (see *Protocol 3*). Transformation frequencies of up to 1000/μg DNA have been reported (13). The transformation of *A. nidulans* allows cloning of genes by complementation (*Protocol 3*), overexpression and underexpression of genes (Section 3.2), molecular disruption of genes (Section 3.3), and the phenotypic analysis of mutations made *in vitro* to be studied *in vivo*.

---

**Protocol 3.** Transformation of *Aspergillus nidulans* using *pyr*G to complement *pyr*G89

*Materials*

- Conidia
- YG: 5 g yeast extract, 10 g glucose per 1000 ml
- 10 mM uridine
- 10 mM uracil
- Trace elements (14)
- Lytic mixture (filter sterilized): 1 mg/ml BSA, 2.5 mg/ml Novozym 234, 0.4 M ammonium sulfate, 10 mM MgSO$_4$, 50 mM citrate buffer (pH 6.0)
- Wash solution: 0.4 M ammonium sulfate, 1% sucrose, 50 mM citrate buffer (pH 6.0)
- Protoplast buffer: 0.6 M KCl, 50 mM CaCl$_2$, 10 mM MES (2-[*N*-morpholino]ethanesulfonic acid) (pH 6.0)
- PEG solution: 25% PEG 6000 (polyethylene glycol), 100 mM CaCl$_2$, 0.6 M KCl, 10 mM Tris (pH 7.5)
- YAG–sucrose (1000 ml): 5 g yeast extract, 10 g glucose, 342 g sucrose, 8 g agar, 1 ml trace elements (14), 2 ml of 1 M MgCl$_2$
- Plasmid carrying *pyr*G

*Method*

1. Inoculate $10^9$ fresh conidia into 50 ml of YG plus 10 mM uridine, 10 mM uracil, and trace elements (14). Incubate for 5.5 h at 32 °C, shaking at

**Protocol 3.** *Continued*

   200 r.p.m. or until the conidia have swollen but not extended a germ tube (see *Figure 1*).
2. Harvest the germlings by centrifugation at 2000 $g$ for 2 min at 25 °C and resuspend in 40 ml sterile lytic mixture. Incubate at 32 °C for 2.5 h.
3. Harvest the protoplasts by centrifugation at 2000 $g$ for 2 min and wash twice with 50 ml ice-cold wash solution.
4. Resuspend the protoplasts in 1 ml protoplast buffer.
5. Add 100 µl of protoplast suspension to 4 µg of a plasmid carrying *pyr*G (15) followed by 50 µl PEG solution, leave at 4 °C for 20 min.
6. Add a further 1 ml of PEG solution and incubate the mixture at room temperature for 20 min.
7. Plate portions (10 µl, 100 µl, 2 × 500 µl) of the transformation mixture in 3 ml YAG–sucrose on to YAG–sucrose plates solidified with 2% agar. Incubate for 3 days at 32 °C to allow transformants to grow.

## 3.2 Generation of conditional mutations using the *alc*A promoter

Expression vectors based on the *alc*A promotor of *A. nidulans* have been constructed (16) and used to overexpress cell-cycle specific genes (17). As the *alc*A promoter is strongly repressible by growth on glucose, it is also possible to use this promoter to turn off the expression of cloned genes in *A. nidulans* (18). This is done by placing expression of a 3′ truncated version of the gene under the control of the *alc*A promoter in a plasmid vector such as pAL3 (16). Upon homologous integration, two versions of the gene are generated: a 3′ truncated non-functional copy and a full-length copy under the control of *alc*A. Growth on media that allow expression from the *alc*A promoter allows expression of the full-length, functional copy of the gene, but growth on *alc*A repressing media prevents expression. If the gene is essential and no functional protein is produced by the truncated 5′ portion of the gene, growth will be prevented on repressing media (18).

**Protocol 4.** Generation of conditional mutations in *A. nidulans* using the *alc*A promoter

*Materials*
- pAL3 or similar vector
- *pyr*G89-containing haploid strain (*Protocol 3*)
- Minimal medium A: 10 mM urea, 0.6 M KCl, 2 mM $MgCl_2$, 12 mM $KH_2PO_4$ (adjust pH to 6.5 with KOH), trace elements, 1.5% agar

- Minimal medium B: 10 mM urea, 7 M KCl, 2 mM MgCl$_2$, 12 mM KH$_2$PO$_4$ (adjust pH to 6.5 with KOH), trace elements, 1.5% agar
- Glycerol
- Glucose
- Ethanol
- Reagents and equipment for Southern blot analysis

*Method*

1. Construct a plasmid such that the 5' region of the gene is under control of the *alc*A promoter in vector pAL3.
2. Transform a *pyr*G89-containing haploid strain as described in *Protocol 3*.
3. Plate transformation mixture on to minimal medium A containing 50 mM glycerol; incubate at 37 °C for 2 days.
4. Test the transformants by replica plating from the original transformation plates on to minimal medium B containing either 20 mM glucose (repressing), or 50 mM glycerol (non-inducing, non-repressing), or 1% ethanol (inducing) to identify strains that potentially have the gene under control of *alc*A.
5. Streak putative alcohol-induced strains to single colony three times and retest their phenotypes.
6. Analyse strains that only grow in the presence of glycerol or ethanol but not on glucose media by Southern blot analysis to confirm integration of the plasmid by homologous recombination.

## 3.3 Molecular disruption of essential genes and phenotypic analysis from heterokaryons

The heterokaryon gene disruption procedure developed by Osmani *et al.* (19) is remarkably valuable for studying the function of essential genes such as those required for the cell cycle. This procedure (*Protocol 5* and see *Figure 3*) allows the creation of a recessive lethal gene disruption in a heterokaryon and determination of the phenotypes caused by the disruption of conidia produced from the heterokaryon.

To explain the procedure we will use the disruption of the *mip*A (γ-tubulin) gene as an example (20), although the procedure should be usable for most genes. As *Figure 3.1* shows the *mip*A gene of strain G191 was disrupted by transformation with a plasmid, PLO12, which carries the *pyr*4 gene of *N. crassa* and an internal fragment of the *mip*A gene. The *pyr*4 complements the *A. nidulans* mutation *pyr*G89 (carried by G191) and, thus, permits growth on media lacking uridine or uracil. It does not, however, direct integration at *pyr*G. If γ-tubulin is essential for reproduction or

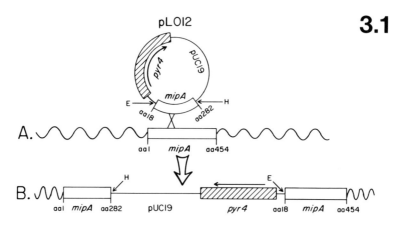

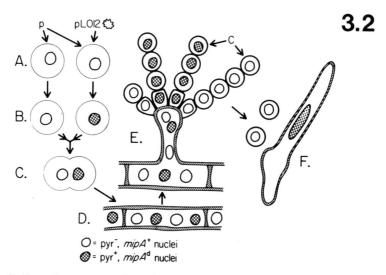

**Figure 3.** Heterokaryon gene disruption in *A. nidulans*. *Figure 3.1* shows the integration of a plasmid (pL012) carrying the *pyr4* gene and an internal fragment of the *mipA* (γ-tubulin) gene into the chromosomal *mipA* gene resulting in the disruption of the *mipA* gene. *Figure 3.2* shows the maintenance of nuclei carrying the disruption in a heterokaryon and determination of the phenotype caused by the disruption in conidia produced by the heterokaryon. As shown in *Figure 3.1*, transformation with pL012 results in disruption of *mipA*. Some protoplasts with nuclei carrying the disruption fuse with untransformed protoplasts (*Figure 3.2C*). In the resultant heterokaryon (*Figure 3.2D*), untransformed nuclei produce γ-tubulin and transformed nuclei produce the *pyr4* product which allows growth in the absence of uridine and uracil. The conidia produced by the heterokaryon are uninucleate and thus contain untransformed or disruptant nuclei (*Figure 3.2E*). If the conidia are incubated in a medium lacking uridine and uracil, only conidia containing disruptant nuclei germinate and they exhibit the phenotype caused by the absence or deficiency of γ-tubulin (*Figure 3.2F*). Reproduced by permission from ref. 20.

## 6: Cell-cycle analysis using A. nidulans

viability, the *mip*A disruption will be lethal. During transformation, however, heterokaryons often form (19, 20), one way in which they can form (fusion of protoplasts carrying parental and transformed [disruptant] nuclei) is shown in *Figure 3.2*. In the heterokaryons (*Figure 3.2 D*) orotidine-5'-phosphate decarboxylase is provided by the *pyr*4 gene in the disruptant nuclei and γ-tubulin is provided by the parental nuclei. Such heterokaryons can thus grow on selective medium lacking uridine or uracil. The conidia produced by the heterokaryon (*Figure 3.2 E, F*) are uninucleate, however, and have either parental nuclei, which do not support growth on selective medium, or nuclei carrying the *mip*A disruption. If γ-tubulin is essential, spores carrying the *mip*A disruption will not be viable. Thus, the failure of spores from the heterokaryon to grow on selective medium indicates that the *mip*A gene is essential. Finally, although the spores carrying the disruption are not viable, they do germinate and exhibit a phenotype caused by the absence of functional γ-tubulin (*Figure 3.2 F*).

---

**Protocol 5.** Heterokaryon gene disruption[a]

1. Construct a disrupter plasmid carrying the *pyr*4 gene.

2. Transform strain G191 as described in *Protocol 3*. Select for transformants on YAG–sucrose (*Protocol 3*) or YAG (5 g yeast extract, 10 g glucose, 20 g agar, distilled water to 1000 ml) osmotically balanced with 0.6 M KCl. Although YAG contains some uridine and uracil, it does not contain enough to support the growth of G191.

3. Test the conidia from transformants for growth on YAG and YAG supplemented with 10 mM uridine and 10 mM uracil (YAGUU). If the gene is not essential, the gene disruption will not be lethal and the spores from virtually all transformant colonies will grow on YAG. Colonies that produce spores that grow on YAGUU but not YAG are putative disruptant heterokaryons. These heterokaryons can not be propagated from spores, but they can be propagated by cutting portions of the colonies from the transformation selection plates and transferring them to fresh YAG plates.

4. Analyse putative disruptant heterokaryons by Southern blotting. Identify heterokaryons that carry the desired gene disruption.

5. Determine the phenotype caused by the disruption by germinating spores from the heterokaryon in YG medium and examining microscopically (*Protocols 1* and *2*).

[a] Strains other than G191 and selective markers other than *pyrG89* can be used, but the ability to complement *pyrG89* with *pyr4* and the easy selection for transformants on osmotically balanced YAG make these particularly useful.

## 4. Cell-cycle synchrony methods

### 4.1 Hydroxyurea block–release

It is possible to generate populations of *A. nidulans* cells at particular points in the cell cycle using drugs or mutants that arrest cells at different points in the cycle. By then releasing these arrested cells, it is possible to allow the cell cycle to continue synchronously. *Protocol 6* describes the use of hydroxyurea in such an approach.

---

**Protocol 6.** Cell-cycle synchrony induced by hydroxyurea block–release

1. Inoculate conidia to $2 \times 10^7$/ml into 500 ml YG containing 0.2% Tween-80 and 20 mM hydroxyurea in a 2 litre conical flask; incubate at 37 °C at 250 r.p.m. for 5 h.

2. Pellet the cells for 2 min, 2000 *g*, 37 °C and decant the medium.

3. Resuspend the cells in YG medium lacking hydroxyurea and pellet as in step 2. Repeat the resuspension and pellet again; then continue incubating at 37 °C with shaking.

4. Take 20 ml samples at 10 min intervals for RNA isolation (or 70 ml for protein isolation) and cool rapidly by the addition of ice. Briefly spin at 2000 *g* to concentrate the cells and harvest by filtration through Miracloth (Calbiochem cat. 475855). Freeze the cells in liquid nitrogen and store at −80 °C.

5. At the same time take samples for biochemical analysis, remove 0.5 ml and fix and stain (according to *Protocol 1*) to determine the mitotic index.

---

### 4.2 Cell-cycle synchrony using conditional mutations

One drawback to cell-cycle analysis using synchronous cultures is the low yield of cells obtained which can be limiting for some types of biochemical analysis. For this reason it is sometimes more suitable to arrest cultures of cells at a particular point in the cell cycle using cell-cycle specific mutations. For studying the transition from $G_2$ through mitosis the reversibility of the temperature sensitive *nim*T23 mutation has been used (21); see *Protocol 7*.

**Protocol 7.** Large-scale isolation of cell-cycle staged cultures using the *nim*T23 mutation

1. Inoculate 4 × 2000 ml flasks with conidia of a *nim*T23-containing strain to 2 × 10$^6$/ml in 1000 ml YG and incubate with shaking at 250 r.p.m. at 20 °C for about 12 h.

2. Monitor the stage of growth by sampling the packed cell volume (PCV) of 10 ml of the culture after centrifugation at 2000 *g* for 2 min.

3. When the PCV has reached 0.3 ml, transfer three flasks to a waterbath set at 55 °C. Allow the temperature of the cultures to rise to 42 °C and then transfer them to a shaking waterbath at 42 °C.

4. After 2 h and 50 min incubation at 42 °C, add benomyl[a] from a stock solution of 20 mg/ml in ethanol, or DMSO, to a final concentration of 5 μg/ml to one flask and continue incubation for 10 min more.

5. At 3 h incubation time, harvest the flask left at 20 °C by filtering under vacuum through Miracloth and washing the mycelial mat with 100 ml ice-cold stop buffer containing 0.9% NaCl, 1 mM NaN$_3$, 10 mM EDTA (pH 7.2), and 50 mM NaF. Remove the mycelial mat on to a paper towel and briefly press dry before freezing in liquid nitrogen. This is the random cell-cycle sample.

6. Harvest the culture at 42 °C that contains no benomyl (G$_2$ arrested) as described in step 5.

7. Transfer the flask containing benomyl and the fourth remaining flask from 42 °C to ice-cold water. Take the temperature rapidly down to 25 °C and continue the incubation for 30–40 min.

8. Harvest both samples as described in Step 5. The sample containing benomyl will be arrested in mitosis and the other cells will be in G$_1$/S.

[a] Benomyl is produced by DuPont and sold commercially as a fungicide. DuPont has been generous in supplying pure benomyl to researchers, but we know of no commercial source for pure benomyl. If it is difficult to obtain benomyl, it should be possible to use the related antimicrotubule agent nocodazole (which is available from several sources) at approximately 2.5 μg/ml. Nocodazole blocks mitosis in *A. nidulans*, but we have not used nocodazole for synchronization experiments.

# Acknowledgements

This work was supported by National Institutes of Health grants GM31837 to B.R.O. and GM42564 to S.A.O.

# References

1. Morris, N. R. and Enos, A. P. (1992). *Trends Genet.*, **8**, 32.
2. Timberlake, W. E. and Marshall, M. A. (1988). *Trends Genet.*, **4**, 162.
3. Bergen, L. G. and Morris, N. R. (1983). *J. Bacteriol.*, **156**, 155.
4. Morris, N. R. (1976). *Genet. Res. Camb.*, **26**, 237.
5. Robinow, C. F. and Caten, C. E. (1969). *J. Cell Sci.*, **5**, 403.
6. Morris, N. R., Kirsch, D. R., and Oakley, B. R. (1982). In *Methods in cell biology* (ed. L. Wilson), Vol. 25, pp. 107–130. Academic Press, New York.
7. Oakley, B. R. and Rinehart, J. E. (1985). *J. Cell Biol.*, **101**, 2392.
8. Kilmartin, J. V., Wright, B. and Milstein, C. (1982). *J. Cell Biol.*, **93**, 576.
9. Blose, S. H., Meltzer, D. I., and Feramisco, J. R. (1984). *J. Cell Biol.*, **98**, 847.
10. Roncal, T., Ugalde, U. O., Barnes, J., and Pitt, D. (1991). *J. Gen. Microbiol.*, **137**, 1647.
11. Harlow, E. and Lane, D. (ed.) (1988). *Antibodies: a laboratory manual*. Cold Spring Harbor Press, Cold Spring Harbor, New York.
12. May, G. S. (1992). In *Applied molecular genetics of filamentous fungi* (ed. G. Turner and J. Kinghorn), p. 1. Blackie and Son Ltd., Glasgow.
13. Osmani, S. A., May, G. S., and Morris, N. R. (1987). *J. Cell Biol.*, **104**, 1495.
14. Cove, D. J. (1966). *Biochim. Biophys. Acta*, **113**, 51.
15. Oakley, B. R., Rinehart, J. E., Mitchell, B. L., Oakley, C. E., Carmona, C., Gray, G. L., and May, G. S. (1987). *Gene*, **61**, 385.
16. Waring, R. B., May, G. S., and Morris, N. R. (1989). *Gene*, **79**, 119.
17. Osmani, S. A., Pu, R. T., and Morris, N. R. (1988). *Cell*, **53**, 237.
18. Lu, K. P., Rasmussen, C. D., May, G. S., and Means, A. R. (1992). *Mol. End.*, **6**, 365.
19. Osmani, S. A., Engle, D. B., Doonan, J. H., and Morris, N. R. (1988). *Cell*, **52**, 241.
20. Oakley, B. R., Oakley, C. E., Yoon, Y., and Jung, M. K. (1990). *Cell*, **61**, 1289.
21. Osmani, A. H., McGuire, S. L., and Osmani, S. A., (1992). *Cell*, **67**, 283.

# 7

# Techniques for studying mitosis in *Drosophila*

CAYETANO GONZALEZ and DAVID M. GLOVER

## 1. Introduction

Of all metazoans, *Drosophila melanogaster* is perhaps best suited for a genetic analysis of cell division. In this chapter, we cannot hope to cover all the experimental techniques needed to study *Drosophila*. However, we aim to supplement existing texts by concentrating on techniques to characterize mitotic mutants. Ashburner has recently undertaken the Herculean task of producing an excellent laboratory handbook (1) which together with a laboratory manual (2) provide an essential reference guide for *Drosophila* research. In addition, we refer to another volume in this series *Drosophila: a practical approach*, edited by David Roberts (3).

### 1.1 Cell-cycle control during *Drosophila* development

In the first two hours of its development the fertilized *Drosophila* egg becomes populated with several thousand diploid nuclei as a result of 13 rounds of nuclear division within a common syncytium. These mitotic cycles consist of alternating S and M phases which take place at approximately 10 minute intervals (4, 5). In cycles 8 and 9, nuclei migrate to the cortex of the embryo, and those that reach the posterior pole become incorporated into the pole cells, the germ-cell precursors. The remaining nuclei undergo four further rounds of divisions within the syncytial body of the embryo before they undergo cellularization following cycle 13. The gastrulation movements of the embryo commence during cycle 14, which differs from the previous division cycles in that a $G_2$ phase is introduced. Divisions no longer take place synchronously through the whole of the embryo, but rather in a series of 25 mitotic domains, the spatial and temporal patterns of which have been carefully mapped by Foe (6). There are 3–4 further rounds of cell division during remaining embryonic development. A broad description of the patterns of these divisions may be found in the text by Campos-Ortega and

Hartenstein (7). A $G_1$ phase is introduced into the 17th cell cycle and is seen in all subsequent cell divisions during larval and pupal development. Techniques for studying mitosis in syncytial and cellularized embryos are given in Section 4.3.

A number of cells enter cycles of polytenization during embryonic development. Cycles of endoreduplication of DNA without nuclear division are first seen in the yolk nuclei which are left behind in the interior of the syncytial embryo when the majority of nuclei migrate to the surface. The generation of polyploid cells during later embryonic development has been described by Smith and Orr-Weaver (8). Much of subsequent larval development consists of cell growth and polytenization. However, cells within the imaginal tissues and the central nervous system retain their potential for proliferation, and do so in the late larval and pupal stages in order to produce adult structures. Diploid cells within the larval brain are particularly suited for the analysis of mitotic events (see Section 5). A description of cell proliferation within the developing central nervous system may be found in White and Kankel (9) and Truman and Bate (10).

A developmentally regulated pattern of specialized cell divisions is seen during gametogenesis. The course of spermatogenesis and spermiogeneis has been reviewed by Lindsley and Tokuyasu (11). The stem cells within the apex of the testes undergo divisions to produce gonial precursor cells. Each of these precursors undergoes a series of four rounds of division to produce a cyst of 16 cells that are interconnected through cytoplasmic bridges. These cells then undergo a period of growth to produce cysts of primary spermatocytes. Following the two meiotic divisions, these become cysts of 64 spermatids which then undergo differentiation to sperm. Oogenesis has been described in detail by King (12) and by Mahowald and Kambysellis (13). The divisions within the germarium of the ovary are similar to those in spermatogenesis in that they lead to the formation of a cyst of 16 cells. One cell becomes designated the future oocyte, and the processes of homologue pairing and meiotic recombination then begin. The remaining 15 cells become the highly polyploid nurse cells that will eventually discharge their contents into the developing oocyte and subsequently shrink and then totally disappear. When the oocyte is fully mature its nucleus is arrested in the first meiotic metaphase. The completion of the meiotic divisions is triggered as the egg passes down the oviduct.

## 1.2 General phenotypes of cell-division mutants

The methods used for screening for mitotic mutants in *Drosophila* have been previously reviewed (14–16). Early studies identified mutations affecting mitosis using clonal analysis of genetically marked cells. By using flies heterozygous for two visible recessive markers it is possible to record events

## 7: Mitosis in Drosophila

that permit the expression of the otherwise recessive phenotype. Barker *et al.* (17) described ways to distinguish marked clones arising from somatic recombination, chromosome breakage and fragment loss, mutation, chromosome loss, and single or double non-disjunction events. This approach has been applied to screen collections of meiotic mutants (17), mutagen-sensitive mutants (18), and temperature-sensitive lethal mutants (19), and has identified a large number of genes essential for mitosis. Such techniques still have a place in the analysis of mitotic mutants of *Drosophila*. A detailed description of the methodology for studying marked clones (generated by X-irradiation) is given by Lawrence *et al.* (20).

In a further search for mitotic mutants, Gatti and his co-workers (21, 22), discovered a rich source in collections of late larval or early pupal lethal mutants, such as the collection generated by Shearn *et al.* (23). The lethal phase is seen so late in development since the heterozygous mother of a homozygous mutant larva will have provided her embryos with their needs for the rapid syncytial divisions and for subsequent embryogenesis. As larval development consists primarily of cell growth and polytenization, the next major demand for proteins essential for mitosis comes when the imaginal tissues proliferate to form adult structures in the late larval and pupal stages of development.

The maternal provision of mitotic products to the embryo is reflected by the maternal effect phenotype of another group of mitotic mutants. A female homozygous for a mutation in such a gene lays an egg in which a protein essential for mitosis is either missing or defective (see ref. 14 for review). Screens for recessive female steriles, such as the one carried out by Schupbach and Wieschaus (24), have been a useful source not only of mutations affecting embryonic morphogenesis but also as a source of mutations affecting mitosis. Many genes essential for mitosis also affect the meiotic divisions and can reduce fertility not only in the female but also in the male. Screens of male sterile mutations are also proving to be valuable sources of mutations affecting the mitotic process (Gatti, personal communication).

Although the effective lethal phase of the zygotic mutants affecting mitosis is only seen in late larval development, some mutations lead to a failure of mitotic divisions in the embryo that immediately follow cellularization. These are mutations in genes whose RNA or protein products are unstable and, therefore, have to be repeatedly replenished. One such example is *string*, one of the two *Drosophila* homologues of the fission yeast gene *cdc25*, which is completely degraded at cellularization and has to be expressed *de novo* at cycle 14 in order to drive the entry of individual cells in mitosis (25). *String* is one member of a large number of mutations that lead to defective cuticle patterns in the early embryo, identified in a major screen for embryonic lethals (26).

## 2. Genetic approaches

### 2.1 Generating *Drosophila* mutants

General schemes for mutagenesis in *Drosophila* have been described by both Roberts (27) and Grigliatti (28). Breeding programmes for the detection and recovery of mutations are also extensively discussed by Ashburner (1) who also discusses the methodology for both chemical and X-ray mutagenesis.

More recently, hybrid dysgenesis has been used as a means for generating mutations tagged with the mobile P-element. The initial screens carried out using this method of mutagenesis involved the mobilization of a large number of defective P-elements. This method has the disadvantage that the mutated gene is only one of several that show an association with P-elements in the mutant stock. A scheme for mobilizing single P-elements using a P-element derivative known as *jump-starter* (*J-S*) was described by Cooley *et al.* (29). *J-S* is essentially a stable P-element that produces functional transposase and is used to mobilize a single defective P-element transposon to a new position. The *J-S* element is then removed and a stable stock established containing the single defective P-element. An alternative stable P-element that can be used to drive the transposition of other P-insertions is $p[ry^+\Delta 2 - 3]$, a defective P-element that encodes transposase but is itself unable to jump, and which is inserted into region 99B (30). Several laboratories have utilized this stable P-element as a source of transposase to drive the transposition of marked defective P-elements.

If these elements carry the *lacZ* gene of *Escherichia coli* downstream of the P-element promoter, they permit the identification of elements of *Drosophila* DNA able to stimulate the expression of beta-galactosidase with specific spatial–temporal patterns (31). The major drawback of this 'enhancer-trap' technique is that, as the enhancer elements can exert their effect over a relatively long range, it may prove difficult to identify the gene under their control. This necessitates that *in situ* hybridization is performed to determine the expression pattern for all the genes in the vicinity of the transposon to ascertain which of them is expressed in a pattern correlating with that of *lac-Z*. Enhancer trap lines have been very useful in identifying genes whose expression is restricted to a particular tissue or a particular subset of cells. In fact one of the applications of these lines is to mark cells that can hardly be visualized by conventional histology. Enhancer traps are also invaluable as markers for specific chromosomes, such as balancers, in embryonic development. This is particularly useful for identifying homozygous, balancer-free, embryos in order to study early zygotic phenotypes using such techniques as immunostaining. *Protocol 1* gives a method for detecting the expression of *lac-Z* in *Drosophila* tissues.

> **Protocol 1.** Staining to reveal *lac-Z* expression
>
> *Materials*
> - Phosphate–citrate buffer: 9 vol. of 0.2 M $Na_2HPO_4$ to 1 vol. of 0.1 M citric acid to 10 vol. of water, the pH should be between 7.0 and 8.0
> - X-gal solution: phosphate–citrate buffer containing 5 mM $K_3Fe(CN_6)$, 5 mM $K_4Fe(CN_6)$, 0.02% of Triton X-100, and a very small amount of X-gal (5-bromo-4-chloro-3-indolyl-$\beta$-D-galactoside) to give saturation, make sure that there is some undissolved X-gal
>
> *Method*
> 1. Dissect out soft tissues surrounded by permeable membranes such as the internal organs of larvae and adults, for direct staining. Dissect the tissue in phosphate–citrate buffer on a siliconized slide. (Embryos must first be devitelinized as described in Section 4.3).
> 2. Fix the tissue for 15 min in 3.7% formaldehyde in phosphate–citrate buffer.
> 3. Rinse the tissue in phosphate–citrate buffer.
> 4. Add saturated X-gal solution.
> 5. Incubate at 30 °C until the colour develops. This may take anything between 2 h and 2 days to appear depending on the level of expression in the tissue being stained.

As with any other method of P-element based mutagenesis, the lines produced by the enhancer trap technique can be used to generate deficiencies in the region of the insertion by imprecise excision of the P-element, thus facilitating the genetic analysis of that region.

## 2.2 Mapping mutants

Whether the mutation has been produced by conventional mutagenesis, using physical or chemical agents, or by mobilization of single or multiple P-elements, the first step in its characterization is its mapping by meiotic recombination. To this end, females that carry the mutant chromosome over a multiply-marked chromosome are mated to the appropriate males and the recombinant offspring are isolated and balanced. The absence or presence of the mutant phenotype in each class of recombinant indicates the location of the gene, i.e. proximal or distal to each marker. The additional function served by meiotic recombination is to clean the chromosome of unwanted mutations which might have been produced during the course of the mutagenesis and that can potentially interfere with the characterization of the mutation under study.

Once a locus has been mapped using recombination, it is desirable to determine if mutant alleles can be uncovered by a chromosome deficiency. This can often be carried out relatively easily, since there are collections of deficiencies that uncover up to 80% of the genome. This allows one to identify the cytological interval in which the gene lies by reference to the extent of the chromosome deficiency on Bridges' maps of the salivary gland polytene chromosomes. The extent of a deficiency can be determined by examining a squashed preparation of polytene chromosomes from larvae heterozygous for the deficiency (*Protocol 2*). The technique will also reveal any major rearrangements in the mutated chromosome under study.

---

**Protocol 2.** Preparation of polytene chromosomes for banding

1. Dissect out the salivary glands from a climbing third instar larva in 0.7% NaCl.
2. Fix the glands in a drop of fresh ethanol:propionic acid (3:1) for at least 30 sec.
3. Place the glands in a drop of 1% orcein–45% propionic acid on a microscope slide and stain for 3–5 min (check).
4. Cover with a siliconized coverslip.
5. Squash between two sheets of blotting paper by tapping with the blunt end of a pair of tweezers.
6. Observe the state of the chromosomes under a microscope, and if the chromosome arms are satisfactorily spread, seal around the edges of the coverslip with nail varnish.

---

The availability of an uncovering deficiency also permits the study of the phenotypic effects of a single copy of the mutant gene, and so classify the allele according to Muller's classical terminology (32). Recessive mutations are usually hypomorphic, and so a stronger phenotype would be seen when the mutant allele is placed against a deficiency. Knowledge of the cytological position of the gene greatly simplifies the task of determining possible allelism to existing mutants. This is achieved by carrying out complementation tests with all the mutants present within the region. Care must be taken to study every possible phenotype as allelic series in which partial complementation takes place are not uncommon, especially with mitotic mutants in which differing alleles can show either zygotic or maternal effect lethal phenotypes.

In those instances in which mutagenesis has been carried out by mobilizing a single P-element, mapping can also be carried out by *in situ* hybridization

## 7: Mitosis in Drosophila

(*Protocols 3–7*) using a P-element probe. In the case of P-element induced mutagenesis, using either multiple or single P-elements, it is advisable to confirm the association of the mutation under study with the presence of a particular P-element by examining phenotypes of recombinants and revertants. *In situ* hybridization is, of course, an extremely valuable technique for localizing *Drosophila* genes once cloned, with reference to chromosomal rearrangements. Strategies for cloning *Drosophila* genes have been discussed by Pirrotta (33). The techniques described in his article are still largely applicable, except that PCR is now used routinely following microdissection of DNA from polytene chromosomes rather than microcloning (34, 35).

---

**Protocol 3.** Preparation of polytene chromosomes for *in situ* hybridization

*Materials*
- 2 × SSC: 0.3 M NaCl, 0.03 M sodium citrate, pH 7.0

*Method*
1. Dissect out the salivary glands from a climbing third instar larva in 0.7% NaCl.
2. Transfer the glands to a drop of 45% acetic acid for 30 sec.
3. Fix the glands in a 1:2:3 mixture of lactic acid:glacial acetic acid:water for 5 min.
4. Cover with a siliconized coverslip and sandwich between two sheets of blotting paper, tapping with the blunt end of a pair of tweezers to squash the cells.
5. Monitor the squashing procedure by examining the chromosomes by phase-contrast microscopy. When the chromosome arms are well enough spread, leave to fix overnight.
6. Immerse the end of the slide carrying the coverslip in liquid nitrogen to freeze the preparation.
7. When the nitrogen ceases to boil, remove the slide and lever off the coverslip with the flick of a scalpel.
8. Place the slide in successive jars containing in 70% ethanol for 3 min; followed by 100% ethanol for 3 min; and allow to dry in the air.
9. Soak the slide in a jar containing 2 × SSC at 65 °C for 30 min; followed by two immersions in 70% ethanol for 5 min; and finally in 100% ethanol for 5 min.
10. Allow to dry in the air. Slides can be stored at room temperature once this stage has been reached.

**Protocol 4.** Labelling nucleic acids for *in situ* hybridization

*Materials*

- 5 × oligolabelling buffer: 250 mM Tris–HCl (pH 8.0), 25 mM $MgCl_2$, 5 mM ß-mercaptoethanol, 2 mM each of dATP, dGTP, dCTP, 1 M Hepes (adjusted to pH 6.6 with 4 M NaOH), 1 mg/ml random primers
- 20 × SSC: 3 M NaCl, 0.3 M sodium citrate, pH 7.0
- 100 × Denhardt's solution: 2% Ficoll, 2% polyvinyl pyrolidone, 2% bovine serum albumin

*Method*

1. Make up the following reaction mixture:
   - 5 µl 5 × oligolabelling buffer minus dTTP
   - 1 µl biotin-16-UTP (Boehringer cat. no. 1093070)
   - approximately 100 ng DNA
   - 1 µl Klenow fragment of DNA polymerase (1 unit)
   - make up to a final volume of 25 µl with water
2. Incubate overnight at room temperature.
3. Remove unincorporated nucleotides using the spun column technique (36). Add 100 µg of tRNA to the excluded material, and precipitate with ethanol using standard conditions (36).
4. Concentrate the precipitate by centrifugation (10 min, room temperature) in a microcentrifuge and resuspend the pellet in 50 µl water. Add 50 µl 8 × SSC, 2 × Denhardt's solution, 20% dextran sulfate, 0.4% denatured salmon sperm DNA.

**Protocol 5.** *In situ* hybridization

*Materials*

- 70 mM NaOH, must be made fresh for use
- 2 × hybridization buffer: 8 × SSC, 2 × Denhardt's solution (see materials section of *Protocol 4*), 20% Dextran sulfate, 0.04% denatured salmon sperm DNA

*Method*

1. Place the slide carrying the chromosome preparation in 70 mM NaOH for 2 min in order to denature the chromosomal DNA.
2. Dehydrate as in Step 8 of *Protocol 3*.

# 7: Mitosis in Drosophila

3. Make up the following mixture in a microcentrifuge tube:
   - 25 μl of biotinylated probe prepared as in *Protocol 4*
   - 25 μl of water
   - 50 μl 2 × hybridization buffer
4. Place the tube in a boiling waterbath for 3 min and then cool on ice.
5. Place 20 μl of the denatured probe on a clean siliconized coverslip, and pick it up with the slide such that it covers the region occupied by the chromosomes.
6. Incubate the slide in a humid chamber overnight at 58 °C.
7. Proceed with the post-hybridization wash and staining using either the conventional peroxidase method (*Protocol 6*) or a fluorescence method (*Protocol 7*).

---

**Protocol 6.** Staining chromosomes using peroxidase following *in situ* hybridization

*NB* **DAB is a carcinogen: carry out manipulations in a fume hood and wear gloves. DAB may be inactivated with 50% bleach before disposal.**

*Materials*
- 20 × SSC: 3 M NaCl, 0.3 M sodium citrate, pH 7.0
- PBS: dissolve 8.0 g NaCl, 0.2 g KCl, 1.44 g $Na_2HPO_4$, 0.24 g $KH_2PO_4$ in water to 1 litre, adjust pH to 7.2
- Triton X-100
- Streptavidin–horseradish peroxidase
- Enzo dilution buffer (Enzo Diagnostics Inc.)
- DAB (3,3-diaminobenzidine)
- Hydrogen peroxide
- Giemsa stain
- Sodium phosphate buffer, pH 6.8
- DPX mountant (Fluka Chemicals Ltd)

*Method*
1. Carry out the following post-hybridization washes in sequence:
   - three times in 2 × SCC at 53 °C for 20 min
   - twice in PBS at room temperature for 5 min

**Protocol 6.** *Continued*

- once in PBS containing 0.1% Triton X-100 at room temperature for 2 min
- briefly rinse in PBS at room temperature

2. Place 100 µl of a 1:250 dilution of streptavidin–horseradish peroxidase in Enzo dilution buffer, in PBS in the region of the slide where the chromosomes are to be found.
3. Cover the solution with a 22 × 40 mm coverslip.
4. Incubate the slide for 30 min at 37 °C in a humid chamber.
5. Remove the coverslip by dipping the slide into a beaker containing PBS.
6. Wash the slide twice in PBS for 5 min at room temperature, and for a further 2 min in PBS containing 0.1% Triton X-100.
7. Stand the slides in PBS prior to staining.
8. Add 100 µl of a 0.5 mg/ml solution of DAB, 0.01% $H_2O_2$ in PBS, and cover with a 22 × 40 mm coverslip.
9. Incubate at room temperature until the signal develops. This should take between 1–10 min, although the slides can be left to develop for longer.
10. Rinse the slides in water followed by PBS.
11. Examine the slides for a brown–black hybridization signal by phase-contrast microscopy. If the signal is absent, or too weak, staining can be resumed by adding more DAB solution as in Step 8.
12. Stain the chromosomes with 5% Giemsa solution in sodium phosphate buffer pH 6.8 for 1 min.
13. Rinse the slides in water and air dry.
14. Check the staining and if satisfactory mount the chromosomes under a siliconized coverslip, in DPX mountant.

---

**Protocol 7.** Staining chromosomes to detect *in situ* hybridization signal by fluorescence

*Materials*

- 2 × SSC (see *Protocol 6*)
- 4 × SSC (see *Protocol 6*)
- Triton X-100
- FITC–avidin (Vector Lab.)
- Propidium iodide

7: *Mitosis in Drosophila*

- Mounting medium: 2.5% *n*-propyl gallate (Sigma cat. no. P3130) in 85% glycerol

*Method*
1. Carry out the following post-hybridization washes in sequence:
   - once in 2 × SSC at 53 °C for 2 min
   - once in 4 × SSC at room temperature for 5 min
   - once in 4 × SSC containing 0.1% Triton X-100 at room temperature for 5 min
   - once in 4 × SSC at room temperature for 5 min
2. Dip the slides in a solution containing 2% (v/v) FITC–avidin in PBS for 30 min.
3. Carry out the following successive washes:
   - once in 4 × SSC at room temperature for 5 min
   - once in 4 × SSC containing 0.1% Triton X-100 at room temperature for 5 min
   - once in 4 × SSC at room temperature for 5 min
4. Mount the preparations under a coverslip using a solution of 1 µg/ml propidium iodide (to stain DNA) in mounting medium.
5. Seal the edges of the coverslip with nail varnish, and allow to dry.
6. Examine using either conventional fluorescence or confocal microscopy.

## 3. General approaches

### 3.1 Localization of transcripts

As cell division proteins are required at specific developmental stages (see the introduction to this chapter) it is often informative to examine the localization of the transcripts of a cloned mitotic gene throughout development. The method of Tautz and Pfeifle (37) (*Protocol 8*) of *in situ* hybridization is generally applicable to most developmental stages, although differing approaches are used for the preparation of specific tissues:

(a) *Testes, brains and imaginal discs* can be dissected from larvae or adults as appropriate, in 0.7% NaCl at room temperature. They are then placed in the fixative PP (see *Protocol 8*). After 1 h in fixative, they are given two 5 min washes in PBT (see *Protocol 8*).

(b) *Ovaries* are dissected from adult females in 0.7% NaCl at room temperature, and fixed for 1 h in 2 ml PP plus 2 ml heptane, before being given two 5 min washes in PBT.

(c) *Embryos* are dechorionated in 50% household bleach (sodium hypohlorite) or Chloros for 3 min and then washed extensively in PBT (see also Section 4.1). They are fixed for 1–4 h in 4 ml of freshly prepared PP containing an equal volume of heptane. The vitelline membrane is then removed by placing the embryos in 8 ml of a mixture of equal parts of fresh heptane and methanol and subjecting them to vigorous shaking. If the embryos are then allowed to settle, the methanol:heptane mixture can then be removed and replaced with 5 ml of 90% methanol, 10% 0.5 M EGTA (ME). Following this treatment it is necessary to rehydrate the embryos by passing them through a series of solutions containing 700 μl ME:300 μl PP; 500 μl ME:500 μl PP; and 300 μl ME:700 μl PP for periods of 5 min. They are then incubated in PP for a further 20–40 min and washed twice in PBT.

If it is necessary to store any of the above tissues before carrying out the hybridization reaction, they can be dehydrated through a series of 30%–50%–70% ethanol for 3 min each and stored at −20 °C.

---

**Protocol 8.** Localization of transcripts by *in situ* hybridization

*Materials*

- Tissue (see text)
- Proteinase K
- PBT: 0.2% Triton X-100 in PBS
- Glycine
- PP: 4% paraformaldehyde dissolved in PBS (see Section 4.3)
- HS: 50% formamide, 5 × SSC (see *Protocol 6*), 100 μg/ml salmon sperm DNA, 0.1% Tween 20
- Probe: prepare using either BCL digoxigenin DNA (cat. no. 11750999) or RNA (cat. no. 1175025) labelling kits
- Wash buffer: 100 mM Tris (pH 9.5), 100 mM NaCl, 50 mM $MgCl_2$, 1 mM levamisol (Sigma), 0.1% Tween 20
- NBT (nitro-blue-tetrazolium salt from BCL kit)
- X-phosphate (5-bromo, 4-chloro, 3-indolyl-phosphate from the BCL detection kit)
- Series of ethanols
- Canada balsam
- Methyl salicylate
- Hoechst 33258
- *n*-propyl gallate (*Protocol 7*)
- Glycerol

## 7: Mitosis in Drosophila

*Method*

1. Prepare the tissue as indicated in the text, and incubate in 50 µg/ml proteinase K in PBT for 3.5 min (embryos) or 8 min (testes, brains, imaginal discs, and ovaries). Use 100 µl solution for 10 brains or equivalent amount of other tissues.

2. Stop the protease digestion by carrying out the following washes prior to hybridization:
   - 2 mg/ml glycine in PBT for 2 min
   - PBT twice for 5 min
   - PP for 20 min to refix the preparation
   - PBT three times for 5 min
   - PBT:HS (1:1) for 10 min
   - HS for 10 min
   - HS for 1 h at 47 °C

3. Remove most of the HS and carry out hybridization at 47 °C for 14–16 h using 5 µl of the appropriate probe per 1–200 embryos.

4. After hybridization, carry out the following successive 15 min washes:
   - HS at 45 °C
   - HS:PBT (4:1) at 45 °C
   - HS:PBT (3:2) at 45 °C
   - HS:PBT (2:3) at 45 °C
   - HS:PBT (1:4) at 45 °C
   - twice in PBT at room temperature

5. Incubate the tissue for 1 h in a 1/1000 dilution of alkaline phosphatase conjugated anti-digoxigenin antibody (BCL).

6. Subject the tissue to four 15 min washes in PBT, followed by three 5 min washes in wash buffer. Then incubate in 1 ml PBT containing 4.5 µl of NBT and 3.5 µl of X-phosphate (BCL). Monitor and stop the colour reaction by washing twice for 5 min with PBT. The incubation should take between 3–30 min depending upon the tissue and the abundance of the transcript.

7. Dehydrate the tissue by passing it through series of 30%, 50%, 70%, 90%, and 100% ethanol for 2 min each step.

8. Mount the preparation in a 1.6 g/ml (w/v) solution of Canada balsam in methyl salicylate. To enable the simultaneous visualization of DNA by fluorescence microscopy, incubate for 5 min in 10 µg/ml Hoechst 33258 and mount in 2.5% *n*-propyl gallate (*Protocol 7*) in 85% glycerol.

## 3.2 Immunolocalization studies

### 3.2.1 General considerations about immunostaining

Immunostaining is the most valuable technique for examining mitotic phenotypes. The organization of the mitotic spindle can be examined using antibodies such as YL1/2 (38), which recognizes microtubules from a variety of species. Centrosomes can be visualized at most developmental stages using antibodies against the 190 kDa centrosome associated antigen, Bx63 (39). Additional data may be gained by staining preparations with antibodies that recognize the mitotic cyclins (40, 41). As with all immunostaining, the methods used for tissue fixation are particularly important. This is especially true of *Drosophila* microtubules which appear to be particularly unstable. We prefer to use formaldehyde as a fixative simultaneously with a brief treatment with the microtubule stabilizing drug taxol (5). We have been concerned that taxol might lead to artefactual structures since it is known to promote nucleation and elongation of microtubules. The simultaneous incubation of the tissue preparations with taxol and formaldehyde does, however, appear to fix the tissue before artefacts can arise. Real-time studies on the early embryonic cycles using fluorescently labelled tubulin (42) or fluorescently labelled anti-tubulin antibodies (43) indicate that the images observed when embryos are fixed using a formaldehyde/taxol procedure are reasonably accurate. Moreover, the protocols which we describe give reproducible staining patterns in several tissues for a large number of mitotic mutants. The other commonly used fixative is methanol, usually applied in the presence of EGTA (39). We find this less reproducible than formaldehyde when used with embryos, but it gives excellent results when used on testes as suggested by Gatti (see below).

The choice of fluorochromes with which secondary antibodies are labelled will depend upon the light microscope being used to make the observations. If a conventional fluorescence microscope is being used, then assuming appropriate filter sets are available, it is usually possible to carry out double immunolabelling using primary antibodies obtained from different species such as mouse and rabbit. The secondary antibodies should be specific to the two species, and conjugated with either rhodamine or Texas red on the one hand and fluorescein on the other. We have found secondary antibodies purchased from Jackson Immuno Research Inc. to be particularly reliable. We usually use either mouse monoclonal antibodies or polyclonal sera from rabbits as the primary antibody. The concentration at which a primary antibody should be used will vary, but as a rule of thumb, hybridoma supernatants can usually be used with minimal (4–5-fold) or no dilution. The immunoglobulin concentration within such supernatants is usually in the order of 25 μg/ml. Monoclonal antibodies in ascites fluid should be diluted (by $10^2$–$10^4$) to around this concentration. Sera can be much more variable, but are usually diluted about 500–1000-fold. The reader is referred to Harlow and

## 7: Mitosis in Drosophila

Lane (44) for a discussion of methods for affinity purification of antisera, and of artefacts that can arise during immunostaining.

Simultaneous labelling of DNA can be carried out using either Hoechst 33258 or DAPI which can be visualized by fluorescence in the ultraviolet. If preparations are to be observed by confocal microscopy using an instrument such as the Biorad MRC600 in its standard configuration, then propidium iodide is our preferred means of visualizing DNA. In this case, the preparation must first be treated by adding RNase to the primary antibody incubation (*Protocol 9*) since propidium iodide also binds RNA. Propidium iodide is added at a concentration of 1 µg/ml after the preparation has had its final wash to remove secondary antibody, or may be added to the mounting medium at a concentration of 1 µg/ml. Alternatively if DAPI or Hoechst are to be used for staining DNA, these dyes can be added at a concentration of 0.3 µg/ml.

---

**Protocol 9.** Indirect immunofluorescence

1. Prepare and fix the tissue under study as described in Sections 4–6.

2. Block any residual fixative by incubating the embryos in 10% fetal calf serum, 0.3% Tween in PBS (*Protocol 6*) for 1 h. If RNase treatment is to be carried out (see text), an aliquot of a boiled solution of the enzyme should be added at this time at a concentration of 2 mg/ml.

3. Incubate the preparation with the primary antibody in 10% fetal calf serum, 0.1% Tween[a] in PBS either at 4 °C overnight, or for 4 h at room temperature[b].

4. Wash several times with 0.1% Tween in PBS over a 1 h period. Best results may be achieved for thick specimens by more extensive washing: i.e. eight 1 h incubations.

5. Incubate with the secondary antibody in 10% fetal calf serum, 0.1% Tween in PBS for 2 h at room temperature or overnight at 4 °C.

6. Wash as in Step 4, but use PBS alone for the last two or three washes.

7. Mount the preparation as described in the text, carrying out DNA staining if desired.

[a] Triton X-100 is an alternative detergent and can be used at a comparable concentration.
[b] This length of incubation is appropriate for thick specimens. The incubation time can be reduced to 45 min at room temperature for squashed cytological preparations on microscope slides.

---

In order to mount thick preparations, such as whole mounts of embryos or larval brains, first place a drop of approximately 50 µl of 2.5% *n*-propyl gallate in 85% glycerol on a clean siliconized coverslip. The amount of mounting fluid is quite critical, and is best judged with experience. If too much is used, the coverslip will move around and it is hard to focus on the specimen; if too little, it can leave dry areas, and pressure will be exerted on the embryo by the coverslip. Remove excess washing buffer from around the specimen, and transfer it into the mounting medium. It is best to pick up embryos using a fine slightly wet brush, and then removing the embryos from the brush with a pair of fine forceps, for transferring to the mounting medium. Take a clean microscope slide and gently lower it on to the drop. Seal the preparation with nail varnish. An alternative way of staining a preparation with propidium iodide is to add it at a concentration of 1 µg/ml to the 2.5 $n$-propyl gallate in 85% glycerol mounting medium. This is particularly effective for thin preparations such as squashed tissue.

Histochemical methods can also be used for immunostaining if the specimen is to be observed by bright-field microscopy. Double labelling is also possible using this approach. If for example, two antigens are to be detected, one using a mouse monoclonal antibody, and the other a rabbit polyclonal antiserum, then the specimen can be incubated with the primary antibodies in the usual way. Following washing, the specimen can be incubated with peroxidase-conjugated goat anti-mouse secondary antibody, washed and the first distribution visualized histochemically. This entails adding 10 µl of 5 mg/ml DAB (diaminobenzidine), and 1 µl $H_2O_2$ to 1 ml 10 mM $CoCl_2$ dissolved in PBT. $CoCl_2$ is used to modify the colour of the DAB derivatives produced by peroxidase, thus enabling double immunolabelling experiments. Similar effects can be achieved with nickel ions. When the colour has developed to the required extent, the specimen can be washed several times with PBT buffer before being incubated with peroxidase conjugated goal anti-rabbit second antibody. Following this incubation and subsequent washing, the second pattern can be revealed by the same histochemical reaction but omitting $CoCl_2$ from the developing buffer.

## 4. Embryos

### 4.1 Collection and dechorionation

There are basically two ways to handle the embryos for dechorionation. Flies are allowed to lay their eggs upon 3% agar plates containing 4% fruit (apple or grape) juice. The embryos are brushed from the collecting trays in 0.7% saline and placed either on a nylon gauze within a Millipore filtration funnel or in a basket of similar sized wire mesh of such a size as to retain the embryos, whilst permitting the chorions to be washed away. Commercial bleach is passed over the embryos for about 3 minutes or until they become transparent as the reflective chorion is dissolved. They are then thoroughly washed with water and poured on to a dry plastic Petri dish, and covered with water. The embryos can be kept attached to the surface of the dish, under water, where they will stay alive and develop, and can be easily observed with a dissection microscope. This facilitates the selection of those which are appropriate for further manipulation. Alternatively, if they are to be subjected to immunostaining (*Protocols 11* and *12*) they can be transferred directly into a vessel for permeabilization.

## 4.2 Introducing reagents into embryos

The vitelline membrane presents an impermeable barrier through which it is difficult to introduce drugs. This can be overcome either by using treatments that permeabilize the membrane or by direct microinjection. A variety of drugs can be used to study the embryonic cell cycle using such techniques. Studies from this laboratory of the effects of microinjecting aphidicolin, for example, have shown that the rapid mitotic cycles of the syncytial embryo can take place without the check point of requiring DNA synthesis to be completed (45). The approach is amenable to studying the effects of microtubule destabilizing drugs such as colchicine, microtubule stabilizing drugs such as taxol, phosphatase inhibitors such as okadaic acid, kinase inhibitors such as staurosporine, and many other drugs that might affect cell-cycle regulation. The techniques also permit the introduction of chemicals such as bromodeoxyuridine (BrdU) as a label to follow DNA synthesis. In this particular case, the incorporation of the drug into DNA can be visualized by immunostaining (see Section 4.2.2.).

### 4.2.1 Microinjection

It is helpful if microinjection of embryos can be carried out at 18 °C in order to slow down the rate of development. Embryos can be collected from fruit juice agar plates smeared with yeast paste placed under culture bottles. Embryos at the appropriate stage are transferred to double-sided Scotch tape stuck to a microscope slide. They are rolled using fine forceps to remove the chorion and then transferred to a line of glue placed close to one edge of a 22 × 40 mm coverslip on a microscope slide. The glue is prepared by dissolving double-sided sticky tape in *n*-heptane. The embryos are allowed to desiccate for a few minutes in a closed box continuing some silica gel in a dish. The line of embryos is covered with halocarbon oil (Voltelef, 10S grade: Atochem UK Ltd), the slide placed on an inverted microscope, and the embryos injected using a pulled capillary held in a micromanipulator. The embryos are then allowed to develop as required on a 3 cm plastic dish, after which time the Voltelef oil can be removed with 3 washes of heptane. If the embryos are to be immunostained, they can be dislodged from the coverslip and transferred to fixative and processed as described below. A detailed description of how to carry out microinjection is given by Santamaria (46).

### 4.2.2 Permeabilization

The procedure given in *Protocol 10* is suitable for permeabilizing embryos for incubation in the presence of drugs.

---

**Protocol 10.** Permeabilization of embryos with octane for drug treatment

*Materials*
- TCM: 5 vol. Schneider's Drosophila Medium Revised (Gibco cat. no. 041–01720 H) to 4 vol. MEM 25 mM Hepes with Hank's salts without L-glutamine (Gibco cat. no. 041–02370 H)
- Octane: we recommend octane from Aldrich (cat. no. 29,698–8), 100 ml anhydrous 99+% Gold Label

*Method*
1. Collect and dechorionate embryos at a suitable stage of development using either a pair of sharp forceps or a micropipette. Place them on a small piece of wet, black blotting paper on a Petri dish.
2. Dip the blotting paper into the organic phase of a mixture of 1 ml TCM, 1 ml octane (previously saturated in TCM) in a 10 ml glass vial. The embryos will fall on to the interface where they will clump together forming a circle.
3. Place the vial on a flat horizontal surface and gently shake with a rotatory movement.
4. After precisely 2 min aspirate out the embryos with a Pasteur pipette. They will form a clump within the pipette at the interface.
5. Discard as much TCM as possible and drop the embryos into a vial containing 1 ml of TCM containing the appropriate amount of the drug. The embryos will stay on the surface covered by some octane.
6. Gently blow air upon the embryos observing them under the dissecting microscope until the octane has evaporated and they become reflectant.
7. After incubation with the drug add a few drops of heptane saturated in TCM in order to reform an aqueous/organic interface. Collect the embryos with a Pasteur pipette as before and drop them into a vial for fixation (*Protocol 11*).

---

## 4.3 Preparation of embryos for immunostaining

The procedure used for devitellinization and fixation can be varied according

## 7: Mitosis in Drosophila

to how many embryos are being prepared. A procedure is given in *Protocol 11* for egg collections from a population cage or a mutant stock that lays good quantities of eggs. *Protocol 12* will give better recoveries for small numbers of embryos.

Paraformaldehyde (3.7%) is our preferred fixative, but the quality is important in order to preserve some subcellular structures. The fixative is best made up fresh. Transfer a weighed quantity of EM grade paraformaldehyde (Polysciences Ltd. cat. no. 0380) into a graduated Pyrex tube and make it up to volume with 15 mM NaCl, 45 mM KCl, 10 mM potassium phosphate, pH 6.8. In a fume hood, heat the solution to 60 °C and maintain at this temperature, with some stirring. Once the paraformaldehyde has gone into solution (about 15–20 min), filter through a Millipore filter and check that the pH is 6.8. The fixative is stored on ice until needed.

If staining microtubules or if it is suspected that the distribution of the protein under study is dependent upon microtubules, steps must be taken to preserve the microtubule structure. This can be achieved by using 37% formaldehyde (instead of 3.7%) in *Protocol 11*, omitting the use of taxol. Alternatively, methanol fixation can be used as outlined below. We have obtained variable results with both of these procedures and prefer the procedures described in *Protocols 11* and *12*.

---

**Protocol 11.** Fixing embryos[a]

1. If microtubules are to be stabilized with taxol[b] (see text), resuspend dechorionated embryos in 1 ml 15 mM NaCl, 45 mM KCl, 10 mM potassium phosphate, pH 6.8: 5 ml heptane. Add 5 μl 1 mM taxol to the aqueous phase, and shake gently for 30 sec[c].
2. Quickly remove the aqueous phase using a drawn-out Pasteur pipette, carefully avoiding the embryos.
3. Add 3 ml fresh 3.7% formaldehyde prepared as described in the text and shake gently for 1–2 h. Alternatively if taxol is not to be used, transfer the dechorionated embryos directly into a glass vial containing 3 ml fresh 3.7% formaldehyde: 5 ml heptane, shaking for 1–2 h.
4. Completely remove the aqueous layer followed by most of the heptane.
5. Add 3 ml of fresh heptane followed by 3 ml methanol.
6. Shake vigorously for 1 min.
7. Leave the phases to separate. Those embryos that sink to the bottom have lost their vitelline membrane, those that remain at the interface have not.
8. Aspirate carefully to remove the heptane and most of the methanol.

> **Protocol 11.** Continued
>
> 9. Add 8 ml fresh methanol and leave for 10 min.
> 10. Aspirate to remove the methanol, and resuspend the embryos in PBS containing 0.1% Tween 20. The embryos can now be stored for up to 1 week at 4 °C.
>
> [a] Alternative protocols for fixation are discussed in the text.
> [b] Taxol is from Molecular Probes Inc. (cat. no. T-3456).
> [c] It is important not to leave the embryos in taxol for longer than 1 min as this can induce artefactual polymerization of microtubules (see ref. 5 for discussion).

The vitelline membrane can be removed from small numbers of embryos by hand if 100% recovery of embryos is required. Alternatively, the following method works very well, and is the *only* approach that works with some mitotic mutants.

> **Protocol 12.** Removing the vitelline membranes from embryos
>
> 1. Place the embryos (up to 50) into a microcentrifuge tube containing 500 µl methanol, 500 µl heptane. Invert the tube several times. Most embryos will immediately sink to the bottom of the vial and the others will follow in a matter of seconds.
> 2. Remove all heptane and as much methanol as possible.
> 3. Wash twice with methanol. Embryos can be kept at this stage at 4 °C for weeks.

Methanol fixation has been described as an alternative method for preserving microtubules, although we find it gives variable results. Dechorionated embryos are placed in a 1:1 mixture of heptane: 97% methanol, 3% 0.5 M $Na_3$–EGTA, pH 7.6. The devitellinized embryos fall to the bottom, and as much liquid as possible is then removed. After two further washes with 97% methanol and 3% 0.5 M $Na_3$–EGTA, pH 7.6, the embryos are incubated for at least 4 h at room temperature or overnight at 4 °C. Finally, they are rehydrated by passing through a series of 20, 40, 60, and 80% PBS in methanol, and finally PBS alone.

Immunostaining can be carried out by taking the preparation of fixed embryos, blocking any residual fixative in protein containing buffer (*Protocol 9*). If the embryos have been incubated with BrdU, then their chromosomal DNA has to be denatured in order to make the antigen accessible to anti-BrdU antibodies. This is achieved by first incubating them for 30 min in 2 M

HCl. This is followed by several rinses with PBS containing 0.3% Tween (or Triton X-100), and finally incubating for at least 60 min in PBS containing 0.3% Tween and 10% fetal calf serum. Immunostaining can then be carried out following *Protocol 9*. A suitable anti-BrdU antibody can be purchased from Becton Dickinson.

## 5. Brains

### 5.1 Procedures for making squashed preparations

The larval brain is the tissue of choice for the study of mitotic chromosomes during late development. Consequently most mitotic mutants of *Drosophila* have been identified by screening squashed preparations of brains stained with orcein. Several protocols have been described for making such preparations, but we have had best success with the one given in *Protocol 13*.

Brains are obtained from third instar larvae by pulling from the mouthparts with a pair of tweezers while holding the larvae by their middle. The brain—ventral ganglion plus optical lobes—usually comes out together with the salivary glands and other internal tissues. Carefully remove these as brain preparations that are free from other tissues are essential to achieve good squashes. For most purposes, dissection is carried out in saline (0.7% NaCl) which is also appropriate for short-term culture; neuroblasts will keep dividing in saline for at least a couple of hours. If more controlled conditions are desired one can incubate the brains in TCM (see *Protocol 10*), but this is not usually required. Short-term culture permits the study of the effects of drugs on the cell cycle, as discussed above (Section 4.2) for embryos. As with embryos, if BrdU is used to follow DNA synthesis, the chromosomal DNA must be denatured using the procedure given in Section 4.3 before immunostaining with an anti-BrdU antibody can be effective.

Depending on the purpose of the study, brains can either be treated with colchicine and hypotonic shock, or left untreated. Colchicine and hypotonic shock render the tissue ideal for the observation of chromosomes. Colchicine arrests cells before anaphase and the hypotonic shock swells the cells, thus allowing sister chromatids and chromosomes to disentangle and spread. This treatment is, therefore, ideal for the purpose of karyotyping to estimate ploidy, to identify chromosome rearrangements, etc. Nevertheless, the method is not suitable for studying many aspects of cell division which are impeded or altered by the presence of colchicine, such as chromosome condensation, the formation of the metaphase plate, the processes of anaphase, etc. The analysis of these processes requires the study of untreated brains. Unfortunately, these are harder to study, compared to colchicine-treated tissue, their overall quality tends to be rather low, and they show fewer mitotic figures. Nevertheless, they provide a much more realistic view of those features of cell division.

**Protocol 13.** Acid–orcein stained preparation

*Materials*
- acetic–orcein stain: this solution may be made by adding 3 g Gurr's synthetic orcein to 100 ml of the appropriate concentration of acetic acid (45 or 60%) and boiling with refluxing for 30 min. Filter through filter paper and store at 4 °C. The solution keeps for many years. Centrifuge in a microcentrifuge before use.

*Method*
1. Place the dissected brain on a microscope slide and add a drop of 45% acetic acid. Leave for 30 sec.
2. After 30 sec, remove the liquid using tissue paper.
3. Add a drop of 3% orcein in 45% acetic acid and leave for 3 min.
4. Remove the liquid using tissue paper.
5. Add a drop of 60% acetic acid and then immediately remove the liquid.
6. Add a drop of 3% orcein in 60% acetic acid, and immediately cover with a coverslip.
7. Squash between two sheets of blotting paper by pressing with two fingers on opposite corners. Keep pressing for at least 10 sec and release pressure gently. Repeat, pressing on the other two corners.
8. Seal around the edges of coverslip immediately with nail varnish.

If mitotic chromosomes are to be used for *in situ* hybridization, then an unstained preparation can be made as described in *Protocol 14*. Chromosomes can be subsequently visualized using a variety of DNA-specific dyes. The procedure can be carried out either on untreated brains or on brains that have been incubated with colchicine followed by subjection to hypotonic shock.

**Protocol 14.** Unstained preparation
1. Place the dissected brains on a microscope slide and add a drop of 45% acetic acid, and leave for 30 sec.
2. Remove the liquid using tissue paper.
3. Add a drop of 60% acetic acid.
4. Cover with an 18 × 18 mm coverslip. Do not squash yet, but wait for 3 min.
5. Squash as in Step 7 of *Protocol 13*.

# 7: Mitosis in Drosophila

> 6. Immerse the end of the slide carrying the coverslip in liquid nitrogen to freeze the preparation.
> 7. When the nitrogen ceases to boil, remove the slide and lever off the coverslip with the flick of a scalpel.
> 8. Dehydrate by successive immersion of the slide in:
>    - 70% ethanol for 3 min
>    - 100% ethanol for 3 min
> 9. Allow to dry in air before use.

---

> **Protocol 15.** Staining squashed preparations of larval brains with Hoescht 33258
>
> 1. Take the preparation from Step 9 of *Protocol 14*, and rehydrate by soaking for 5 min in 150 mM NaCl, 30 mM KCl, sodium phosphate buffer, pH 7.0.
> 2. Stain in 0.5 µg/ml Hoechst 33258 (Sigma) for 10 min.
> 3. Wash briefly in 150 mM NaCl, 30 mM KCl, sodium phosphate buffer, pH 7.0.
> 4. Mount in 0.16 M $Na_2HPO_4$, 0.04 M sodium citrate, pH 7.0.

*Protocol 15* describes how to stain a squash with Hoechst 33258. The preparations can also be stained with propidium iodide (*Figure 1*). This dye gives strong DNA staining, but also binds RNA. Consequently it is necessary to treat the preparation with RNase A if only DNA is to be visualized. To stain a squashed preparation such as one produced by Step 9 of *Protocol 14* it is necessary to add a drop of 2.5% *n*-propyl gallate in 85% glycerol containing 1 µg/ml propidium iodide to the preparation before mounting. The preparation can then be covered with a coverslip which can be sealed around the edges with nail varnish.

DAPI and Hoechst 33258 bind preferentially to AT-rich constitutive heterochromatin. A number of alternative dyes can be used to stain DNA. These include chromomycin $A_3$ which binds preferentially to GC-rich constitutive heterochromatin. It does not penetrate living cells, unlike 7-amino actinomycin-D, a DNA-specific dye that can be observed using the same filter sets as for propidium. This dye shows preferential binding to GC-rich regions of DNA, and has some advantage in not binding to RNA.

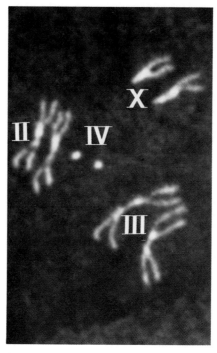

**Figure 1.** Karyotype of a neuroblast from a female larva following the staining of a squashed brain preparation with propidium iodide. The two X (V-shaped) and fourth (dot-like) chromosomes are distinctive. The second and third pair of autosomes can be distinguished because the latter are usually longer and lack the secondary constriction.

## 5.2 Procedures for immunostaining larval neuroblasts

We have developed two methods for immunostaining cells from the larval central nervous system. The first uses a squashed preparation, whereas in the second the whole brain is prepared for immunostaining *in toto*.

*Protocol 16* describes an approach suitable for immunostaining a preparation to reveal subcellular structure such as microtubules, with counterstaining to reveal DNA. The choice of fluorescent labels particularly suited to subsequent observation using a confocal microscope such as the Biorad MRC-600 was discussed in Section 3.2.1. The procedure may be adapted for use with other fluorochromes, or for double-antibody staining.

---

**Protocol 16.** Immunostaining squashed preparations of larval brains

1. Dissect out larval brains in 0.7% NaCl as described in Section 4.2.

2. Place the brains in a microcentrifuge tube containing 3.7% formaldehyde (see Section 4.3) in PBS, and incubate at room temperature for 1 h.
3. Carefully remove the liquid and replace it with 10% fetal calf serum (FCS), 0.3% Triton X-100 in PBS (PBS, see *Protocol 6*) and incubate for 1 h at room temperature.
4. Follow Steps 1–9 of *Protocol 14*.
5. Carry out immunostaining as described in *Protocol 9*.

The procedure that we have developed for immunostaining intact preparations of larval brain is given in *Protocol 17*. This procedure was developed with the prospects of viewing the resulting preparation by scanning confocal microscopy. This facilitates optical sectioning of thick specimens, thus permitting a study of the three-dimensional relationships of components of the mitotic cell.

The brains should be dissected as described in Section 5.1. In handling the brains it is helpful to use a drawn-out Pasteur pipette that has been wetted with saliva to avoid the preparation sticking to the glass.

As the specimen to be stained is very thick, it is important to take special care in premeabilizing the tissue to facilitate antibody penetration, and to wash thoroughly after incubation with antibodies to reduce background staining. This is helped if Step 2 of *Protocol 9* is followed carefully. The RNase treatment used in this step also reduces background staining of RNA so that DNA can be easily seen following propidium iodide staining.

We have found the best results when taxol is included in the buffers as described in *Protocol 17*. We were concerned that artefacts might arise from the use of this drug, but found that scaling-down the time that the tissue is incubated with the drug to only 30 seconds gave identical results. Moreover, the technique give highly reproducible staining patterns characteristic of a number of mitotic mutants (see discussion in ref. 47). We have been able to visualize mitotic spindles in wild-type brains using the above procedure and omitting taxol from the buffers. This results in spindles that have similar shape to the wild-type spindles observed when taxol is included, but the staining is extremely fuzzy. Furthermore, whereas the use of taxol as a stabilizing agent allows us to see many spindles per wild-type brain, without taxol we can detect only a few per brain. We have also attempted methanol fixation in the presence of EGTA following the conditions described by Kellogg *et al.* (48) for *Drosophila* embryos. In our hands this is even less efficient at stabilizing mitotic spindles in neuroblast cells, and does not permit the examination of a sufficient number of structures to make satisfactory conclusions as to the reproducibility of mitotic defects.

Nevertheless, the rare structures that have been observed correspond to those seen with formaldehyde/taxol fixation.

The thickness of the third instar larval brain means that special care has to be taken in mounting the specimen for microscopy. It is important to mount brains in a position that facilitates observation of the region to be studied. For general purposes, neuroblasts located on the thoracic region of the ventral side of the ganglion provide the best material. To observe these cells, the brains should be mounted with their ventral sides upwards bringing the neuroblasts as close as possible to the microscope objective. After the correct positioning of the brain has been achieved, and the specimen has been mounted and sealed, preparations can be kept at −20 °C for several months without any noticeable loss in their quality for microscopic analysis.

---

**Protocol 17.** Immunostaining whole-mount preparations of larval brain

1. Dissect out the brain and rinse in 0.7% NaCl. If microtubules are to be observed by immunostaining, add taxol to the dissection buffer at a final concentration of 1 μM taxol[a].
2. Fix for 30 min in 3.7% formaldehyde, 0.7% NaCl.
3. Rinse in 0.7% NaCl.
4. Carry out immunostaining as described in *Protocol 9*.

[a] See text for a discussion on the use of taxol.

---

## 5.3 *In situ* hybridization to cells of the larval brain

*In situ* hybridization to diploid cells provides the only route towards mapping the pericentromeric sequences of *Drosophila* which are both under-represented and aggregated in polytene cells. As with the mapping of euchromatic sequences on polytene chromosomes, the mapping of heterochromatic sequences on diploid chromosomes can be facilitated through the use of genetically characterized chromosomal rearrangements. These are useful as a simple means of distinguishing the two major autosomes which appear very similar, as well as defining landmarks within the heterochomatic regions, thus increasing the level of resolution. Moreover, *in situ* hybridization provides a means of following the behaviour of heterochromatic sequences in mutations that affect mitosis through effects upon somatic pairing and chromosome positioning during the different cell-cycle stages.

Radioactive methods for the detection of probes have now been superseded by fluorescence-based protocols that have allowed a dramatic increase in resolution, a noticeable decrease in the time required, and a reduction in the hazard associated with the earlier protocols (*Protocol 18*). Probes are prepared using the same procedures as described for *in situ* hybridization to polytene chromosomes (*Protocol 4*).

---

**Protocol 18.** *In situ* hybridization to diploid cells

1. Dissect out larval brains in 0.7% NaCl as described in Section 4.2.
2. Incubate in 0.5 μg/ml colchicine (Sigma cat. no. C 9754) in 0.7% NaCl in a dark, humid chamber for 2 h.
3. Wash in 0.5% trisodium citrate.
4. Fix for 30 sec in 45% acetic acid.
5. Remove the liquid using blotting paper and replace with 60% acetic acid.
6. After 3 min place the brains in a drop of 60% acetic acid on a microscope slide and cover with a siliconized coverslip.
7. Follow Steps 5–9 of *Protocol 14*.
8. Bake the slide at 58 °C for 1 h in a dry oven.
9. Denature the chromosomal DNA by treating with 70% formamide in 2 × SSC (*Protocol 6*), 70 °C for 2 min.
10. Dehydrate by successive immersion of the slide in:
    - 70% ethanol for 3 min
    - 100% ethanol for 3 min
11. Apply 10 μl of denatured probe and carry out the hybridization following Steps 5 and 6 of *Protocol 5*.
12. Following hybridization, carry out the washing and FITC staining as described in *Protocol 7*.

---

# 6. Mutations that also affect male meiosis

## 6.1 Cytological studies on testes squashes

The second meiotic division is analogous to mitotic divisions and as such might be expected to share many characteristics. In structure, the meiotic spindle in the *Drosophila* male appears superficially similar to the mitotic spindle, although one might expect to see significant differences between these two structures when their detailed organization is understood. Mutations, such as *asp* (43) and *polo* (44), are known to affect the meiotic spindle in addition to their mitotic phenotype. Abnormal spindle structure can be seen in the testes either directly by phase-contrast microscopy (*Figure 2*) or by

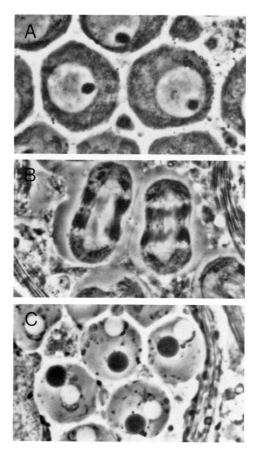

**Figure 2.** Primary spermatocytes (A), meiotic spindles (B), and early spermatids (C) in unfixed, unstained squashed preparations of testes viewed with phase-contrast optics. The large spherical body within each spermatocyte is the nucleus, and the small dark body within it is the nucleolus. Meiotic spindles are decorated with mitochondria which bind tightly to the spindle fibres. Each spermatid contains a nucleus (the light spherical body) associated with a Nebenkern (the dark body), a mitochondrial derivative. In some cells the acrosome can be seen as a ribbon-like structure associated with the periphery of the nucleus.

immunostaining. If the mutation results in chromosome non-disjunction, then this can be seen cytologically by the appearance of spermatid nuclei of varying rather than uniform size. If some male fertility persists, then non-disjunction can also be followed genetically.

Testes squashes can be carried out following *Protocol 19*. Examples of the appearance of meitoic spindles and spermatids in a well-squashed preparation are shown in *Figure 2*. Squashing is achieved by withdrawing saline from

## 7: Mitosis in Drosophila

underneath a coverslip that covers the preparation on a microscope slide. The degree of squashing can be controlled by how much liquid is withdrawn. If, for instance, the number of spermatids within a cyst is to be counted, then a mild squash is required to preserve the cyst intact. On the other hand, if a particular structure such as the acrosome is to be studied, more extreme squashing might be required. In any event, the right degree of squashing is always attained by removing the liquid slowly.

---

**Protocol 19.** Aqueous squashed preparations of testes

1. Dissect out the testes from either third instar larvae, pupae, or adults in 0.7% NaCl.
2. Place the dissected material in a drop of approximately 100 µl of 0.7% NaCl on a siliconized coverslip.
3. Cut each spiral testis at about one-third from its apical end using tungsten needles[a].
4. Cover with non-siliconized coverslips and squash by applying a small piece of blotting paper to one of the edges of the coverslip, following the process using a 40 × dry phase-contrast objective.
5. When the appropriate degree of squashing has been attained, remove the blotting paper.

[a] Tungsten needles can be prepared as described in ref. 2.

---

Squashing can also be carried out by substituting 45% acetic acid for saline. This permits examination of both chromosomes and sperm heads, the only structures that withstand this treatment. Most other structures, including nuclei with non-condensed DNA, are damaged under these conditions.

If a preparation is to be made for the purposes of immunostaining, it is possible to fix the tissue either using formaldehyde (*Protocol 20*), or methanol (*Protocol 21*). In order to preserve microtubules, it is necessary to add taxol to the buffers when using formaldehyde as fixative. However, *Protocol 21* devised in Maurizio Gatti's laboratory gives excellent microtubule preservation.

---

**Protocol 20.** Immunostaining testes preparations—formaldehyde fixation

1. Dissect testes in 183 mM KCl, 47 mM NaCl, 10 mM Tris–HCl, pH 6.8. If microtubules are to be studied, add 0.5 µM taxol.

**Protocol 20.** *Continued*

2. Carry out the squashing procedure as described in *Protocol 19* and in the text.
3. When the preparation is at a suitable stage of squashing, dip the end of the slide carrying the coverslip into liquid nitrogen.
4. Remove the coverslip from the slide using a razor blade or scalpel.
5. Dip the slide immediately into 70% ethanol, and leave there for 5 min.
6. Transfer the slide to 100% ethanol for 5 min.
7. Transfer into 3.7% formaldehyde in 1 × PBS (*Protocol 6*) for 10 min.
8. Wash three times, each for 5 min, with PBS.
9. Incubate in 10% FCS, 0.1% Triton X-100, in PBS, for 30 min.
10. Carry out antibody incubations, washes, and mounting of the preparation as described in *Protocol 9*.

---

**Protocol 21.** Immunostaining testes preparations—methanol fixation

*Materials*

- Buffer 1: 183 mM KCl, 47 mM NaCl, 10 mM Tris–HCl, 1 mM EDTA, 1 mM PMSF, pH 6.8 at room temperature; *NB* PMSF should be made up fresh for each use
- Buffer 2: 1 mM $CaCl_2$, 2.6 mM KCl, 1.5 mM $KH_2PO_4$, 0.5 mM $MgCl_2$, 137 mM NaCl, 8.1 mM $Na_2HPO4$

*Method*

1. Dissect larval or adult testes in buffer 1.
2. Carry out the squash in the same buffer following the considerations outlined in the text.
3. When the preparation is at a suitable stage of squashing, dip the end of the slide carrying the coverslip into liquid nitrogen.
4. Remove coverslip from the slide using a razor blade or scalpel.
5. Immerse the slide immediately into methanol at −20 °C and leave for 5 min.
6. Transfer the slide into acetone at −20 °C, and leave for 1–2 min.
7. Transfer the slide into buffer 2 containing 1% Triton X-100 and 0.05% acetic acid. Leave for 10 min at room temperature.

> 8. Give the slides three 5 min washes in the above buffer without the Triton X-100 and acetic acid and proceed to immunostaining as described in *Protocol 9*.

## 6.2 Assays for non-disjunction in male meiosis

The genetic analysis of meiotic non-disjunction is a very powerful means of estimating chromosome stability during cell division. It enables the identification of chromosome misbehaviour occurring at very low frequencies, which would be otherwise impossible by cytological techniques. Secondly, the analysis of meiotic non-disjunction provides additional information about the role of mitotically active genes in meiosis, thus offering a way to identify common, and most likely, essential functions governing cell division. Finally, the judicious choice of markers allows one to distinguish between failure of chromosome segregation at either the first or second meiotic division. This is particularly relevant as the second meiotic division is very similar to mitosis.

The simplest test for meiotic non-disjunction is based on the use of compound chromosomes, i.e. two homologues sharing a common centromere. Females carrying such a chromosome produce aneuploid eggs that can be either 'diplo' or 'nullo'—that is to say they carry either two copies or none—of that particular homologue. In the case of the X chromosome, exceptional progeny such as patroclinous females and Y-chromosome bearing males would indicate non-disjunction during the second and first meiotic divisions, respectively. When compound autosomes are used, such as *C(2)EN* or *C(3)EN* (49) the situation is somewhat different in as much as that in the absence of non-disjunction in the male, the gametes laid by these females are inviable. Consequently, meiotic non-disjunction is revealed here by the simple presence of offspring. By using the appropriate transheterozygous combination of recessive markers (i.e. *cn bw/b* x *C(2)EN c bw*) it is also possible to determine whether non-disjunction took place during the first division (indicated by *cn bw/b* descendants) or the second division (indicated by *cn bw/cn bw* or *b/b* descendants).

The crosses *st sr/th e* males × *C(3)EN th st* females, and *ci/ey* males × *C(4)RM spa$^{pol}$* females would also be appropriate to study non-disjunction of the third and fourth chromosomes, respectively.

Whichever test is used it is important to carry out this kind of analysis by examining the offspring of single-pair matings so that any oddities due to a particular individual, that might not have anything to do with the mutation under study, can be observed. It is important to set up sufficient single-pair crosses of both mutant and control individuals to achieve statistical significance.

## Acknowledgements

The work in our laboratory is supported by grants, fellowships, and studentships from the Cancer Research Campaign, the Medical Research Council, the Wellcome Trust, the European Community, and the Human Frontiers Science Programme. We thank Hiroyuki Ohkura, Daryl Henderson, and Helen White-Cooper for their comments on the manuscript.

## References

1. Ashburner, M. (1989). *Drosophila: a laboratory handbook*. Cold Spring Harbor Laboratory Press, NY.
2. Ashburner, M. (1989). *Drosophila: a laboratory manual*. Cold Spring Harbor Laboratory Press, NY.
3. Roberts, D. (ed.) (1986). *Drosophila: a practical approach*. IRL Press, Oxford.
4. Foe, V. E. and Alberts, B. M. (1983). *Journal of Cell Science*, **61**, 31.
5. Karr, T. L. and Alberts, B. M. (1986). *Journal of Cell Biology*, **102**, 1494.
6. Foe, V. E. (1989). *Development*, **107**, 1.
7. Campos-Ortega, J. A. and Hartenstein, V. (1985). *The embryonic development of* Drosophila melanogaster. Springer-Verlag, Berlin.
8. Smith, A. and Orr-Weaver, T. (1991). *Development*, **112**, 997.
9. White, K. and Kankel, D. R. (1978). *Developmental Biology*, **65**, 296.
10. Truman, J. W. and Bate, C. M. (1988). *Developmental Biology*, **125**, 145.
11. Lindsley, D. L. and Tokuyasu, K. T. (1980). In *The genetics and biology of* Drosophila, Vol. 2d (ed. M. Ashburner and T. R. F. Wright), p. 225. Academic Press, London.
12. King, R. C. (1970). *Ovarian development in* Drosophila melanogaster. Academic Press, London.
13. Mahowald, A. and Kambysellis, M. P. (1980). In *The genetics and biology of* Drosophila, Vol. 2d (ed. M. Ashburner and T. R. F. Wright), p. 141. Academic Press, London.
14. Glover, D. M. (1989). *Journal of Cell Science*, **92**, 137.
15. Ripoll, P., Casal, J., and Gonzalez, C. (1987). *Bioessays*, **77**, 204.
16. Gatti, M. and Goldberg, M. L. (1991). In *Methods in cell biology*, Vol. 35 (ed. L. Wilson), p. 543.
17. Baker, B. S., Carpenter, A. T. C., and Ripoll, P. (1978). *Genetics*, **90**, 531.
18. Baker, B. S. and Smith, D. A. (1979). *Genetics*, **92**, 883.
19. Smith, D. A., Baker, B. S., and Gatti, M. (1985). *Genetics*, **110**, 647.
20. Lawrence, P. A., Johnston, P., and Moratta, G. (1986). In *Drosophila: a practical approach* (ed. D. B. Roberts), IRL Press, Oxford.
21. Gatti, M., Pimpinelli, S., Bove, C., Baker, B. S., Smith, D. A., Carpenter, A. T. C., and Ripoll, P. (1983). In *Proceedings of the XV International Congress of Genetics*, Vol. III (ed. V. L. Chopra, D. C. Joshi, R. P. Sharma, and H. C. Bansal), p. 193. Oxford and IBH Publishing Company, Dehli.
22. Gatti, M. and Baker, B. S. (1989). *Genes and Development*, **3**, 438.

23. Shearn, A., Rice, T., Garen, A., and Gehring, W. (1971). *Proc. Natl Acad. Sci USA*, **68**, 2594.
24. Schupbach, T. and Wieschaus, E. (1989). *Genetics*, **121**, 101.
25. Edgar, B. A. and O'Farrell, P. H. (1989). *Cell*, **57**, 177.
26. Jurgens, G., Wieschaus, E., Nusslein-Volhard, C., and Kluding, H. *Wilhelm-Roux's Archives Developmental Biology*, **193**, 283.
27. Roberts, D. B. (1986). In *Drosophila: a practical approach* (ed. D. B. Roberts), p. 1. IRL Press, Oxford.
28. Grigliatti, T. (1986). In *Drosophila: a practical approach* (ed. D. B. Roberts), p. 39. IRL Press, Oxford.
29. Cooley, L., Kelly, R., and Spradling, A. C. (1988). *Science*, **239**, 1121.
30. Robertson, H. M., Preston, C. R., Phillis, R. W., Johnson-Schlitz, D., Benz, W. K., and Engels, W. R. (1988). *Genetics*, **118**, 461.
31. O'Kane, C. and Gehring, W. J. (1987). *Proceedings of the National Academy of Sciences*, **84**, 9123.
32. Muller, H. J. (1932). In *Proceedings of the Sixth International Congress of Genetics*, Vol. 1, p. 213. Ithaca, New York.
33. Pirrotta, V. (1986). In *Drosophila: a practical approach* (ed. D. B. Roberts), p. 83. IRL Press, Oxford.
34. Saunders, R. D. C., Glover, D. M., Ashburner, M., Siden-Kiamos, I., Louis, C., Monastrioti, M., Savakis, C., and Kafatos, F. (1989). *Nucleic Acids Research*, **17**, 9027.
35. Johnston, D. H. (1991). In *PCR: a practical approach* (ed. M. J. McPherson, P. Quirke, and G. R. Taylor), p. 121. IRL Press, Oxford.
36. Sambrook, J., Fritsch, E. F., and Maniatis, T. (1989). *Molecular cloning: a laboratory manual*, (2nd edn). Cold Spring Harbor Laboratory Press, NY.
37. Tautz, D. and Pfeifle, C. (1989). *Chromosoma*, **98**, 91.
38. Kilmartin, J. V., Wright, B., and Milstein, C. (1982). *Journal of Cell Biology*, **93**, 576.
39. Whitfield, W. G. F., Millar, S. E., Saumweber, H., Frasch, M., and Glover, D. M. (1988). *Journal of Cell Science*, **89**, 467.
40. Lehner, C. F. and O'Farrell (1989). *Cell*, **56**, 957.
41. Whitfield, W. G. F., Gonzalez, C., Maldonado-Codina, G., and Glover, D. M. (1990). *EMBO Journal*, **9**, 2563.
42. Kellogg, D. R., Mitchison, T. J., and Alberts, B. M. (1988). *Development*, **103**, 675.
43. Warn, R. M., Flegg, L., and Warn, A. (1987). *Journal of Cell Biology*, **105**, 1721.
44. Harlow, E. and Lane, D. P. (1988). *Antibodies: a laboratory manual*. Cold Spring Harbor Laboratory Press, NY.
45. Raff, J. W. and Glover, D. M. (1988). *Journal of Cell Biology*, **107**, 2009.
46. Santamaria, P. (1986). In *Drosophila: a practical approach* (ed. D. B. Roberts), p. 159. IRL Press, Oxford.
47. Gonzalez, C., Saunders, R. D. C., Casal, J., Molina, I., Carmena, M., Ripoll, P., and Glover, D. M. (1990). *Journal of Cell Science*, **96**, 605.
48. Sunkel, C. E. and Glover, D. M. (1988). *Journal of Cell Science*, **89**, 25.
49. Novitski, E., Grace, D., and Strommen, C. (1981). *Genetics*, **98**, 257.

# 8

# The use of cell-free extracts of *Xenopus* eggs for studying DNA replication *in vitro*

C. J. HUTCHISON

## 1. Introduction

Cell-free extracts of *Xenopus* eggs provide the only available systems that support the initiation, elongation, and completion of eukaryotic chromosomal replication *in vitro* (1, 2). These extracts have been used extensively to study the control of entry into S phase (3, 4), the dependency upon completion of S phase as a requirement for entry into mitosis (5, 6), and the role of the nuclear envelope in the initiation and cell-cycle control of S phase (7–10). Although the exploitation of these extracts reflects the current interest in cell-cycle control mechanisms, it also reflects their limitations. The extracts lend themselves to analysis of periodic enzyme activities and to immunochemical dissection. However, they are difficult to fractionate. This has restricted those of us who are interested in the enzymology of DNA synthesis to testing for the involvement of previously identified proteins at replication forks (11, 12). Nevertheless, these extracts still represent powerful experimental tools, whose preparation and manipulation will be reviewed.

## 2. Maintaining a *Xenopus* colony

The starting point for the preparation for egg extracts is to obtain a reliable source of eggs. Although they are considerably easier to maintain than most amphibians, *Xenopus laevis* still display idiosyncrasies which prohibit experimentation at certain times of year (usually early to mid-summer; this season does not of course exist in Scotland and presumably accounts for our colony's year round reliability). Talking to other workers in the field, we appear to have fewer problems than most (excepting those who are able to offer their animals a Californian climate), so I will outline our husbandry procedures.

Our animals are purchased as small mature females either from a national supplier (our preference is Philip Harris, Bristol) or direct from Snake farm, South Africa, which requires an import licence. The quality of the animals is very variable from either source and we recommend that they are allowed to 'fatten up' for at least 2 months before being used (in the wild, *Xenopus* populations often live on the verge of starvation and are often heavily infested with parasites). The animals are kept at a density of 16 per cage, in tanks which are 660 × 450 × 220 mm. The tanks are half-filled with tap water which has stood for 48 h before use. The colony is maintained at a constant room temperature of 21 °C with a light/dark cycle of 12 h (this contributes significantly to reducing seasonal variations in egg laying). The animals are fed twice a week on chopped ox heart which can be obtained cheaply from any abattoir. Although preparing a frog's lunch is time consuming, the animals much prefer fresh meat to the alternative of trout pellet and are as a consequence healthier. Tanks are usually cleaned 24 h after feeding. A useful size for a colony is 160 females and 16 males.

Each animal is used to obtain eggs once every 3 months over a period of 3 years. We normally inject four animals at a time to ensure the collection of a mimimum volume of 15 ml of dejellied eggs. The animals are injected into their dorsal lymph sacs with 650 i.u. of human chorionic gonadotrophin (trade name Chorulon, Intervet UK Ltd) 16 hours before the eggs are required. To collect the eggs the frogs are placed in individual tanks which are fitted with a perforated false floor (tank dimensions: 350 × 235 × 160 mm) and half-filled with saline tap-water (110 mM NaCl) equilibrated to 20 °C.

## 3. Preparation of egg extracts

Selection of the eggs which are collected is critical to the subsequent preparation of extracts. A healthy animal will lay eggs which have uniformly pigmented hemispheres and which are separate from each other. Eggs which are laid in strings are usually necrotic and should be discarded. To use the eggs the jelly coats must first be removed. This can easily be achieved using mild reducing reagents such as cysteine or by using DTT. Most authors prefer cysteine (110 mM NaCl, 20 mM Tris–HCl at pH 8.0, 2% cysteine, prepared fresh) as this is commonly used by embryologists who perform *in vitro* fertilizations. However, for the purposes of producing extracts DTT is perfectly good and is at hand in most laboratories. Eggs are dejellied by incubating them in a solution containing 110 mM NaCl; 20 mM Tris–HCl, pH 8.5; 1 mM DTT, and a few granules of phenol red (this solution, called dejelly solution, is prepared just before use). By gently rocking the eggs, the jelly coats detach within 5 minutes (longer incubations are deleterious). Removal of the jelly coat can be ascertained by observing the packing of the eggs (eggs with a jelly coat do not contact each other). Once the eggs are dejellied they are washed extensively in saline tap-water. At this point necrotic eggs are removed. These are easily identifiable as they are white and swollen.

## 8: Cell-free extracts of Xenopus eggs and DNA replication

Unfertilized eggs are arrested at metaphase of second meiosis. Some authors prefer to mimic the fertilization reaction by either electric shock or treatment with a calcium ionopohore (1). As disruption of the eggs normally achieves this end we routinely avoid activation. The dejellied eggs are transferred to a 100 ml glass beaker and washed three times with ice-cold extraction buffer (*Protocol 1*). The eggs are then packed into centrifuge tubes and excess buffer is removed. We have used Beckman SW50.1 rotors, Sorvall HB-4 rotors, and MSE 3 × 5 ml rotors with equal success. Eggs which are placed in 4 ml or 5 ml centrifuge tubes will settle quickly so that buffer can be removed from the surface. However, some authors use a prespin step (120 $g$ for 2 min in a bench top centrifuge) to ensure maximum packing of the eggs and removal of buffer (20). Centrifugation is performed at 10 000 $g$ for 10 min at 4 °C. This results in a stratified extract consisting of a yolk pellet, an ooplasmic layer, and a lipid cap (*Figure 1*). The ooplasmic layer is removed

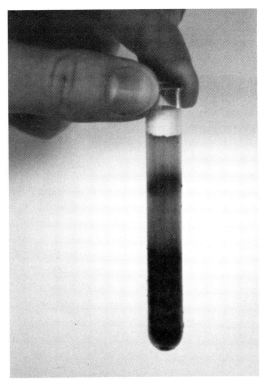

**Figure 1.** *Xenopus* egg extract after the first centrifugation step, illustrating the lipid cap (top of the tube), ooplasmic layer (just under the lipid cap), and yolk pellet (bottom of the tube). Only the ooplasmic layer is retained. In some preparations, as here, the ooplasmic layer is divided in two by a band of pigment granules. These are removed by the second centrifugation step.

and its volume measured. Cytochalasin B is then added to a final concentration of 50 μg/ml to increase the fluidity of the extract. Centrifugation is then repeated (10 000 g for 10 min at 4 °C) to remove debris. The volume of the final extract is measured and aprotonin is added to a final concentration of 10 μg/ml. (We have systematically tested a range of protease inhibitors in the extract including leupeptin, PMSF, pepstatin A, and iodoacetamide. All have a range of inhibitory effects either on protein synthesis or DNA replication. Even aprotonin varies from batch to batch, so that new batch numbers must be tested by addition to frozen extracts which have been characterized for replicative capacity). By adding aprotonin variation between extracts is reduced.

---

**Protocol 1.** Steps in the preparation of egg extracts

*Materials*

- Dejelly solution: 110 mM NaCl, 20 mM Tris–HCl (pH 8.5), 1 mM DTT, few granules of phenol red
- Extraction buffer (EB): 100 mM KCl, 5 mm MgCl$_2$, 20 mM Hepes (pH 7.5), 2 mM 2-mercaptoethanol
- Cytochalasin B (prepared in ethanol at 5 mg/ml and stored at −20 °C)
- Aprotinin (1 mg/ml stock)
- Glycerol (optional)

*Method*

1. Combine eggs from several females, ensuring the quality of each batch, and remove debris including faeces.
2. Drain off existing saline tap-water and replace with 500 ml of dejelly solution. Rock gently for a maximum for 5 min.
3. Rinse dejellied eggs three times in large volumes (∼ 2 litres) of saline tap-water.
4. Observe dejellied eggs and remove necrotic ones.
5. Transfer eggs to a glass beaker on ice and decant off saline tap-water. Rinse three times with ice-cold extraction buffer and transfer to centrifuge tubes, allow eggs to settle, and pipette off excess buffer.
6. Spin at 10 000 *g* for 10 min at 4 °C.
7. Remove the ooplasmic layer (*Figure 1*) with a cold Pasteur pipette and transfer to a 10 ml measuring cylinder (20 ml of eggs should yield approximately 6 ml of extract at this stage). When the volume has been determined add 5 mg/ml cyctochalasin B to give a final concentration of 50 μg/ml.

8: Cell-free extracts of Xenopus eggs and DNA replication

> 8. Transfer to fresh centrifuge tubes and spin at 10 000 g for a further 10 min at 4 °C.
>
> 9. Add aprotonin to 10 μg/ml and use immediately **or** add aprotonin and glycerol (5%) and freeze in 15 μl droplets in liquid nitrogen, storing 10 droplets per cryogenic vial.

## 4. Using egg extracts

### 4.1 Fresh vs frozen

Extracts can be used fresh or can be frozen for future use. Fresh extracts recapitulate all of the events associated with DNA replication and nuclear division in early embryos. The temporal sequence of these events, which occur synchronously in large populations of added nuclei, is maintained (5). However, repeated experiments on the same extracts are precluded. In contrast, extracts which have been frozen and thawed only support the initiation and completion of DNA replication. Furthermore, populations of nuclei within a common cytoplasmic environment initiate DNA replication asynchronously over a 6 h period, while individual nuclei take approximately 1 h to double their DNA content. However, repeated experiments can be performed using individual extracts, which makes them useful for some fractionation procedures (1, 2, 8).

Fresh extracts are straightforward to use. They do not require the addition of either ATP or an energy regenerating system and will assemble nuclei and replicate DNA periodically for up to 6 h. To maintain a cell-cycle periodicity equivalent to that of an early embryo, additions to the extract (of DNA templates or deoxynucleotide analogues) should be limited to 10% of the total volume of the extract. We have tested a range of buffers in dilution assays and find SUNaSp (*Protocol 2*) to be best. Extracts can be diluted by up to 50% in depletion and rescue experiments, but under these conditions, only nuclear assembly and DNA replication are maintained. Once an extract is prepared it should be used within 1 h. Most extracts are less stable on ice than at 21 °C. So for experiments which require a preincubation step, e.g. immuno-depletion, we normally carry these out at room temperature (RT). As the behaviour of *Xenopus* eggs is sensitive to DNA concentration, we normally limit the amount of DNA template we use to $10^5$ sperm heads (300 ng DNA) per 100 μl of extract. This provides a good signal in replication studies but maintains a 30 minute cell cycle.

Frozen extracts require careful peparation and use. They vary considerably in their efficiency and the only sure way of obtaining good extracts is to make several extracts from different batches of eggs, test each one and discard up to

50% of them (a standard test includes an estimation of the efficiency of nuclear assembly and DNA replication, see below). In their original protocol, Blow and Laskey (1) froze extracts after the addition of 1 mM ATP and 10% glycerol. This methodology was subsequently refined and most workers now agree that the best results are achieved using glycerol at 3–5%.

Clarified extracts are supplemented with 10 µg/ml aprotonin and 5% glycerol. After careful mixing, the extract is frozen in liquid nitrogen. This is done by filling a cryogenic tube with liquid nitrogen, then pipetting 15 µl aliquots of extract into it. The aliquots freeze as small droplets and, for convenience, the equivalent of 150 µl of droplets are stored in a single tube. We have kept extracts in liquid nitrogen for up to 1 year without noticeable deterioration. By vigorously shaking a tube immediately after freezing, the droplets separate and can be individually removed from the tubes using precooled forceps. For use the extracts are thawed rapidly at 21 °C then placed on ice. They are supplemented with an energy regenerating system consisting of 150 µg/ml creatine phosphokinase and 60 mM creatine phosphate (final concentration) and used immediately. As these extracts do not maintain normal cell-cycle dynamics nuclear concentration is less important. Nuclear assembly and DNA replication is efficient at nuclear concentrations of up to $10^6$/100 µl extract which provides enough material for biochemical analysis. Although some authors have reported dilutions of 60% without loss of activity (13) our experience is that additions to a maximum of 20% of the original volume is the upper limit for most procedures. Again we recommend that preincubation steps are performed at RT and that experiments are run at 21 °C.

## 4.2 Choice of DNA template

*Xenopus* embryos are unusual in their capacity to assemble nuclear structures and initiate semi-conservative DNA replication on a range of DNA templates (14, 15). Similarly, egg extracts are also able to use a range of DNA templates which facilitates studies on chromatin organization (16), the role of the nuclear envelope in cell-cycle control (17), and nuclear envelope assembly (18).

### 4.2.1 Demembranated sperm heads

For most purposes demembranated sperm heads (*Protocol 2*) are the most convenient template. These can be obtained in very large quantities and are used efficiently by egg extracts. Sperm are released by suspending freshly isolated testes in 3 ml of Barth X (*Protocol 2*) and homogenizing gently with a loose-fitting glass pestle. The homogenate is poured into a conical glass centrifuge tube and particulate material allowed to settle to the bottom. The sperm suspension is then removed and made 10% with respect to both dimethyl sulfoxide and new-born calf serum and snap frozen in 0.5 ml aliquots ($0.5 \times 10^7$ sperm) in liquid nitrogen. For replication and nuclear assembly assays, single 0.5 ml aliquots of sperm are removed from liquid nitrogen and thawed rapidly and demembranated using lysolecithin (phosphatidylcholine).

---

**Protocol 2.** Preparation of demembranated sperm heads

*Materials*
- Barth X: 88 mM NaCl, 2 mM KCl, 0.33 mM Ca(NO$_3$)$_2$, 0.41 mM CaCl$_2$, 0.82 mM MgSO$_4$
- SUNaSp: 0.25 M sucrose, 75 mM NaCl, 0.5 mM spermidine, 0.15 mM spermine (27)
- Lysolecithin
- BSA

*Method*
1. Thaw frozen sperm ($5 \times 10^6$ in 500 μl Barth X) rapidly by incubating at 37 °C.
2. Dilute the freshly thawed sperm to 3 ml in SUNaSp and recover by centrifugation at 3000 $g$ for 15 minutes at room temperature in a bench top centrifuge.
3. Resuspend the sperm in 200 μl of SUNaSp and determine their number using a haemocytometer.
4. Add 40 μl of 1 mg/ml lysolecithin and gently agitate the sperm suspension at room temperature for 90 sec.
5. Terminate the reaction by adding 3 ml of ice-cold SUNaSp containing 3 mg/ml BSA and recover the demembranated sperm by centrifugation for 15 min at 3000 $g$ in a bench top centrifuge.
6. Resuspend the demembranated sperm in SUNaSp at a final concentration of $10^5$/μl.

---

### 4.2.2 Double-stranded DNA plasmids

*Xenopus* egg extracts will initiate semi-conservative DNA replication on a range of double-stranded DNA plasmids. The efficiency of replication varies between templates (1), large plasmids replicating more efficiently than small plasmids. For this reason most authors now use lambda DNA for studies involving nuclear assembly or DNA replication (18, 19). To prepare lambda

DNA as a substrate for nuclear reconstruction, solutions containing plasmid are heat-treated to produce non-covalently ligated multimers of > 800 000 bp which form nuclei large enough to observe using a light microscope (20).

---

**Protocol 3.** Preparation of lambda DNA

*Materials*

- Potassium phosphate buffer: 50 mM NaCl, 0.1 mM EDTA, 10 mM $K_2HPO_4$, pH 7.2

*Method*

1. Grow and prepare Bacteriophage lambda DNA as described by Sambrook *et al.* (21).
2. Suspend DNA at 200 μg/ml in potassium phosphate buffer.
3. Heat 10 μl of DNA suspension for 5 min at 65 °C.
4. Cool the suspension to 21 °C in a waterbath and add 7 μl to 100 μl of egg extract for assay.
5. Triturate the extract carefully, but extensively, to ensure even distribution of the DNA.

---

### 4.2.3 Somatic cell nuclei

Somatic cell nuclei have been used in egg extracts to study the role of the nuclear envelope in cell-cycle control (17) as well as changes in the organization of replication centres (12). For the specific purpose of investigating the role of the nuclear envelope in the $G_2$ block to reinitiation Laskey and his colleagues have used an isolation procedure which disrupts the cell membrane but not the nuclear membrane. Intact nuclei with unpermeabilized nuclear membranes can be prepared by treatment either with the bacterial exotoxin SLO (Murex Diagnostics Ltd, formerly Wellcome-Diagnostics) or with the detergent digitonin (Calbiochem). Cell pellets obtained from monolayer cultures are resuspended in ice-cold Pipes buffer (50 mM KCl; 5 mM $MgCP_2$; 2 mM EGTA; 50 mM K–Pipes, pH 7.0) at a concentration of $5 \times 10^5$/ml. An equal volume of Pipes buffer containing 1.5 μg/ml SLO is added and the suspension incubated on ice for 10–30 min. During this period the tubes are inverted at 5 minute intervals and samples removed to test the integrity of the plasma membrane using an IgG exclusion assay (see below). Leno *et al.*, (17) report that < 1% of cells include IgG during this incubation. SLO is then removed by pelleting the cells at 600 *g* for 5 min in a benchtop centrifuge and washing the pellet twice in Pipes buffer.

# 8: Cell-free extracts of Xenopus eggs and DNA replication

During the washing procedure it is essential to keep the rotor buckets at 0 °C as permeabilization is subsequently achieved by resuspending the pellet in Pipes buffer and raising the temperature to 23 °C. Samples removed at this point should display 90% inclusion of IgG.

If it is not important to maintain the integrity of the nuclear membrane, a more convenient method of nuclear isolation is described by Kill *et al.* (12). In this procedure (*Protocol 4*) nuclei are released from cells by gentle homogenization. For convenience we use cultures of human fibroblasts which have been made quiescent by growth in low (0.5%) serum for 7 days. Nuclei isolated from these cultures are generally separated from each other and between 90–95% will replicate when added to egg extracts.

---

**Protocol 4.** Isolation of quiescent human nuclei

*Materials*

- Nuclear isolation buffer (NIB): 10 mM NaCl, 3 mM $MgCl_2$, 10 mM Tris–HCl (pH 7.6) 0.5% Nonidet P-40
- Complete medium: DMEM supplemented with 10% new-born calf serum (NCS)
- Low-serum medium: DMEM supplemented with 0.5% NCS
- SUNaSp (*Protocol 2*)
- Dounce homogenizer with loose-fitting pestle

*Method*

1. Establish cultures of human fibroblasts by seeding 90 mm Petri dishes with $2 \times 10^5$ cells and growing in complete medium for 2 days.
2. Replace the medium with low-serum medium and leave the cultures for a further 7 days.
3. Release the cells from the dish by incubating in versene/trypsin at RT.
4. Quench the trypsin by adding 10 ml of complete medium, count the cells using a haemocytometer, and recover by centrifugation for 10 min at 1000 *g* at room temperature.
5. Resuspend the cells in ice-cold nuclear isolation buffer (NIB) at a concentration of $0.5 \times 10^5$/ml and allow to swell for 15 min.
6. Rupture the cells by placing the suspension in a 2 ml Dounce homogenizer and giving 5 strokes with a loose-fitting pestle. Remove a sample and observe using a phase-contrast microscope. If a significant number of nuclei remain attached to cytoplasmic debris, repeat the homogenization.

> **Protocol 4.** *Continued*
> 7. Repeat the process until > 95% of the nuclei are released.
> 8. Dilute the nuclei and cell debris in 10 ml of NIB and gently layer over 4 ml of 30% sucrose (w/v in NIB).
> 9. Centrifuge at 2400 r.p.m. for 15 min in a bench top centrifuge. Nuclei are recovered in the pellet fraction.
> 10. Resuspend isolated nuclei at $10^4/\mu l$ in SUNaSp.

The timing of initiation events as well as the duration of S phase in individual nuclei varies considerably with different DNA templates. In our hands, demembraned sperm heads initiate DNA replication after 30 minutes incubation in fresh extracts and 60 minutes incubation in frozen extracts. In fresh extracts, they act as a synchronously replicating population and the duration of S phase is approximately 20 minutes (22). In frozen extracts, individual nuclei will initiate replication at different times, but each one then takes only 30 minutes to double its DNA content (2). In contrast, somatic nuclei act as asynchronous populations in fresh and frozen extracts. Initiation only occurs after incubation for at least 90 minutes in the extracts and each nucleus takes approximately 90 minutes to complete its synthesis (12). Replication of lambda DNA is also slower than sperm heads, presumably reflecting the slower rate of nuclear assembly.

## 4.3 Labelling protocols

Three methods of detecting and quantifying DNA replication have been commonly used, namely: [$^{32}$P]dCTP labelling followed by agarose gel electrophoresis and autoradiography; BrdU density substitution in conjunction with [$^{32}$P]dCTP labelling followed by caesium chloride gradient centrifugation; and biotin-11-dUTP labelling followed by fluorescence microscopy. [$^{32}$P]dCTP labelling is fast but has been criticized as not distinguishing between replication and repair processes. Density substitution is tedious and requires considerable care, but results are unequivocal. Biotin-11-dUTP labelling provides a direct estimate of the number of nuclei synthesizing DNA but can give false positives since avidin conjugates can bind non-specifically to DNA.

### 4.3.1 [$^{32}$P]dCTP labelling

[$^{32}$P]dCTP labelling can be performed essentially as described in Hutchison *et al.* (23). Either continuous labelling (*Protocol 5*) or pulse labelling (*Protocol 6*) can be carried out with the same quantities of radioactivity. [$^{32}$P]dCTP (3000 Ci/mmol) is preferred and incorporation is estimated following agarose gel electrophoresis and autoradiography (*Protocol 7*).

# 8: Cell-free extracts of Xenopus eggs and DNA replication

To obtain an estimate of total incorporation of [$^{32}$P]dCTP into DNA, labelled bands are excised from the dried agarose gels and counted in a scintillation counter. For comparisons between incubations it is important to retain some of the original sample in order to estimate the total number of c.p.m. added to the tube. We and others (1, 23) have estimated the nucleotide pool size in typical egg extracts at 50 µM. Thus the amount of incorporation as a fraction of the total of input radioactivity can be used to estimate the total amount of DNA synthesized, which is usually between 80–100% for sperm templates.

---

**Protocol 5.** Continuous labelling

*Materials*

- [α-$^{32}$P]dCTP (2 mCi/ml, 3000 Ci/mmol, Amersham)
- Stop buffer: 8 mM EDTA, 80 mM Tris–HCl (pH 8.0), 0.13% phosphoric acid, 10% Ficoll, 5% SDS, 0.2% bromophenol blue

*Method*

1. Pipette 2 µCi of [$^{32}$P]dCTP into a microcentrifuge tube and dry down under vacuum.
2. Pipette 40 µl of extract containing 10$^5$ sperm into the micro-centrifuge tube and mix briefly by trituration.
3. Incubate the sample in a Perspex block at 21 °C.
4. Terminate the reaction by adding 40 µl of stop buffer and store sample frozen at −20 °C.

---

**Protocol 6.** Steps in pulse labelling

1. Dry down [$^{32}$P]dCTP under vacuum and redissolve at 2 µCi/µl in SUNaSp (*Protocol 2*).
2. Pipette 1 µl of [$^{32}$P]dCTP into 40 µl extract containing 10$^5$ sperm and incubate in a Perspex block.
3. Terminate the reaction by adding 40 µl of stop solution (*Protocol 5*) and store the sample frozen.
4. Label the next sample in the time sequence and repeat the procedure.

**Protocol 7.** Processing labelled samples

1. Thaw frozen samples, supplement with proteinase K (500 μg/ml final concentration) and incubate for 1 h at 37 °C.
2. Pipette 30 μl of each sample into each pocket of a 0.8% agarose horizontal gel and resolve at 60 V (130 mA) for 4 h.
3. Rinse resolved gels extensively by incubation in three changes of ~ 2 litres of water (15 min per change).
4. Dry the gels under vacuum and autoradiograph overnight.
5. Develop the autoradiographs and realign with the gels. Mark the positions of the labelled bands and excise from the gels. Immerse excised material in scintillant and count.

**Protocol 8.** Density substitution and CsCl gradient centrifugation

*Materials*

- TE buffer: 50 mM Tris–HCl (pH 8.0), 5 mM EDTA
- Caesium chloride
- Beckman Ti 75 rotor or equivalent
- Pyrophosphate buffer: 0.1 m $Na_4P_2O_7.10\ H_2O$, 1 mM EDTA, 500 μg/ml herring sperm DNA
- 10% TCA
- Whatman GF/C1 filter discs
- Wetting solution: 0.1 m $Na_4P_2O_7.10\ H_2O$, 1 mM EDTA, 100 μg/ml thymidine monophosphate

*Method*

Density substitution is performed using a modification of the method originally described by Harland and Laskey (ref. 15).

1. Mix 100 μl of egg extract containing sperm heads with 5 μCi of [$^{32}$P]dCTP (3000 Ci/mMol) and 5 μl of 20 mM bromodeoxyuridine triphosphate (BrdUTP). Incubate this mixture in Perspex blocks at 21 °C.
2. Terminate the reaction by adding 2 volumes of TE and then placing the sample on dry ice/ethanol.

3. Prepare the sample for caesium chloride gradient analysis by adding 40 µl of calf thymus DNA (4 mg/ml) and 36 µl of 10% SDS. Next add 15 µl of proteinase K (10 mg/ml) and incubate the sample for 1 h at 37 °C.

4. Extract the DNA by adding an equal volume of phenol. Mix thoroughly and centrifuge to separate the two phases. Remove and retain the aqueous phase.

5. Precipitate the DNA from the aqueous phase by adding two volumes of ethanol and incubating for 2 h at −20 °C. Recover the precipitable material by centrifugation.

6. Resuspend the precipitable material in 40 µl of TE.

7. Prepare CsCl at an initial density of 1.74 g/ml by dissolving 10.44 g of solid CsCl in 4 ml of $H_2O$. When the CsCl is fully dissolved measure the increase in volume. Make up the volume of the CsCl solution to 6 ml by adding the sample (40 µl) and $H_2O$. Pipette the solution into 13 ml polyallomer centrifuge tubes and overlay with mineral oil. Balance the tubes and seal or cap them.

8. Create an equilibrium gradient (~ 1.83–1.68 g/ml) by centrifugation for 60 h at 44 000 rpm at 21 °C in a Beckman Ti75 rotor or its equivalent.

9. Recover material from the gradient by puncturing the bottom of the centrifuge tube and collecting drops in glass tubes (for a 6 ml gradient, 0.2 ml fractions are collected).

10. To measure the gradient, remove 20 µl samples from every fifth fraction and determine its refractive index using a refractometer. The density of CsCl can be estimated from the refractive index by preparing standard solutions of CsCl.

11. To determine the position to which unreplicated DNA migrates, measure the O.D. of each fraction at 260 nm. Since carrier DNA is the only material present that can be detected at an absorbance of 1.0, its position should be clearly identifiable in a maximum of four fractions towards the top of the gradient.

12. After measuring the O.D. and refractive index of various fractions, dilute each fraction with 2 vol. of ice-cold pyrophosphate buffer and 3 vol. of 10% TCA. Incubate the fractions at 4 °C for 5 min.

13. Collect the precipitable material on Whatman GF/C 1 filter discs which have been pre-soaked in wetting solution. Wash each filter three times with 5% TCA and once with ice-cold methanol. Dry the filters before scintillation counting.

A typical gradient is illustrated in *Figure 2*.

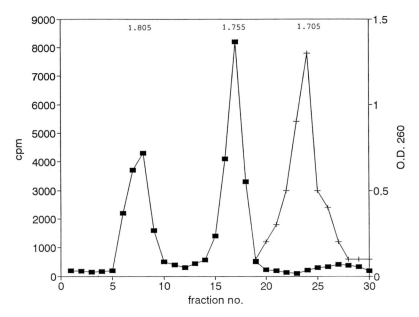

**Figure 2.** Migration of re-replicated DNA on CsCl gradients. Density substitution was performed for 4 h in cycling extracts. Labelled DNA was isolated and separated on CsCl gradients as described (*Protocol 8*). Fractions were collected and the position of labelled DNA determined (■) and compared to the position of unreplicated (carrier DNA), which was measured by OD 260 (+). The density of CsCl (g/ml) at each peak fraction is indicated. These correspond to the expected densities of heavy heavy, heavy light, and light light DNA, respectively.

### 4.3.2 Biotin-11-dUTP labelling

Biotinylated nucleotides are available from a number of companies, although for practical purposes the 11 carbon side-chain linked analogue supplied by Enzo Biochemicals is most convenient as it is relatively cheap and available in mM quantities. Biotin labelling combined with Texas red-conjugated streptavidin detection was first used in flow cytometric analyses of replication in populations of nuclei (2), but is more conveniently used as a method for determining the timing of replication in nuclei which are examined by fluorescence microscopy (22). Continuous labelling and pulse labelling with biotin-11-dUTP is straightforward, although we prefer pulse labelling as the fraction of nuclei which are labelled over long periods of time declines, indicating that the biotinylated nucleotide is removed by repair processes. Pulse labelling is normally carries out for periods of time varying from 1 min to 1 h. One microlitre of 40 µM biotin-11-dUTP is added to 10 µl samples of extract containing sperm and incubated at 21 °C. At the end of the incubation period the nuclei are fixed for fluorescence microscopy by adding 200 µl of 1 mM ethylene glycol bis-(succinic acid *N*-hydroxysuccinimide ester) (EGS) in one-third strength extraction buffer (*Protocol 1*) and incubated for 30 min at 37 °C. Slides are prepared for fluorescence microscopy as described below (*Protocol 9*). Texas red–streptavidin labelling can be performed at RT for 10 min or for 4 h at 4 °C. Dilutions are made according to the manufacturer's information (Amersham International plc).

## 5. Microscopic analysis of nuclear assembly

It is now well established that DNA replication in cell-free extracts of *Xenopus* eggs is dependent upon nuclear envelope assembly (2, 8, 9), therefore it is essential, in any experiment in which DNA replication is inhibited, to examine nuclear envelope assembly. The simplest method is to make temporary slides of extracts, containing nuclei, in formaldehyde. These can be examined under phase-contrast optics and give straightforward results. Alternatively, slides can be prepared for indirect immunofluorescence and stained with anti-lamin antibodies. This assay is convenient as lamina assembly appears to be the final step in nuclear envelope assembly (9, 24) and can be used in conjunction with biotin-11-dUTP labelling.

### 5.1 Phase-contrast microscopy

Temporary slides for phase-contrast microscopy are easy to prepare. Four microlitre samples are removed from incubations of extracts with sperm heads and pipetted into an equal volume of DAPI fixative (15 mM Pipes, pH 7.2, 80 mM KCl, 5 mM EDTA, 15 mM NaCl, 10 μg/ml DAPI, 3.3% formaldehyde) which has been spread on a glass slide. A 22 mm × 22 mm glass coverslip is dropped over the mixture and its edges sealed with nail varnish. Sperm heads or pronuclei are located using UV fluorescence. These are then examined by phase-contrast optics. Pronuclei are characterized as being kidney-shaped and surrounded by a distinct phase-dark boundary, the nuclear envelope, enclosing a non-refractive nucleoplasm. By contrast sperm-heads which have failed to assemble into pronuclei are S-shaped and indistinct under phase optics.

### 5.2 Immunofluorescence

Indirect immunofluorescence is performed on nuclei which have been fixed with EGS and recovered from egg extracts on to glass coverslips. Ten microlitre samples of extracts containing $10^4$ pronuclei are fixed by suspension in 200 μl of EGS. The suspension is incubated for 30 min at 37 °C then layered over a 30% glycerol cushion prepared in extraction buffer (*Protocol 1*) covering a 22 mm diameter glass coverslip. After centrifugation at 1500 *g* for 15 min at room temperature in a bench top centrifuge the nuclei have migrated through the glycerol on to the coverslip, but the bulk of the cytoplasm remains above the cushion. The supernatant is discarded and the coverslips air dried for 10 min. The coverslips can now be incubated with primary and secondary antibody. We normally perform primary antibody incubations at 4 °C overnight and secondary antibody incubations at 4 °C for 4 h.

---

**Protocol 9.** Preparation of nuclei for immunofluorescence

1. Fix nuclei in EGS by diluting 10 µl of extract containing $10^4$ nuclei in 200 µl of 1 mM EGS (EGS is preferred as it is a bifunctional crosslinking fixative with a long carbon spacer. This is especially important for preserving antigens associated with the nuclear envelope). Fixation is complete after incubation for 30 min at 37 °C and fixed samples can be accumulated at 4 °C.

2. Prepare glycerol cushions in cytology chambers as follows. Glass coverslips (22 mm diameter) are adhered to 25 mm diameter cytology chambers by painting molten histology wax on to the bottom of the chamber and dropping the coverslip on to the wax. The chamber is constructed from Perspex, has a depth of 1 cm and has a centrally located hole (diameter 8 mm) running through it. The coverslip forms a seal on the bottom of the chamber and the hole has a capacity of 0.4 ml. This is three-quarters filled with 30% glycerol and 100 µl of sample is then carefully layered on top.

3. For centrifugation, place a perspex support in the bottom of a 26 mm diameter 50 ml centrifuge tube and stack five histology chambers on top. Centrifuge at 1500 $g$ for 15 min at RT in a bench top centrifuge.

4. Remove the histology chambers from the centrifuge tube by lifting the Perspex support from the bottom with a stiff wire rod inserted through a small hole in the bottom of the tube. Aspirate off the sample and glycerol and remove the coverslips by placing the histology chambers face up on a heated table.

5. Air dry the coverslips face up for 10 min to ensure that the nuclei adhere properly, then transfer to humid chambers in which antibody incubations are carried out. (Humid chambers can be constructed simply by placing Whatman 3MM chromatography paper in the bottom of a Petri dish, soaking this in water then partially draining with tissue paper. The coverslips can be placed directly on top of the chromatography paper and the chamber will remain moist and prevent evaporation of antibody solution when closed.)

## 5.3 IgG exclusion assay

The most straightforward method for determining nuclear membrane integrity is IgG exclusion (25). This is important when performing experiments which might lead to non-specific inhibition of nuclear envelope assembly. Unfixed samples of extract (~ 3 μl) containing nuclei are placed on a coverslip and mixed with 0.5 μl of 50 μg/ml Hoechst 33258 and 0.5 μl of undiluted fluorescein-tagged IgG (all diluted in extraction buffer, see *Protocol 1*). A coverslip is dropped over the mixture which is observed with UV optics to locate nuclei and FITC optics to monitor IgG exclusion. Nuclei excluding IgG will appear dark against a bright fluorescent background. However, it should be noted that unfixed nuclei degenerate within 10 minutes so results should be recorded rapidly.

## 6. Fractionation of extracts

### 6.1 High-speed centrifugation

Egg extracts can be readily fractionated into cytosolic and membrane components by high-speed ultracentrifugation. Neither fraction is capable, on its own, of nuclear assembly, but when recombined nuclear assembly can proceed. This fractionation step is useful for immunodepletion experiments or for partial fractionation of nuclear membrane precursors.

Extracts are placed in centrifuge tubes (for convenience we use a Beckman SW50 rotor fitted with inserts for 0.8 ml tubes) and spun at 200 000 $g$ for 1 h at 4 °C. This yields a stratifed extract consisting of a pellet of polyribosomes, a loose layer of membrane, a clear cytosolic layer, and a small lipid cap. The cytosolic layer is carefully removed and spun at 200 000 $g$ for 30 min at 4 °C to remove any contaminating membrane. The membranes are washed by resuspension in extraction buffer (*Protocol 1*) followed by centrifugation at 200 000 $g$ for 30 min at 4 °C. Each fraction can be stored by freezing in liquid nitrogen in the presence of 5% glycerol. The fractions are recombined at a ratio of 1 part membrane to 10 parts cytosol for assays.

### 6.2 Immunodepletion

As stated previously, egg extracts have proved difficult to fractionate but can be depleted of specific proteins using antibodies. This approach has been heavily exploited to investigate both nuclear assembly and the role of cell-cycle proteins in DNA replication (3, 4, 9, 18, 24). Several methods have been described, but these essentially involve either inoculating extracts with saturating amounts of antibody or using antibody-affinity matrices to remove antibody:protein complexes. Removal of antibody:protein complexes has the advantage that rescue experiments can be attempted. However, this requires the fractionation of extracts into soluble and insoluble components by high-speed centrifugation. Proteins are then removed from the soluble fraction before the consequences of removal are assessed by recombining the two fractions. The disadvantage of this approach is that the experimental manipulation will often inactivate the extract. Inoculation of extracts with saturating amounts of antibody is straightforward and has no detectable consequences to nuclear assembly. However, as antibody:protein complexes remain in the extract rescue experiments are difficult to perform. Recently, we have developed a method for immunodepletion which utilizes antibody complexes attached to magnetic polystyrene beads (Dynabeads). As these can be removed from the extracts without centrifugation, fractionation of the extracts is not required and most survive the manipulation.

Dynabeads are a product of the Dynal company and comprise 2.8 μm supramagnetic polystyrene beads which are conjugated to sheep secondary antibodies. Ten microlitres of the beads will bind approximately 1 μg of IgG. The beads are prepared as described in *Protocol 10*.

**Protocol 10.** Immunodepletion with Dynabeads

1. Wash 50 μl of beads three times with 500 μl of PBS/Tween (0.1%) and twice with 500 μl PBS by centrifugation. Resuspend the beads in 1 ml of a solution containing 50 μg (a 100 fold excess) of affinity-purified primary antibody (in PBS containing 1% BSA) and rotate overnight at 4 °C.

2. Wash the Dynabeads three times with 500 μl of PBS/Tween, twice with 500 μl of PBS, and twice 500 μl of SUNaSp (*Protocol 2*). After the final wash, resuspend the Dynabeads in 50 μl of SUNaSp. Buffer is removed at each step by placing on a magnet.

3. Add the Dynabeads to two volumes of extract and incubate for 30 min at RT on a rotating wheel. Remove the Dynabeads by placing the tube containing the extract on a magnet and pipetting off the extract. Mix the extract with fresh Dynabeads and repeat the process.

4. After the second depletion, retain a small fraction of the extract for immunoblotting analysis and use the remainder for nuclear assembly and DNA replication assays.

Using this procedure we are able to remove all detectable lamin Liii from egg extract using a monoclonal antibody L6 8A7. However, the procedure does not prevent nuclear membrane assembly in the depleted extracts although DNA replication is not detected (Holman, Hutchison and Stick, unpublished data).

Depleted proteins can be recovered from the Dynabeads after extensive washing of the beads in PBS and elution with either SDS–sample buffer (for immunoblotting analysis) or by salt elution from rescue experiments. Extracts depleted of proteins using Dynabeads can be rescued either by adding protein expressed from vectors in *Escherichia coli* or by re-adding depleted proteins which have been eluted from the Dynabeads.

# References

1. Blow, J. J. and Laskey, R. A. (1986). *Cell*, **47**, 577.
2. Blow, J. J. and Watson, J. V. (1987). *EMBO J.*, **6**, 1997.
3. Blow, J. J. and Nurse, P. (1990). *Cell*, **62**, 855.
4. Fang, F. and Newport, J. W. (1991). *Cell*, **66**, 731.
5. Hutchison, C. J., Brill, D., Cox, R., Gilbert, J., Kill, I. R., and Ford, C. C. (1989). *J. Cell Sci. (Suppl)*, **12** 197.
6. Dasso, M. and Newport, J. W. (1991). *Cell*, **61**, 811.
7. Blow, J. J. and Laskey, R. A. (1988). *Nature*, **332**, 546.
8. Sheehan, M. A., Mills, A. D., Sleeman, R. A., Laskey, R. A., and Blow, J. J. (1988). *J. Cell Biol.*, **106** 1.
9. Meier, J., Campbell, K. H. S., Ford, C. C., Stick, R., and Hutchison, C. J. (1991). *J. Cell Sci.*, **98**, 271.
10. Leno, G. H. and Laskey, R. A. (1991). *J. Cell Biol.*, **12**, 557.
11. Hutchison, C. J. and Kill, I. R. (1989). *J. Cell Sci.*, **93**, 605.
12. Kill, I. R., Bridger, J. M., Campbell, K. H. S., Moldonado-Codina, G., and Hutchison, C. J. (1991). *J. Cell Sci.*, **100**, 869.
13. Cox, L. S. and Leno, G. H. (1990). *J. Cell Sci.*, **97**, 177.
14. Forbes, D., Kirschner, M., and Newport, J. W. (1983). *Cell*, **34**, 13.
15. Harland, R. M. and Laskey, R. A. (1980). *Cell*, **21**, 761.
16. Philpott, A. and Leno, G. H. (1992). *Cell*, **69**, 759.
17. Leno, G. H., Downes, C. S., and Laskey, R. A. (1992). *Cell*, 69, 151.
18. Dabauvalle, M-C., Loos, K., Merkert, H., and Scheer, U. (1991). *J. Cell Biol.*, **112**, 1073.
19. Cox, L. S. and Laskey, R. A. (1991). *Cell*, **66**, 271.
20. Newport, J. W. (1987). *Cell*, **48**, 205.
21. Sambrook, J., Fritsch, E. F., and Maniatis, T. (1989). *Molecular cloning: a laboratory manual*, p. 2.60–2.81. Cold Spring Harbor Laboratory Press, NY.
22. Hutchison, C. J., Cox, R., and Ford, C. C. (1988). *Development*, **103**, 553.
23. Hutchison, C. J., Cox, R., Drepaul, R. S., Gomperts, M., and Ford, C. C. (1987). *EMBO J.*, **6**, 2001.
24. Newport, J. W., Wilson, K., and Dunphy (1990). *J. Cell Biol.*, **111**, 2247.
25. Blow, J. J. and Sleeman, A. M. (1990). *J. Cell Sci.*, **95**, 383.
26. Gurdon, J. B. (1976). *J. Embryol. Exp. Morphol.*, **36**, 523.

# 9

# In vitro SV40 DNA replication

MARK K. KENNY

## 1. Introduction

Simian virus 40 (SV40) has served as an excellent model for various cellular processes including transformation, transcription, RNA processing, and DNA replication (1). Replication of the genome of this small DNA tumour virus very closely resembles the replication of cellular DNA. SV40 has a 5 kb circular duplex genome which exists in a chromatin structure indistinguishable from that of the cell. Replication initiates at a single 64 bp core origin of replication and proceeds bidirectionally in a semi-conservative manner. SV40 uses only a single viral protein for its replication, the large T antigen, albeit it serves several functions. All of the other replication proteins are provided by the host cell. There is a species specificity to SV40 DNA replication. It will replicate in primate cells but not in rodent cells.

The establishment by Li and Kelly of an efficient and reproducible cell-free system for the replication of SV40 DNA has led to a major advancement in our understanding of both SV40 and cellular DNA replication (2–5). Most of the proteins and the basic mechanism of SV40 DNA replication have now been elucidated, and this has been the subject of some excellent reviews (6–9). The initiation of replication is the likely target of any cell-cycle controls and is briefly described here.

The first step in the initiation of SV40 DNA replication is the ATP-dependent binding of T antigen to GAGGC recognition motifs within the origin (10–12). T antigen forms a double hexamer around the origin, and the DNA within this nucleoprotein complex is partially melted (13–16). By virtue of its DNA helicase activity, T antigen is able to extensively unwind origin-containing DNA using the hydrolysis of ATP (17–20). This DNA unwinding reaction also requires a single-stranded DNA binding protein (eukaryotic SSB = RPA) to stabilize the single-stranded DNA, and a topoisomerase (eukaryotic topo I or II) to relieve the torsional stress ahead of the advancing helicase. In addition to the aforementioned proteins, the actual initiation reaction (i.e. the synthesis of the first RNA primers) requires DNA polymerase α–primase complex (polα–primase) (21). Protein–protein interactions between T antigen, SSB, and polα–primase are likely to play an important role in the initiation process (22, 23). Although the basic framework of SV40 replication is reasonably well understood, there will likely be additional factors and mechanisms uncovered which are involved in processes such as chromatin assembly and replication, coordination of DNA unwinding and DNA synthesis, coordination of leading and lagging strand synthesis, and cell-cycle control.

Two major questions regarding eukaryotic DNA replication and cell-cycle control are (a) what triggers the entry of cells into S phase and (b) how does the cell ensure that all of its DNA is replicated 'once and only once' per cell cycle (24). When SV40 infects a cell it must induce the cell to enter S phase before the viral DNA can be replicated. Thus, even though SV40 uses T antigen rather than a cellular initiator protein, it is reasonable to expect that some of the same controls affecting initiation of cellular DNA synthesis will also affect the initiation of SV40 DNA synthesis. Unlike the cell, however, SV40 does not obey 'once and only once' replication control. It goes through many rounds of replication without an intervening mitosis. Hence SV40 may not be a good model for this aspect of cell-cycle control. Better *in vitro* DNA replication models for this type of cell-cycle control may be the *Xenopus* system (see Chapter 8) or the recently established bovine papilloma virus (BPV) system (25). *Xenopus* has the obvious advantage of being totally dependent on cellular functions, but it is not as easy to biochemically manipulate as the SV40 system. At least under certain conditions, BPV replicates 'once and only once' per cell cycle, though it is not known whether this control feature is reproduced in the *in vitro* assay.

The strengths of the *in vitro* SV40 DNA replication system are that it relies heavily on cellular DNA replication proteins and mechanisms and it is very amenable to biochemical analysis. Now that the major proteins of the replication machinery are known and their genes are being cloned, molecular genetic manipulations will also be possible. By using extracts from different phases of the cell cycle, it is possible to identify and purify factors which influence SV40 replication in a cell-cycle dependent manner. Furthermore, because of the ability to break down the replication process into individual reactions and to use various combinations of purified proteins, one can ask which protein or step is being affected by putative cell-cycle control factors (e.g. protein kinases, protein phosphatases, tumour suppressor proteins).

The use of the *in vitro* SV40 DNA replication system to study the cell cycle is in its early days, and yet some very interesting observations have already been made (24). Extracts from S- or $G_2$-phase cells have been shown to be considerably more active in supporting SV40 DNA replication and DNA unwinding than $G_1$ extracts (26). Furthermore, the cell-cycle regulator $p34^{cdc2}$ protein kinase can activate $G_1$ extracts for DNA replication and unwinding (27, 28). The three key proteins involved in initiation of SV40 DNA replication (T antigen, SSB and polα–primase) have all been shown to be phosphoproteins and are *in vitro* substrates for p34$^{cdc2}$ kinase. Phosphorylation of threonine 124 of T antigen can be accomplished by p34$^{cdc2}$ *in vitro* and is required for efficient origin binding and replication (29). In addition to threonine 124, T antigen is also phosphorylated *in vivo* on other (mainly serine) residues. Hyperphosphorylation of these serines inhibits origin binding and replication (30). Additionally, protein phosphatase 2A can partially activate $G_1$ extracts, particularly when using hyperphosphorylated T antigen (31). Although the protein phosphatase 2A catalytic subunit is present throughout the cell cycle, a T antigen specific phosphatase activity has been detected which is selectively active is S-phase extracts (32). Both human and yeast SSBs are phosphorylated in S and $G_2$ cells but not in $G_1$ cells (33). It has been suggested (though not proven) that the activation of $G_1$ extracts by p34$^{cdc2}$ is due to phosphorylation of SSB (28). Lastly, the tumour suppressor gene product p53 has been shown to inhibit SV40 DNA replication both *in vivo* and *in vitro* (34–37). The relevance of these observations to the SV40 lytic cycle and cell-cycle control remains to be determined (38, 39).

When setting up the *in vitro* SV40 DNA replication system, there are three reagents required that are not commercially available:

- SV40 origin-containing plasmid DNA
- SV40 large T antigen
- replication-competent extract from human or monkey cells

## 2. T-antigen purification

Immunoaffinity purification of SV40 large T antigen was first reported by Simanis and Lane (40) and by Dixon and Nathans (41). It is now invariably purified by one of these methods or modifications thereof. T antigen has been purified from three major over-expression systems: SV40 *cs*1085-infected monkey cells (42), recombinant adenovirus-infected human cells (4), or recombinant baculovirus-infected insect cells (43). The T antigen-expressing baculovirus system is becoming the most popular because of the high yields and the finding that the T antigen produced is favourably phosphorylated for replication reactions (31, 43, 44). The following procedures for column preparation and immunoaffinity purification can be scaled up or down according to one's needs.

### 2.1 Column preparation

The immunoaffinity column is prepared by crosslinking the T antigen monoclonal antibody PAb419 (45) to protein A–Sepharose (see *Protocol 1*). Binding to protein A–Sepharose rather than other matrices has the advantage that the antibody is favourably oriented such that the constant region is bound to the matrix while the antigen-combining region is free to interact with the T antigen.

---

**Protocol 1.** Column preparation[a]

1. Mix PAb419 antibody with 1 ml (packed volume) of protein A–Sepharose fast flow (Pharmacia). Ascites fluid normally contains about 1–10 mg/ml of antibody; hybridoma supernatant, about 1–10 mg/litre. Aim to bind about 5 mg of antibody to 1 ml of protein A–Sepharose. Preclear antibody solutions of cellular debris and particulate matter by centrifugation and/or filtration through an $\sim$ 8 µm filter (Millipore) before binding to the protein A–Sepharose. Follow (a) **or** (b).

    (a) If using ascites, add $\geq$ 1 ml of PAb419 ascites to 1 ml of protein A–Sepharose in a 15 ml conical screw-cap tube and add phosphate-buffered saline (PBS) to 15 ml. Incubate the mixture for 1 h at room temperature (or overnight at 4 °C) on a rotating wheel or rocker. Wash the Sepharose beads 3–4 times with 15 ml 0.1 M sodium borate (pH 9.0) and brief ($\sim$ 1–2 min) centrifugation at about 1000 $g$. Resuspend beads in a final volume of 2 ml (i.e. a 50% suspension).

    (b) If using hybridoma supernatant, place 1 ml of protein A–Sepharose beads in a small column and pass $\geq$ 1 litre of PAb419 hybridoma supernatant through the column overnight at 4 °C. Wash beads extensively with $\sim$ 100 ml of 0.1 M sodium borate (pH 9.0). Transfer beads to a 15 ml conical screw-cap tube and adjust volume to 2 ml (i.e. a 50% suspension).

2. Check amount of antibody on beads. Remove 20 µl (i.e. 10 µl packed volume) of PAb419 protein A–Sepharose solution. (Cut off end of micropipette tip if necessary to prevent beads from clogging tip.) Add 1 ml 0.1 M glycine–HCl (pH 2.0), mix and incubate at $\sim$ 90 °C for $\sim$ 2 min. Centrifuge out beads, determine protein concentration of supernatant, and calculate the amount of antibody on the beads. If there is insufficient antibody, use more ascites or hybridoma supernatant and repeat Steps 1 and 2.

3. Prepare coupling solution. Add 0.5 g dimethyl pimelimidate dihydrochloride (DMP, Sigma) to 40 ml of 0.1 M di-sodium tetraborate. The pH should be between 8.5 and 9.0. Add 10 ml of 0.1 M sodium borate (pH 9.0). Mix DMP solution and PAb419 beads in a 50 ml conical screw-cap tube on a rotating wheel or rocker for 1 h at room temperature (or overnight at 4 °C). **Make up DMP solution fresh and use immediately.**

4. Stop the coupling reaction by incubating the beads in 0.2 M ethanolamine (pH 8.0, 2-aminoethanol, Sigma) for 1 h at room temperature on a rotating wheel or rocker.

5. Check the coupling efficiency by removing a defined aliquot of beads and treating with 0.1 M glycine–HCl (pH 2.0) as above (see Step 2). Coupling should go > 95% to completion.

6. Beads can be stored stably at 4 °C in 0.1 M sodium borate (pH 9.0) for > 1 year. Before using the PAb419 beads for the first time, wash with T antigen Elution Buffer (see *Table 1*) to remove any antibody that was not covalently coupled.

[a] After Harlow and Lane (46)

## 2.2 Immunoaffinity purification

This protocol assumes that cells infected with a T antigen-producing virus are already prepared. Cells can be harvested and used fresh, or frozen as a pellet and stored at −80 °C. The purification procedure described here (see *Protocol 2*) is designed for a lysate from 100 ml of baculovirus-infected insect cells (at $1.0–1.5 \times 10^6$ cells/ml). A good infection can yield approximately 1 mg of T antigen. Scale up or down as required.

The lysate is loaded on to three columns (DEAE, protein A, PAb419) connected in series such that the flow-through from the first column passes directly on to the second column, and so on. The DEAE–Sepharose precolumn removes nucleic acids and any particulate matter not removed by centrifugation. The protein A–Sepharose column removes any protein A binding proteins. A short, squat DEAE column should be used to improve the flow rate which will decrease as the lysate is loaded. If the flow rate and time permit, the flow-through can be reloaded (1 or 2 times) to achieve a more complete depletion of T antigen from the lysate. An aliquot of the lysate and the flow-through should be saved so that the efficiency of immunodepletion can be checked, and if a problem arises with the purification, it can be traced. If desired, the T antigen eluted from the column can be made into separate pools (e.g. early, middle, and late). The specific activity of later fractions tends to be greater than the first couple of fractions.

The purified T antigen is dialysed into buffer containing 50% glycerol and stored at −20 °C. Under these conditions the T antigen does not freeze and is stable for several months. Alternatively, the T antigen can be dialysed against a low glycerol buffer [10 mM Pipes (pH 7.0), 5 mM NaCl, 0.1 mM EDTA, 10% glycerol, 1 mM DTT] and frozen at −80 °C. However, note that repeated freezing and thawing of T antigen aliquots stored in this manner will gradually reduce activity.

**Table 1.** Buffers for T antigen purification

**T Lysis Buffer**
20 mM Tris–HCl, pH 9.0[a]
300 mM NaCl
1mM EDTA[b]
10% glycerol
0.5% Nonidet P-40
0.1 mM DTT[c] (add just before use)

**T Neutralization Buffer**
100 mM Tris–HCl. pH 6.8
300 mM NaCl
1 mM EDTA
10% glycerol
0.5% Nonidet P-40
0.1 mM DTT (add just before use)

**T Loading Buffer**
20 mM Tris–HCl, pH 8.0
300 mM NaCl
1 mM EDTA
10% glycerol
0.5% Nonidet P-40

**T 1 M Wash Buffer**
50 mM Tris–HCl, pH 8.0
1 M NaCl
10% glycerol
1 mM EDTA

**T EG Wash Buffer**
50 mM Tris–HCl, pH 8.5
500 mM NaCl
1 mM EDTA
10% glycerol
10% ethylene glycol

**T Elution Buffer**
20 mM Tris–HCl, pH 8.5
1 M NaCl
10 mM MgCl$_2$
1 mM EDTA
10% glycerol
55% ethylene glycol

**T Dialysis Buffer**
20 mM Tris–HCl, pH 8.0
10 mM Nacl
1 mM EDTA
50% glycerol
1 mM DTT (add just before use)

*Note*: It is recommended that a 'cocktail' of protease inhibitors be added to buffers just before use. Suggested inhibitors include 0.1 mM PMSF, 0.5 μg/ml leupeptin, and 0.5 μg/ml pepstatin (Boehringer–Mannheim). Inhibitors do not need to be added to the dialysis buffer. Store all buffers at 4 °C.

[a] The pH indicated of all buffers is of a 1 M stock solution at room temperature.
[b] ethylenediamine tetraacetic acid
[c] dithiothreitol

---

**Protocol 2.** Immunoaffinity purification of T antigen[a]

*NB* All procedures performed at 4 °C

*Materials*

- 20 mM triethylamine: (Sigma), prepare fresh by adding 56 μl stock solution of triethylamine to 20 ml of distilled water, the pH should be ~ 11

- See *Table 1* for buffer recipes

*Method*

1. Resuspend cell pellet in 5–10 × pellet volume of T Lysis Buffer. Incubate on ice for 15 min with occasional mixing.
2. Centrifuge at ~ 25 000 $g$ for 15 min.
3. Add 0.5 × volume of T Neutralization Buffer to supernatant. Mix. Save a small aliquot.
4. Load neutralized lysate on to three columns (equilibrated with T Loading Buffer) connected in series:
   (a) DEAE–Sepharose precolumn (~ 5 ml)
   (b) protein A–Sepharose precolumn (~ 1 ml)
   (c) PAb419–protein A–Sepharose column (~ 1 ml, ~ 5 mg of PAb419/ml of beads)
   
   Save flow-through.
5. Wash columns with ~ 100 ml T Loading Buffer.
6. Uncouple columns and wash the PAb419 column with ~ 100 ml T 1 M Wash Buffer.
7. Wash with ~ 25 ml T EG Wash Buffer.
8. Elute T antigen with ~ 10 ml T Elution Buffer using a slow flow rate. Collect ~ 0.3 ml fractions.
9. Determine protein concentrations using ~ 10 µl/fraction in Bradford assay (Bio-Rad protein assay) (47). Pool peak fractions.
10. Dialyse overnight against 1 litre of T Dialysis Buffer.
11. (a) Store in aliquots at −20 °C.
    (b) Determine protein concentration.
    (c) Check activity in replication assay.
    (d) Check purity by SDS–PAGE.
12. Regenerate protein A–Sepharose precolumn and PAb419–protein A–Sepharose column:
    (a) Wash columns with ~ 20 ml T Elution Buffer.
    (b) Wash columns with ~ 20 ml 20 mM triethylamine.
    (c) Wash columns with ~ 20 ml 0.1 M sodium borate (pH 9.0). Store at 4 °C.

[a] After Mastrangelo *et al.* (13).

## 3. Extract preparation

Replication-competent extracts are generally prepared from HeLa cells or other primate cell lines capable of growing in spinner culture, although adherent cells have also been used. As *in vivo*, SV40 will not replicate in rodent cell extract (48). The determinant of this species specificity has been shown to be the polα–primase complex. Rodent cell extract supplemented with human polα–primase will support SV40 DNA replication. Alternatively, the closely related polyoma virus will replicate in mouse cells and its DNA can be replicated in mouse cell extract (49). Although the *in vitro* polyoma virus DNA replication system has not been as extensively studied as the SV40 system, they appear to behave in a very similar manner.

The following procedure for extract preparation (see *Protocol 3*) is designed for 10 litres of mid-log phase HeLa cells at 6–15 × $10^5$ cells/ml. Scale up or down accordingly. The ease of lysis and the quality of the extracts may vary amongst different cell lines. Some cells may even lyse simply by incubation in Hypotonic Buffer (*Table 2*). If this is the case, the washing step

---

**Table 2.** Buffers and materials for extract preparation

- Dounce homogenizer with B pestle (Kontes)
- Chilled centrifuges and rotors
- Fresh 1 M DTT
- Large dialysis tubing

- **Phosphate-buffered saline,** pH 7.2 (PBS)
    137 mM NaCl
    3 mM KCl
    10 mM $Na_2HPO_4$
    2 mM $NaH_2PO_4$

- **Hypotonic Buffer**
    20 mM Tris–HCl, pH 7.5
    5 mM KCl
    1.5 mM $MgCl_2$
    1 mM DTT (add just before use)

- **Extract Dialysis Buffer**
    20 mM Tris–HCl, pH 7.5
    50 mM NaCl
    0.1 mM EDTA
    10% glycerol
    1 mM DTT (add just before use)

*Note*: Protease inhibitors can be added to buffers if desired (see *Table 1*). Store all buffers at 4 °C.

in this buffer should be omitted. The release of nuclei should be monitored by microscopy if using a cell line for the first time. Extracts can be prepared from cells at various stages in the cell cycle (see *Chapter 1* for cell-synchrony methods).

Optimally, the extract should have a protein concentration of ⩾ 10 mg/ml. The extract is stable at −80 °C for more than 1 year and retains activity after freeze/thaw cycles.

---

**Protocol 3.** Extract preparation[a]

*NB* All procedures performed at 4 °C

*Materials*
- See *Table 2* for buffer recipes

*Method*
1. Harvest cells (10 litres of mid-log phase HeLa cells at 6–15 × $10^5$ cells/ml).
2. Wash with ∼ 150 ml PBS.
3. Wash with ∼ 100 ml Hypotonic buffer[b].
4. Resuspend in ∼ 30 ml Hypotonic Buffer. Incubate on ice for 10 min to swell cells.
5. Lyse cells by Dounce homogenization (∼ 15 strokes with a B pestle).
6. Measure extract volume and adjust to 0.2 M NaCl (with 5 M NaCl) while stirring.
7. Immediately centrifuge lysate at ∼ 35 000 $g$ for 30 min.
8. Centrifuge supernatant at 100 000 $g$ for 1 h.
9. Dialyse supernatant against 2 × ∼ 1 litre (i.e. one change) of Extract Dialysis Buffer.
10. Centrifuge supernatant at 100 000 $g$ for 1 h.
11. Divide extract into desired aliquots and store at −80 °C.

[a] After Wobbe *et al.* (5).
[b] Omit this step if the cell line used is particularly easily lysed.

---

# 4. Replication reaction

The *in vitro* SV40 DNA replication reaction described below is based on modifications (50) to the original procedure of Li and Kelly (2).

## 4.1 The reagents

The components of the *in vitro* SV40 replication reaction are listed in *Table 3*.

In addition to T antigen and extract, the other specialty reagent required for the replication reaction is DNA containing the SV40 origin of replication (ori$^+$ DNA). There are several suitable plasmids available (e.g. see refs. 2–5, 37, 50). The DNA is prepared by standard methods including ethidium bromide/CsCl density gradient centrifugation (51). The DNA should be predominantly covalently closed circular and free of contaminating RNA and chromosomal DNA. Various plasmid sizes and amounts of DNA have been used in replication assays but 0.3 μg of an approximately 3 kb plasmid (e.g. pSVO1ΔEP, ref. 5) in a 50 μl reaction is suitable for most purposes.

The creatine phosphate–diTris salt (pH 7.7) acts both as a component of the ATP-regenerating system and as the buffer. The replication reaction has a narrow pH optimum and thus the pH of the buffer should be accurately determined. A 0.5 M creatine phosphate–diTris (Sigma; FW 453.4) stock solution can be made up by dissolving 1.0 g in distilled water to give a final volume of 4.41 ml. Addition of ~ 60 μl of 10 M NaOH should yield a buffer of pH 7.7 ± 0.1 when a 50 mM solution is measured at room temperature. Adjust further if necessary.

The DTT should be made fresh monthly or frozen in aliquots at −20 °C. The nucleotides are stable for several months if stored as neutral solutions at −20 °C. A 1 mM stock solution of [$^3$H]dTTP with a specific activity of ~ 500 c.p.m./pmol is usually convenient. Alternatively, $^{32}$P-labelled deoxynucleotides can be used if desired or if analysing products on a gel. The other component of the ATP-regenerating system is the creatine phosphokinase (Sigma). This can be stored at −20 °C at a concentration of 5 mg/ml (~ 200 units/mg) in 25 mM Tris (pH 7.5), 25 mM NaCl, 10% glycerol, 1 mM EDTA, and 1 mM DTT.

The optimal levels of T antigen and extract must be determined by titration. If pure or partially purified proteins are used instead of a crude extract, the replication reactions should be supplemented with 200 μg/ml bovine serum albumin (BSA). A 5 mg/ml stock solution of BSA in 10 mM potassium phosphate (pH 8.0) should be denatured at 100 °C for 5 min. Aliquots can be stored at −20 °C but should only be thawed once and thereafter stored at 4 °C.

## 4.2 The assay

The basic procedure for the *in vitro* SV40 DNA replication assay is summarized in *Protocol 4*.

Replication reactions containing the components listed in *Table 3* are assembled on ice. Generally, a 'reaction mix' of all the components is made which is sufficient for the number of reactions in a given assay. Each of the

**Table 3.** In vitro SV40 DNA replication reaction components

| | Stock solution[a] |
|---|---|
| 40 mM creatine phosphate–diTris salt, pH 7.7 | 500 mM |
| 7 mM MgCl$_2$ | 1M |
| 0.5 mM DTT | 50 mM |
| 4 mM ATP | 100 mM |
| 200 µM each UTP, GTP, CTP | 10 mM each |
| 100 µM each dATP, dGTP, dCTP | 10 mM each |
| 20 µM [$^3$H]dTTP (~ 500 c.p.m./pmol) | 1 mM |
| 20 µg/ml creatine phosphokinase | 5 mg/ml |
| SV40 ori$^+$ DNA (~ 0.3 µg/50 µl reaction) | 150 µg/ml |
| T antigen (0.1–1.0 µg/50 µl reaction) | ≥ 100 µg/ml |
| extract (0.1–0.5 mg/50 µl reaction) | ≥ 10 mg/ml |

[a] Convenient stock concentrations; to be used as a guide.

components can be added individually to the 'reaction mix'. Alternatively, a 5 × 'buffer mix' can be made containing creatine phosphate–diTris salt, MgCl$_2$, DTT, and nucleotides. This 5 × 'buffer mix' can be made in advance and frozen in convenient aliquots at −80 °C. The proteins (creatine phosphokinase, T antigen, and extract), DNA, and 5 × 'buffer mix' are then added individually to the 'reaction mix'.

Reactions with typical volumes of 25–50 µl are incubated at 37 °C usually for ≥ 1 h. The in vitro replication reaction has a sharp temperature optimum and thus should be performed in a good, circulating waterbath that maintains a constant temperature.

It is important when first setting up the SV40 replication system to establish that the synthesis observed is dependent on T antigen and a functional SV40 origin of replication (ori$^+$ DNA). Synthesis in the absence of T antigen, with a negative control plasmid (ori$^-$ DNA) or with no DNA should be < 4% and can be < 1% of that observed in a complete reaction. The total level of synthesis is another indication that the reaction is working properly. A 50 µl reaction containing 0.3 µg of ori$^+$ DNA (~ 3 kb plasmid, e.g. pSVO1ΔEP) and optimal levels of T antigen and extract, should incorporate 20–150 pmol of dTMP into DNA in 1 h.

Reactions are terminated by placing on ice and adding of one drop (about 25–50 µl) of 0.15 M sodium pyrophosphate, one drop of 1 mg/ml carrier DNA, and 1 ml of 10% trichloroacetic acid (TCA)—carrier DNA can be made by dissolving DNA (e.g. salmon sperm DNA) in 10 mM EDTA, heating the solution at 100 °C for 5 min, and cooling on ice. The incorporation of radioactively labelled nucleotides into DNA is detected by acid precipitation of nucleic acid. After a short incubation (about 10 min) on ice, the reaction mixtures are filtered through prewetted, 25 mm glass fibre filters (ENZO Diagnostics, Inc.) using a filter holder (Hoefer). The filters are washed sequentially with 1% TCA and 95% ethanol and then allowed to air-dry (drying can be accelerated by incubating filters in a drying oven or under a heating lamp). The amount of radioactivity on the filters is determined by scintillation counting using fluor. The specific activity of the [$^3$H]dTTP can be determined by spotting a known quantity of the 'reaction mix' or of the [$^3$H]dTTP stock solution on to filters and determining the amount of radioactivity. An alternative to the TCA precipitation assay is the adsorption of nucleic acids to DE-81 filters followed by washing of the filters with 0.5 M sodium phosphate pH 7.0 (51). The products of the replication reaction can also be examined by neutral or alkaline gel electrophoresis (52, 53).

---

**Protocol 4.** Replication assay

1. Assemble reactions on ice.
2. Incubate at 37 °C for 1 h or desired time period.
3. Stop reaction:
    (a) Move tubes to 4 °C
    (b) Add 1 drop of 0.15 M sodium pyrophosphate
    (c) Add 1 drop of 1 mg/ml carrier DNA
    (d) Add 1 ml of 10% TCA
    (e) Incubate for 10 min on ice
4. Filtration:
    (a) Filter TCA precipitates through glass fibre filters
    (b) Wash with 1% TCA
    (c) Wash with 95% ethanol
    (d) Dry filters
5. Quantitation:
    (a) Count filters in scintillation counter
    (b) Determine specific activity
    (c) Calculate incorporation

---

## 5. Further possibilities

In addition to the *in vitro* replication reaction, there are many 'partial' reactions which look at various stages of SV40 DNA replication and involve different subsets of proteins. If an effect of different extracts or putative cell-cycle control factors is observed in the replication reaction, these 'partial' reactions should be useful in identifying the target of these cell-cycle controls. These other assays include origin binding (11, 12, 29) and untwisting by T antigen (14–16), DNA unwinding (18, 20, 28), RNA primer formation (21), and protein–protein interactions between T antigen, SSB, and polα–primase (23).

Lessons learned about cell–cycle control using *in vitro* SV40 DNA replication will eventually have to be tested *in vivo* and by using an *in vitro* cellular DNA replication system when possible. SV40 has previously been found to use many cellular proteins and pathways for its own benefit. It will not be surprising if it also exploits aspects of normal cell-cycle control.

## Acknowledgements

I would like to thank David Lane and Jerry Hurwitz and my colleagues in these two laboratories. The author is supported by the Cancer Research Campaign.

## References

1. Tooze, J. (ed.) (1980). *DNA tumor viruses* (2nd edn). Cold Spring Harbor Laboratory Press, NY.
2. Li, J. J. and Kelly, T. J. (1984). *Proc. Natl Acad. Sci. USA*, **81**, 6973.
3. Li, J. J. and Kelly, T. J. (1985). *Mol. Cell. Biol.*, **5**, 1238.
4. Stillman, B. W. and Gluzman, Y. (1985). *Mol. Cell. Biol.*, **5**, 2051.
5. Wobbe, C. R., Dean, F., Weissbach, L., and Hurwitz, J. (1985). *Proc. Natl Acad. Sci. USA*, **82**, 5710.
6. Stillman, B. (1989). *Annu. Rev. Cell Biol.*, **5**, 197.
7. Challberg, M. D. and Kelly, T. J. (1989). *Annu. Rev. Biochem.*, **58**, 671.
8. Borowiec, J. A., Dean, F. B., Bullock, P. A., and Hurwitz, J. (1990). *Cell*, **60**, 181.
9. Hurwitz, J., Dean, F. B., Kwong, A. D., and Lee, S.-H. (1990). *J. Biol. Chem.*, **265**, 18043.
10. Dean, F. B., Dodson, M., Echols, H., and Hurwitz, J. (1987). *Proc. Natl Acad. Sci. USA*, **84**, 8981.
11. Deb, S. and Tegtmeyer, P. (1987). *J. Virol.*, **61**, 3649.
12. Borowiec, J. A. and Hurwitz, J. (1988). *Proc. Natl Acad. Sci. USA*, **85**, 64.
13. Mastrangelo, I. A., Hough, P. V. C., Wall, J. S., Dodson, M., Dean, F. B., and Hurwitz, J. (1989). *Nature*, **338**, 658.
14. Borowiec, J. A. and Hurwitz, J. (1988). *EMBO J.*, **7**, 3149.
15. Roberts, J. M. (1989). *Proc. Natl Acad. Sci. USA*, **86**, 3939.
16. Dean, F. B. and Hurwitz, J. (1991). *J. Biol. Chem.*, **266**, 5062.

17. Stahl, H., Droge, P., and Knippers, R. (1986). *EMBO J.*, **5**, 1939.
18. Dean, F. B., Bullock, P., Murakami, Y., Wobbe, C. R., Weissbach, L. and Hurwitz, J. (1987). *Proc. Natl Acad. Sci. USA*, **84**, 16.
19. Dodson, M., Dean, F. B., Bullock, P., Echols, H., and Hurwitz, J. (1987). *Science*, **238**, 964.
20. Wold, M. S., Li, J. J., and Kelly, T. J. (1987). *Proc. Natl Acad. Sci. USA*, **84**, 3643.
21. Matsumoto, T., Eki, T., and Hurwitz, J. (1990). *Proc. Natl Acad. Sci. USA*, **87**, 9712.
22. Collins, K. L. and Kelly, T. J. (1991). *Mol. Cell. Biol.*, **11**, 2108.
23. Dornreiter, I., Erdile, L. F., Gilbert, I. U., von Winkler, D., Kelly, T. J., and Fanning, E. (1992). *EMBO J.*, **11**, 769.
24. Huberman, J. A. (1991). *Chromosoma*, **100**, 419.
25. Yang, L., Li, R., Mohr, I. J., Clark, R., and Botchan, M. R. (1991). *Nature*, **353**, 628.
26. Roberts, J. M. and D'Urso, G. (1988). *Science*, **241**, 1486.
27. D'Urso, G., Marraccino, R. L., Marshak, D. R., and Roberts, J. M. (1990). *Science*, **250**, 786.
28. Dutta, A. and Stillman, B. (1992). *EMBO J.*, **11**, 2189.
29. McVey, D., Brizuela, L., Mohr, I., Marshak, D. R., Gluzman, Y., and Beach, D. (1989). *Nature*, **341**, 503.
30. Mohr, I. J., Stillman, B., and Gluzman, Y. (1987). *EMBO J.*, **6**, 153.
31. Virshup, D. M., Kauffman, M. G., and Kelly, T. J. (1989). *EMBO J.*, **8**, 3891.
32. Ludlow, J. W. (1992). *Oncogene*, **7**, 1011.
33. Din, S., Brill, S. J., Fairman, M. P., and Stillman, B. (1990). *Genes and Dev.*, **4**, 968.
34. Gannon, J. V. and Lane, D. P. (1987). *Nature*, **329**, 456.
35. Braithwaite, A. W., Sturzbecher, H.-W., Addison, C., Palmer, C., Rudge, K., and Jenkins, J. R. (1987). *Nature*, **329**, 458.
36. Sturzbecher, H.-W., Brain, R., Maimets, T., Addison, C., Rudge, K., and Jenkins, J. R. (1988). *Oncogene*, **3**, 405.
37. Wang, E. H., Friedman, P. N., and Prives, C. (1989). *Cell*, **57**, 379.
38. Prives, C. (1990). *Cell*, **61**, 735.
39. Fanning, E. (1992). *J. Virol.*, **66**, 1289.
40. Simanis, V. and Lane, D. P. (1985). *Virology*, **144**, 88.
41. Dixon, R. A. F. and Nathans, D. (1985). *J. Virol.*, **53**, 1001.
42. DiMaio, D. and Nathans, D. (1982). *J. Mol. Biol.*, **156**, 531.
43. Hoss, A., Moarefi, I., Scheidtmann, K.-H., Cisek, L. J., Corden, J. L., Dornreiter, I., Arthur, A. K., and Fanning, E. (1990). *J. Virol.*, **64**, 4799.
44. Kenny, M. K., Lee, S.-H., and Hurwitz, J. (1989). *Proc. Natl Acad. Sci. USA*, **86**, 9757.
45. Harlow, E., Crawford, L. V., Pim, D. C., and Williamson, N. M. (1981). *J. Virol.*, **39**, 861.
46. Harlow, E. and Lane, D. (1988). *Antibodies: a laboratory manual*. Cold Spring Harbor Laboratory Press, NY.
47. Bradford, M. (1976). *Anal. Biochem.*, **72**, 248.
48. Murakami, Y., Wobbe, C. W., Weissbach, L., Dean, F. B., and Hurwitz, J. (1986). *Proc. Natl Acad. Sci. USA*, **83**, 2869.

49. Murakami, Y., Eki, T., Yamada, M., Prives, C., and Hurwitz, J. (1986). *Proc. Natl Acad. Sci. USA*, **83**, 6347.
50. Dean, F. B., Borowiec, J. A., Ishimi, Y., Deb, S., Tegtmeyer, P., and Hurwitz, J. (1987). *Proc. Natl Acad. Sci. USA*, **84**, 8267.
51. Sambrook, J., Fritsch, E. F., and Maniatis, T. (1989). *Molecular cloning: a laboratory manual* (2nd edn). Cold Spring Harbor Laboratory Press, NY.
52. Ishimi, Y., Claude, A., Bullock, P., and Hurwitz, J. (1988). *J. Biol. Chem.*, **263**, 19 723.
53. Tsurimoto, T., Fairman, M. P., and Stillman, B. (1989). *Mol. Cell. Biol.*, **9**, 3839.

# 10

# Analysis of DNA replication origins and directions by two-dimensional gel electrophoresis

JOEL A. HUBERMAN

## 1. Introduction

### 1.1 Importance of the two-dimensional (2D) gel methods

Prior to 1987 when the 2D gel electrophoretic 'replicon mapping' techniques were introduced (1, 2), no generally applicable method was available for the identification or characterization of eukaryotic DNA replication origins. Consequently, it was uncertain whether or not specific origins existed. Thanks to new results obtained with the 2D gel techniques (and, more recently, with additional origin mapping methods, refs 3–5), a great deal of information about eukaryotic origins is now available. Several recent reviews discuss the biological importance of these new results (6–9).

### 1.2 Purpose of this chapter

The widespread use of the 2D methods has generated a wealth of information about the technical parameters which are critical for their effective utilization. The purpose of this chapter is to summarize the most important features of this accumulated information and thus to provide the reader with the practical details needed to proficiently run 2D gels.

### 1.3 Historical antecedents

Both of the 2D gel methods 'introduced' in 1987 were essentially identical to previously used techniques. What was new was the employment of these techniques for the specific purpose of identifying replication origins. Since the earlier publications describing the techniques, which are predecessors of the currently used methods, contain information of both historical and practical interest, I shall provide a brief discussion of them here.

### 1.3.1 Neutral/neutral (N/N) electrophoresis

To distinguish replicating DNA molecules from non-replicating ones, gel conditions are required in the second dimension which permit the separation of molecules based on their structure as well as on their size. In 1980, Sundin and Varshavsky (10) demonstrated that increasing the agarose gel concentration between the first and second dimensions would provide sufficient structure-dependent separation to permit resolution of several different forms of replicating SV40 DNA. Three years later, Bell and Byers (11) showed that adding an increased voltage gradient to the increased agarose concentration in the second dimension would permit the separation of branched molecules from linear molecules. The conditions of the Bell and Byers' technique were further modified by Brewer and Fangman (1) so that all the structures anticipated during replication of double-stranded DNA could be distinguished from each other, and restriction fragments containing replication origins or termini could be identified, thus permitting replicon mapping.

### 1.3.2 Neutral/alkaline (N/A) electrophoresis

The 2D gel conditions employed in the N/A replicon mapping technique permit both separation of nascent DNA strands from parental strands and size fractionation of the nascent strands. Critical to the success of this technique is the alkaline buffer used in the second dimension. This buffer was introduced by McDonell *et al.* in 1977; their paper contains an excellent discussion of the potential and limitations of electrophoresis through alkaline agarose gels (12). In 1980, Sundin and Varshavsky demonstrated that N/A electrophoresis is capable of separating and size-fractionating the nascent strands of replicating SV40 DNA (10). The N/A replicon mapping technique was subsequently introduced by Huberman *et al.* (2, 13), who showed that a combination of N/A 2D gel electrophoresis with Southern blotting and short hybridization probes would permit the determination of directions of replication fork movement and locations of origins and termini.

## 2. How the 2D gel replicon mapping methods work

### 2.1 Features common to both the N/N and N/A methods

Both methods are based on 2D gel separations of replicating DNA molecules. Usually the molecules studied are restriction fragments generated by restriction enzyme cleavage of replicating genomic DNA.

For both methods, the purpose of the first dimension of electrophoresis is to separate molecules according to their extent of replication. Consequently, optimal conditions for first dimension electrophoresis are independent of whether the second dimension is intended to be neutral or alkaline.

In both cases, the strongest signal is always produced by the intact non-replicating restriction fragment detected by the hybridization probe employed in the particular experiment (round spots in the bottom diagrams of *Figures 1, 2,* and *4*). Even when effective methods are used to enrich for replicating DNA (see below), non-replicating DNA usually remains present in considerable excess over replicating DNA.

Eukaryotic chromosomal DNA molecules are so large that they are inevitably sheared during purification. If the average size of the DNA fragments in a preparation is 100 kb, and if this DNA is subsequently cut by a restriction enzyme into fragments of 5 kb, then 1/20 of the resulting fragments will be shorter than 5 kb, with one end generated by the restriction enzyme and the other end generated by shearing. These restriction/shear fragments will migrate during 2D gel electrophoresis as a continuous smear from the position of the smallest detectable molecules to the position of the intact restriction fragment. This smear is indicated in the bottom diagrams of *Figures 1, 2,* and *4* as a signal which descends downward and rightward from the spot due to the intact non-replicating restriction fragment.

## 2.2 The N/N replicon mapping technique

The N/N technique is based on Brewer and Fangman's observation (1) that when replicating restriction fragments are fractionated according to the extent of replication in the first dimension and on the basis of structure in the second dimension, different patterns are generated depending on the manner in which the restriction fragment is replicated. Structure-dependent migration in the second dimension is accomplished by the use of a high agarose concentration, high voltage gradient, high ethidium bromide concentration (ethidium bromide stiffens the DNA) and low running temperature (1). Under the recommended second dimension conditions, non-linear molecules migrate more slowly than double-stranded linear molecules of corresponding mass. The degree of retardation is dependent on the extent of deviation from linearity. Thus, in a N/N 2D gel, restriction fragments replicated in different fashions generate arcs of different shapes as illustrated in *Figure 1*. All replicating molecules migrate more slowly than non-replicating double-stranded linear molecules of equivalent mass. Non-replicating double-stranded linear fragments of varying sizes form a fast-migrating arc (*Figure 1A–E*).

DNA molecules in living cells undergo recombination as well as replication, and one type of recombination intermediate, X-shaped molecules with two pairs of equally sized arms, is frequently detected in N/N gel studies. Such molecules have a constant mass (twice unit mass), but their crossover point can occur anywhere from 0 to 50% of their full length. The X-shaped family of recombination intermediates migrates as a nearly vertical line (*Figure 1A–E*) above the position of molecules with 2n mass (1, 14).

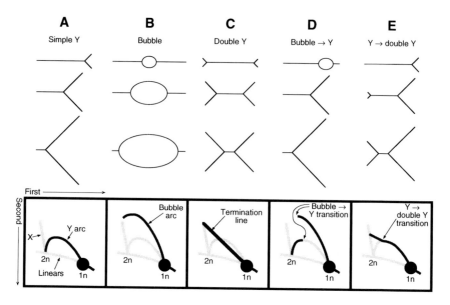

**Figure 1.** N/N 2D gel electrophoresis of various types of replication intermediate. Modified from refs 1 and 14. The arc of linears with sizes larger than 1n and the line of X-shaped structures are frequently not detected, but they are shown in pale grey in all the bottom diagrams for reference. A visible arc of linears can be caused by probe non-specificity or by incompleteness of the restriction digest. Similarly, a Y arc is detected only when the restriction fragment of interest replicates as a simple Y, but a pale Y arc is shown in all the other bottom diagrams for reference. See the text for further explanation.

Restriction fragments which are replicated from one end to the other by replication forks emanating from an external origin (the majority of restriction fragments in eukaryotic genomic DNA) assume a forked or Y-like configuration during replication. Such 'simple Y' molecules are maximally non-linear, and thus maximally retarded, when they are 50% replicated (1, 14). Consequently, when appropriate gel conditions are used, the family of Y-like replicating molecules generates an arc ('Y arc' in *Figure 1A*) which rises up from the arc of linears, peaks at 50% replication, and then returns to the arc of linears at 100% replication.

Replication of restriction fragments containing bidirectional origins at their centres produces a family of molecules containing internal bubbles of increasing size. This family generates a 'bubble arc' (*Figure 1B*), which usually rises higher than a Y arc and does not return to the arc of linears (1).

If replication terminates in the centre of a restriction fragment, then the fragment is replicated by two forks which enter at the ends and converge in the middle. The resulting family of 'double Y' replication intermediates generates a linear signal, as shown in *Figure 1C* (1). The end of the termination line is at the top of the line of X-shaped molecules, at the position corresponding to X-shaped molecules with four equal arms.

If a restriction fragment contains an original which is not centrally located, early replication intermediates are bubble-containing structures, and later intermediates are Y-like structures. As a result, the early portion of a bubble arc and the late portion of a Y arc are generated (*Figure 1D*, ref. 1). The positions of the transition along the bubble and Y arcs provide information about the location of the origin within the restriction fragment.

When termination occurs at a non-central position, the family of replication intermediates generates a signal which begins as a Y arc and finishes as a termination line (14). The end of the termination line is on the line of X-shaped molecules at a position determined by the location of the termination event within the restriction fragment (*Figure 1E*).

Experience has shown that the various arcs and lines diagrammed in *Figure 1* can usually be distinguished from each other in practice. Consequently, restriction fragments containing origins and termini can be detected, and, through analysis of the transition patterns generated by overlapping restriction fragments, can be mapped with an accuracy of several hundred base pairs.

## 2.3 The N/A replicon mapping technique

The N/A technique is best explained by reference to *Figure 2*. The top diagram in this figure shows a replicating DNA molecule consisting of three

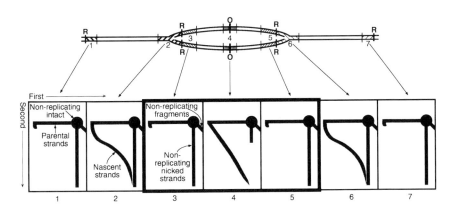

**Figure 2.** Determination of origin location and directions of DNA replication by N/A 2D gel electrophoresis. R indicates restriction enzyme cleavage sites. Hybridization probes are numbered 1–7. See the text for further explanation.

adjacent restriction fragments, with an origin located in the middle of the central fragment. DNA from growing cells is cut with an appropriate restriction enzyme and then subjected to 2D N/A electrophoresis. In the first (neutral) dimension, molecules are fractionated according to extent of replication. In the second (alkaline) dimension, individual strands are separated according to their sizes. The resulting gel is Southern-blotted to a nylon membrane and then probed sequentially with the indicated probes. The lower panels in *Figure 2* show the signals expected for each probe.

Because non-replicating DNA is usually present in considerable excess (see Section 2.1), the nicked strands generated by occasional (usually rare) random nicks within the non-replicating DNA produce a detectable signal, evident as a vertical smear below the spot of non-replicating intact strands. For each nick, two nicked strands are generated, each with one end created by the nick and the other end created by the restriction enzyme. Probes which are located at the ends of restriction fragments (probes 1 and 2, 3 and 5, 6 and 7) detect nicked strands of all sizes. However, internal probes (such as probe 4 in the diagram) detect only those nicked strands which are long enough to extend from one of the restriction sites to the position of the probe. If the probe is centrally located (like probe 4) then only nicked strands which are half size or longer can be detected (notice the shortened vertical streak for probe 4 in the lower panel of *Figure 2*).

Because parental DNA strands have constant size regardless of the extent of replication, they generate horizontal line signals (lower panels, *Figure 2*). However, nascent strands vary in size according to the extent of replication. Consequently, they generate an approximately diagonal signal. Because first dimension electrophoresis is somewhat affected by the shape of the replicating molecule as well as by its size, the different types of replication intermediates shown in *Figure 1* produce slightly different nascent strand signals (13). For example, molecules replicated as simple Ys produce sigmoid nascent strand arcs (probes 2 and 6, *Figure 2*), and molecules with an internal bubble yield straight or upwardly curved nascent strand arcs (probe 4, *Figure 2*).

Probes 1, 3, 5, and 7 (*Figure 2*) are located in origin-distal portions of their restriction fragments. Consequently, these probes detect only the longest nascent strands. In contrast, probes 2, 4 and 6 are in origin-proximal portions of their restriction fragments. These probes detect the full range of nascent strand sizes. In each restriction fragment, the probe closest to the origin detects the shortest nascent strands.

If the data presented in the lower panels of *Figure 2* had been generated in a real experiment, one would have been able to conclude that replication forks move from right to left through the restriction fragment defined by probes 1 and 2 and from left to right through the fragment defined by probes 6 and 7. One would also have been able to conclude that the central fragment contains a replication origin, located at or near probe 4.

The N/A technique is also capable of detecting and mapping sites where replication terminates. Examples are provided in refs 2 and 13.

## 2.4 Advantages and disadvantages of the two techniques

For several reasons, the N/N technique is more sensitive than the N/A technique. First, the entire restriction fragment of interest can be used as a hybridization probe in the N/N technique, whereas two or more smaller probes must be used for each restriction fragment when using the N/A technique. Second, complete replicating intermediates (parental strands plus nascent strands) contribute to the signal arcs detected in the N/N technique. In contrast, only the nascent strands provide information in the N/A technique. Third, the arcs of interest in the N/N technique migrate in a region of the gel which is relatively free of contaminating molecules. The nascent strands of interest in the N/A technique migrate directly below the parental strands, in a position where nicked parental strands also migrate.

The N/A technique offers the advantage that the direction of replication fork movement can be directly determined. It is also possible to determine the direction of fork movement with a modification of the N/N technique (*Protocol 4* and Section 3.6), but this requires a suitable arrangement of restriction sites in the studied fragment plus a possibly difficult in-gel restriction digestion. Another advantage of the N/A technique is that it permits measurements of nascent strand sizes and these measurements can be used to provide a more precise estimate of the origin or terminus position than can be obtained with the N/N technique (2). Characterization of nascent strands may also prove useful in understanding the mechanism of initiation at an origin.

Extensive nicking can generate problems with both techniques. In the case of the N/A technique, nascent strand nicking leads to loss of signal. With regards to the N/N technique, nicking of parental strands at replication-forks converts bubble structures to Y-like structures, and this can lead to a loss or underestimation of critical bubble arcs.

Although both techniques rely on 2D gel electrophoresis of replicating restriction fragments followed by Southern blotting and probing, the rationales for the two techniques are completely different from each other. The two techniques are independent and complementary. In the experience of my laboratory, and in the experience of many other laboratories, use of both techniques together provides the most rapid and most secure route to obtaining satisfactory understanding in unknown situations. In relatively well-understood situations, however (for example, if one is testing a mutated chromosomal origin sequence for activity), the N/N technique by itself is likely to be sufficient. If so, it is the method of choice due to its greater sensitivity and simplicity.

## 3. Methods

### 3.1 DNA preparation

DNA should be prepared from rapidly growing cells or tissues by a technique which minimizes shearing, nicking, and the possibility of branch migration (movement of replication forks due to rewinding of parental strands and extrusion of nascent strands). In cases where a good procedure is available for nuclear matrix preparation, shearing can be minimized by first preparing nuclear matrices with DNA attached (long DNA molecules are resistant to shear while they are attached to the nuclear matrix) and then incubating with a restriction enzyme (restriction fragments are sufficiently short to be shear resistant; refs 15, 16). Because replication forks are attached to the nuclear matrix in many organisms, this approach also permits enrichment for replicating DNA (16). If a nuclear matrix preparation procedure is not available, shearing can also be minimized by agarose embedding (17).

Regardless of the organism one is working with, it is important to:

- work rapidly
- keep samples cold (unless in ethanol or CsCl)
- chelate nuclease-activating metal ions by using ethylenediamine tetraacetic acid (EDTA)
- maintain a salt concentration of 200 mM or higher (except when dissolving ethanol precipitates of DNA)

### 3.2 Incubation of purified DNA with a restriction enzyme

Digestion with a restriction enzyme (or enzymes) usually requires elevated temperatures and may require low ionic strength. Both conditions favour branch migration. Therefore, it is important to keep the time of incubation as short as possible by using a high ratio of enzyme to DNA. We routinely use a ten-fold excess of enzyme over DNA and incubate for one hour.

Do not use spermidine during incubation. It can remain associated with some of the DNA molecules during electrophoresis, reducing their mobility, and causing them to trail during first dimension electrophoresis (B. Brewer, personal communication).

Commercial restriction enzymes may be contaminated with single-strand-specific endonucleases. Because such endonucleases can destroy replicating DNA, it is wise to test new batches of restriction enzyme with an appropriate assay, such as conversion of single-stranded circular DNA to linear form.

If enrichment of replicating molecules by adsorption to benzoylated naphthoylated diethylaminoethyl cellulose (BND–cellulose) is planned, it is helpful to incubate the DNA preparation with RNase at the same time as incubating with the restriction enzyme. RNA can compete with replicating DNA for binding to the BND–cellulose.

## 3.3 Enrichment for replicating molecules

The N/N technique is sufficiently sensitive that enrichment for replicating molecules is frequently not required. Such enrichment is usually required for the N/A technique. Possible enrichment procedures include:

- synchronization to S phase (1)
- attachment to the nuclear matrix (16)
- binding to BND–cellulose (2)

These enrichment procedures are independent of each other. If necessary, they can be combined to produce enhanced enrichments. Combination of these enrichment techniques also permits replicating DNA from a larger amount of starting material to be loaded into the first dimension gel without overloading the gel; this may be important if extremely weak signals are anticipated.

In high salt, BND–cellulose binds selectively to single-stranded DNA. The bound single-stranded DNA can be eluted with caffeine, which competes with the nitrogenous bases in single-stranded DNA for binding to the BND–cellulose. Since replicating DNA is partially single-stranded, this technique permits ~ 20-fold enrichment for replicating molecules. Because the technique is generally useful and easy to carry out, a protocol is provided here (*Protocol 1*). This protocol involves batch adsorption to BND–cellulose and is modified from the earlier batch procedure of Gamper *et al.* (18). Column procedures also work.

---

**Protocol 1.** Enrichment of replicating DNA by adsorption to BND–cellulose

**NB** All procedures carried out at room temperature.

*Materials*
- BND–cellulose (Serva 45025 or Sigma B6385)
- 5 M NaCl
- 1 M NET: 1 M NaCl, 10 mM Tris (pH 8.0), 1 mM EDTA (pH 8.0)
- DNA sample
- Caffeine
- Ethidium bromide

**Protocol 1.** *Continued*
- Isopropanol
- TE buffer: 10 mM Tris (pH 8.0), 1 mM EDTA (pH 8.0)
- 3 M potassium acetate
- 70% ethanol
- Microcentrifuge (Beckman TJ–6)
- Centrifuge
- Electrophoresis equipment

*Method*
1. Prepare commercial BND–cellulose for use as follows:
   (a) Weight out 2 g of BND–cellulose. Place in a 15 ml conical centrifuge tube.
   (b) Add 10 ml of 5 M NaCl and suspend the BND–cellulose, making sure that all particles are wet.
   (c) Centrifuge by accelerating to 850 $g$, then slowing with brake.
   (d) Remove the supernatant, resuspend the pellet in 10 ml of 5 M NaCl, and repeat the centrifugation.
   (e) Repeat Step (d) three more times.
   (f) Wash the pellet once in three pellet volumes of water to reduce the salt concentration.
   (g) Resuspend the pellet in 10 ml of 1 M NET and centrifuge. Repeat.
   (h) Resuspend in a final volume of 10 ml of 1 M NET. Store in the refrigerator until used.
2. Calculate how much BND–cellulose is needed to bind the single-stranded component of the DNA preparation. To make this calculation, assume that 5% of the DNA is single-stranded. The actual amount is far less, but it's better to err on the safe side. About 100 mg (~ 0.2 ml of packed material) of BND–cellulose will bind ~ 10 µg of single-stranded DNA.
3. If the required amount of BND–cellulose is 500 µl packed volume or less, use the microcentrifuge method described below. For larger volumes, use the syringe column method described by Gamper *et al.* (18) or use larger centrifuge tubes (centrifugation at 850 $g$ for 60 sec is usually sufficient to pellet the BND–cellulose in larger tubes).
4. For each DNA sample to be fractionated, prepare a 1.5 ml microcentrifuge tube with the correct amount of packed BND–cellulose in 1 M NET:

(a) Centrifuge a slurry of prepared BND–cellulose at full speed in a microfuge (~ 16 000 $g$) for 20 sec.

(b) Remove the supernatant. If an angle microcentrifuge was used, the surface of the packed BND–cellulose will be slanted, rendering complete removal of the supernatant difficult. This is not a problem; simply remove as much as possible. Add more slurry (or remove some) if necessary to obtain the proper final volume of packed material.

(c) Wash the BND–cellulose in the tubes by resuspending (using a rapid twisting motion with a small spatula with a pointed tip that fits to the bottom of the microfuge tube) in 1 ml of 1 M NET and centrifuging, three times. Remove the final supernatant.

5. Each DNA sample should be in 1 M NET, in a volume about twice that of the packed BND–cellulose volume. Apply the sample to the packed BND–cellulose, resuspend the BND–cellulose in the sample with the aid of a spatula as above, allow about 30 sec for completion of adsorption, then centrifuge for 20 sec. Save the supernatant as a separate fraction (salt-wash fraction #1).

6. Wash the BND–cellulose 5 times with volumes of 1 M NET equal to the volume of the packed BND–cellulose. Each time resuspend the BND–cellulose in the wash, then centrifuge, as above. Save the supernatants as separate fractions (salt-wash fractions #2–6).

7. Similarly, wash the BND–cellulose 5 times with volumes of 1 M NET, containing 1.8% caffeine, equal to the volume of packed BND–cellulose. Save the supernatants as separate fractions (caffeine-wash fractions #1–5).

8. If desired at this point, assay the fractionation results by agarose gel electrophoresis of samples of appropriate volume from each fraction. Use sufficient volume to be able to detect a signal by ethidium bromide fluorescence, but remember that the high salt in each sample may cause electrophoretic artefacts if too large a volume is used. The potential for electrophoretic artefacts can be minimized if the samples are allowed to sit in the sample wells under the overlying electrophoresis buffer for an hour or more before starting electrophoresis, thus permitting the salt to diffuse away from the samples.

9. Pool fractions according to the results of the gel assay in Step 8, or simply pool the first five salt-wash fractions (discard the sixth), and pool all five caffeine-wash fractions. The caffeine-wash DNA is enriched for replicating molecules, but the salt-wash DNA (which is depleted of replicating molecules) frequently provides an excellent control during 2D gel electrophoresis.

**Protocol 1.** *Continued*

10. Centrifuge at 10 000 $g$ in non-siliconized glass tubes or plastic microcentrifuge tubes for 10 min to pellet particles of BND–cellulose (*important*). Transfer the supernatants to fresh centrifuge tubes. Any glass tubes used at this point should be siliconized.
11. Add an equal volume of isopropanol to each tube. Mix thoroughly. Incubate at −20 °C for 30 min or longer.
12. Centrifuge at 10 000 $g$ for 90 min.
13. Remove the supernatant from each tube. Air dry. Dissolve the 'pellet', which may be invisible or may be a thin turbid film spread over a wide area, in a small volume of TE buffer. For example, if a 15 ml centrifuge tube was used, then 150 μl of TE would be appropriate. Transfer to a microfuge tube. Rinse the larger centrifuge tube with two-thirds of the original volume of TE buffer (in the example, use 100 μl), which should be pooled with the original solution in the microfuge tube.
14. To the microcentrifuge tube, add one-ninth volume of 3 M potassium acetate, mix, and then add 2.2 volumes of ethanol. Mix thoroughly.
15. Centrifuge in a microcentrifuge (~ 16 000 $g$) for 30 min. Remove the supernatant. Rinse the pellet in 70% ethanol for 5 min or longer. Air dry. Dissolve the pellet in an appropriate volume of TE buffer (see *Protocol 2*). Store in the refrigerator for use within several days, or in a constant temperature freezer (−20 °C or −80 °C). Pellet is stable at this temperature for at least a year.

## 3.4 First dimension electrophoresis

The first dimension should separate replicating molecules primarily according to mass. Conditions should be chosen to minimize the effects of molecular shape. For this reason, low gel percentages, low voltage gradients, and long run times are recommended. The conditions provided in *Protocol 2* below are appropriate for restriction fragments of ~ 2–7 kb. Modified conditions, suitable for molecules up to 15 kb, have been published (19, 20).

My laboratory uses TAE buffer (40 mM Tris–acetate, 2 mM EDTA, pH 8.5) for first dimension electrophoresis, while Brewer and Fangman (1) use TBE (89 mM Tris–borate, 2 mM EDTA, pH 8.0). Both buffers give satisfactory results. TAE provides a somewhat better resolution of large DNA molecules; TBE provides a better resolution of smaller molecules. My laboratory includes ethidium bromide at low concentration in the first dimension gel; Brewer and Fangman do not (1). In our experience, ethidium bromide at low concentration does not significantly affect the mobility or integrity of replicating molecules. Its use during first dimension electrophoresis

# 10: 2D gel electrophoresis and replicon mapping

allows the progress of electrophoresis to be monitored and obviates the requirement for gel staining between the first and second dimensions.

The following protocol should be considered a guide. The suggested conditions should be modified as appropriate to accommodate variations in the molecular size range of interest, in the number of samples to be run, and in the size of the second dimension gel.

---

**Protocol 2.** First dimension electrophoresis

*Materials*
- 20 × 20 cm 0.4% agarose gel
- TAE buffer: 40 mM Tris–acetate, 2 mM EDTA, pH 8.5
- Ethidium bromide
- DNA samples (*Protocol 1*)
- Ficoll (Type 400, Pharmacia)
- Marker dye
- Electrophoresis equipment
- 360 nm UV light

*Method*

1. Use a 20 × 20 cm 0.4% agarose gel, prepared in TAE buffer supplemented with 0.1 µg/ml ethidium bromide. Use a high-quality comb to prepare sample wells (approximately 5 mm × 2 mm). The gel may be formed as thick as necessary to accommodate the sample volume, up to a thickness of 1 cm.

2. Prepare samples so that the final DNA concentration does not exceed 200 µg/ml. For single-copy yeast sequences, we obtain good signals from N/A gels with 500 ng–2 µg of caffeine wash DNA (see *Protocol 1*). The amount required for other cases can be calculated by comparison of sequence complexities and generation times to those of yeast. Samples should be prepared so that the final ionic composition of the sample is similar or identical to that of the electrophoresis buffer; this will minimize distortion of the electric field in the vicinity of the sample wells. Use 3% Ficoll to increase sample density. Any marker dye can be used.

3. Run multiple samples in separate lanes of the first dimension gel. At least one empty lane should always be placed between sample-containing lanes to prevent cross-contamination. One or more lanes should contain size marker DNAs to cover the size range of interest. One lane should contain the caffeine-wash DNA (or other DNA preparation) intended for the alkaline second dimension gel. Another lane should contain the caffeine-wash DNA (or other DNA preparation)

**Protocol 2.** *Continued*

    intended for the neutral second dimension gel. The amount of DNA required in this lane is about one-quarter the amount required for the alkaline gel if the same short hybridization probe(s) is to be used for both gels. If the N/N gel is to be probed with the complete restriction fragment of interest, about one-tenth of the amount required for the alkaline gel is satisfactory. Several additional lanes should contain samples of salt-wash DNA from the BND–cellulose fractionation (*Protocol 1*), or other control DNAs in amounts approximately equal to the caffeine-wash samples. Because caffeine-wash DNA is recovered in small amounts, it is usually difficult to determine its concentration by direct measurement. To compensate, use multiple salt-wash lanes, each differing from the next by 2- to 4-fold in concentration, to cover the estimated range of possible concentrations of the caffeine-wash samples.

4. Carry out first dimension electrophoresis for sufficient time to provide good resolution in the molecular weight region of interest, but not so long that the region of interest will not fit into the available space in the second dimension gel. We usually employ submerged gel electrophoresis at 0.5–0.8 V/cm for 16–24 h. Under these conditions, molecules of 1 kb migrate about 8 cm from the origin. Electrophoresis is at room temperature, and no buffer recirculation is required.

5. Use 360 nm UV light to examine the first dimension gel. Shorter wavelengths will generate nicks at an unacceptably rapid rate. Once the marker DNAs have migrated the desired distance, stop the electrophoresis and photograph the gel using 360 nm light. Inspect the photograph to determine which salt-wash (or other control) lanes contain amounts of DNA most similar to those in the two caffeine-wash lanes.

6. Using 360 nm light for illumination, cut out the two caffeine-wash lanes and their matching control lanes with a clean razor blade, scalpel, or sharpened spatula, aided by a ruler or other straight edge. Do not use an ordinary spatula because it may produce a rough edge, leading to blurring of the signal. Cut so that little or no non-DNA-containing agarose remains on either side of the excised lane. The length of the excised lane is determined by the width of the second dimension gel. In our case, if the lanes are < 8.5 cm in length, then two of them can be included side by side in a second dimension gel.

## 3.5 Second dimension electrophoresis

The second dimension gels can be prepared while the first dimension gel is running. Once again, these protocols should be considered to be guides and should be modified as appropriate.

---

**Protocol 3.** Neutral second dimension electrophoresis (modified from Brewer and Fangman, ref. 1)

*Materials*
- 20 × 20 cm 1% agarose gel
- TBE buffer: 89 mM Tris–borate, 2 mM EDTA, pH 8.0
- Ethidium bromide
- DNA samples
- DNA size markers in TBE
- Marker dye
- Ficoll (Type 400, Pharmacia)
- Electrophoresis equipment
- 360 nm UV
- Southern blotting equipment

*Method*
1. Use a 20 × 20 cm 1% agarose gel, a few millimetres thicker than the first dimension gel, prepared in 1 × TBE with 0.5 µg/ml ethidium bromide. Form the gel using one or two preparative well combs which produce a long preparative well (~ 17 cm) flanked by two small marker wells (~ 2.5 mm each). Place one comb near the top of the gel. The second comb, if used, should be placed at or just below the middle of the gel.
2. After the second dimension gel has solidified, enlarge the preparative well(s) to permit insertion of the first dimension lanes. Use a sharp instrument to cut away the agarose above the preparative well(s). The cuts from the top well can extend all the way to the top of the gel, leading to the formation of a large rectangular hole (*Figure 3*). A hole of similar size should be cut above the second preparative well at the middle of the gel if a second comb was used. Save the agarose which has been cut away in a small beaker; it will be used to fill in around the first dimension lanes.
3. Carefully transfer the appropriate first dimension lanes into the hole. We find it useful to place one caffeine-wash lane and its matching salt-wash lane into each hole, side by side. Use a convention, such as always placing the caffeine-wash lane on the right and always placing

**Protocol 3.** *Continued*

the high-molecular-weight end of each lane on the left. Place the bottom of each first dimension lane on the exposed surface of the gel trough so that it will also be on the bottom in the second dimension gel. The stacking effect at the 0.4%/1% gel boundary will produce sharp second dimension bands, even from gel lanes that were initially over 5 mm wide. The first dimension lanes should be placed close (about 2 mm) and parallel to the edge of the preparative well, but not against that edge or bubbles may be trapped between the two gels. If two preparative combs were used, then a total of four first dimension lanes can be placed into a single second dimension gel.

4. Use a microwave oven to melt the agarose which was cut away from the hole. Make sure the agarose is thoroughly melted and thoroughly mixed. Let it cool somewhat, to ~ 65 °C, then pour or pipette it quickly, but gently, into the spaces surrounding the first dimension lanes. Add enough agarose to cover the first dimension lanes. Use a spatula or other instrument to remove any bubbles that may be trapped between the first and second dimension gels. At this point, with liquid agarose and no air bubbles between the first and second dimension gels, the first dimension gel may be moved by pressure from a spatula so its bottom edge is abutted against the top edge of the second dimension gel. Work quickly so that the fill-in gel can be completely poured, all bubbles can be removed, and the first dimension lanes can be properly positioned before the gel begins to solidify. Practice these manipulations with some dummy gels before attempting them with gels containing experimental material.

5. After the agarose has solidified, place the second dimension gel into its electrophoresis box in a coldroom and fill the buffer troughs with 1 × TBE buffer, supplemented with 0.5 µg/ml ethidium bromide and pre-equilibrated to coldroom temperature. Add enough buffer to cover the gel and let the gel equilibrate to coldroom temperature (at least 30 min).

6. Add appropriate DNA size markers in TBE to the marker wells, along with a marker dye and 3% Ficoll.

7. The voltage and run time appropriate for second dimension electrophoresis vary according to the sizes of the restriction fragments of interest. We use coldroom temperature and ~ 5 V/cm for ~ 3 h for fragments of 2–7 kb and ~ 3.5 V/cm for ~ 7 h for fragments of 5–12 kb Others recommend an even wider range of voltages and times for varying conditions (1, 19, 20). We find that buffer recirculation is unnecessary, but Brewer and Fangman recommend it (1).

8. Monitor progress of the gel run using 360 nm UV light. The arc of

# 10: 2D gel electrophoresis and replicon mapping

double-stranded linears plus the size markers should be visible. Good results for a wide range of fragment sizes are obtained when a 500 bp marker has migrated ~ 8 cm. After electrophoresis, photograph the gel and then transfer the DNA to a nylon membrane using standard Southern blotting procedures.

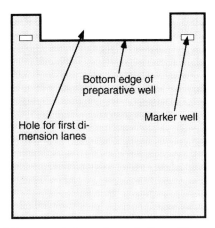

**Figure 3.** Appearance of the second dimension gel after cutting out a space for the first dimension lanes.

**Protocol 4.** Alkaline second dimension electrophoresis

*Materials*

- Pure water
- Alkaline electrophoresis buffer (AEB): 40 mM NaOH, 2 mM EDTA
- Bromocresol green
- Ficoll (Type 400, Pharmacia)
- Electrophoresis and Southern blotting equipment

*Method*

1. Prepare the alkaline second dimension gel in exactly the same way as the neutral second dimension gel (*Protocol 3*, Steps 1–4) with one exception: the alkaline second dimension gel should be prepared in pure water (no TBE, no alkaline buffer, no ethidium bromide). The gel solidifies more readily in water than in alkaline buffer.

**Protocol 4.** *Continued*

2. After the first dimension lanes have been placed and embedded into the second dimension gel, place the gel into its electrophoresis tank and add enough alkaline electrophoresis buffer (AEB) to fill the tanks and cover the gel. Incubate for at least 1 h at room temperature before running the gel.

3. Just before starting electrophoresis, add appropriate DNA size markers plus 0.1% bromocresol green (BCG) and 3% Ficoll (all in AEB) to the marker wells. *Do not* use standard dye markers because they are unstable under alkaline conditions. BCG is stable, and its electrophoretic properties are indistinguishable from those of bromophenol blue.

4. Use 0.5–0.75 V/cm for alkaline gel electrophoresis, at room temperature with no buffer recirculation. To prevent the BCG dye from diffusing out of the gel during the overnight run, cover the gel with a sheet of glass after electrophoresis has started and the BCG has moved into the gel. Keep the voltage low enough to prevent excessive heating. Good resolution of single strands in the range from 200 b to 5 kb is obtained when the BCG marker has migrated about 7 cm. This usually takes 15–20 h.

5. After electrophoresis, transfer the DNA to a nylon membrane using standard Southern blotting procedures. Note that the DNA is already denatured.

## 3.6 Modification of the N/N procedure to permit the determination of replication direction

A modification of the original N/N procedure permits the obtaining of information about the direction of replication fork movement through simple Y restriction fragments (7). It is strongly recommended that this modified N/N procedure or the N/A procedure be used to determine replication fork direction. In several cases, information about the direction of replication fork movement has permitted identification of origins that might have been missed (due to failure to use a sufficiently complete set of overlapping restriction fragments) based on data provided by the conventional N/N technique alone (21, 22). In addition, determination of direction of replication fork movement can help to quantitate the frequency of origin usage. Because nicks at forks can convert bubble arcs to Y arcs, quantitation based on the conventional N/N technique may not be accurate (23, 7).

The modification which permits direction mapping is a second digestion with a restriction enzyme, between the first and second electrophoretic

# 10: 2D gel electrophoresis and replicon mapping

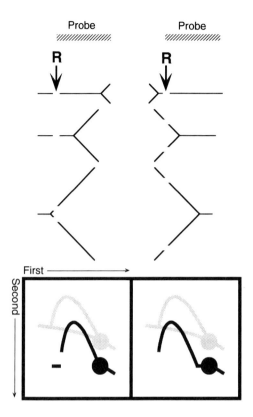

**Figure 4.** Determination of direction of DNA replication by modified N/N 2D gel electrophoresis. R indicates the cutting site for the second restriction enzyme. See the text for further explanation.

dimensions. When the blot of the resulting 2D gel is hybridized with a probe corresponding to just one of the two subfragments, different 2D gel patterns are detected depending on the direction of replication fork movement through the fragment. In the example provided in *Figure 4*, the signals produced by the intact fragment are shown in pale grey. If the second digestion is not complete, these signals will be visible; they can provide helpful orientation. The subfragments resulting from the second digestion migrate faster (lower in the gel) than the intact fragment. In *Figure 4*, the second restriction enzyme is assumed to cut the intact restriction fragment close to its left end. If replication forks move through the intact fragment from right to left, then the large subfragment will replicate before the small subfragment. Consequently, when the blot of the 2D gel is hybridized with a probe corresponding to the large subfragment, a Y arc should arise directly from the spot of non-replicating molecules (left panels in *Figure 4*; the horizontal line to the left of the Y arc is due to the large subfragment being linear once the replication forks enter the small subfragment). If, instead, replication forks move from left to right through the intact fragment, then the Y arc should be displaced to the left of the spot of non-replicating molecules (right panels in *Figure 4*).

The second restriction digestion is unusual in that it is carried out on the DNA in the first dimension lane, after the lane is excised from the first dimension gel. B. Brewer (personal communication) has found that the following enzymes will cut DNA in Beckman LE agarose to completion: *Ava*I, *Bam*HI, *Bgl*II, *Eco*RI, *Eco*RV, *Hin*dIII, *Nco*I, *Pst*I, *PVU*II, *Sac*II, *Sna*BI, *Spe*I, *Stu*I, *Xba*I, and *Xho*I. Many other types of agarose and many other enzymes are also likely to work. Additional information about enzyme/agarose compatibilities is provided in the New England BioLabs 1992 catalog (pp. 189–90). It is recommended that each batch of enzyme and agarose be tested with readily obtainable DNA before being used with experimental DNA

**Protocol 5.** Restriction enzyme digestion of DNA in the first dimension gel (modified from a protocol kindly provided by B. Brewer)

*Materials*

- Disposable pipette reagent reservoir (Costar 4870)
- 10 mM Tris (pH 8.0), 0.1 mM EDTA
- Appropriate restriction enzyme buffer
- BSA
- Stock restriction enzyme

*Method*

1. Excise the first dimension lane as usual (*Protocol 2*), then place it in a disposable pipette reagent reservoir or equivalent trough. Costar plastic troughs have a 12 cm long wedge-shaped reservoir. Fill with 10 mM Tris (pH 8.0), 0.1 mM EDTA. Incubate at room temperature, 30 min, with gentle agitation. Repeat once.

2. Prepare 250 ml of restriction enzyme buffer. Include bovine serum albumin (BSA) (100 µg/ml) but not spermidine.

3. Drain the Tris–EDTA from the first dimension gel lane and fill the trough with restriction buffer. Incubate for 1 h at room temperature, with gentle agitation. Repeat once.

*10: 2D gel electrophoresis and replicon mapping*

4. Drain the restriction buffer. Remove excess buffer with a Pasteur pipette and then blot the gel lane briefly with absorbant paper. Pipette 10–20 μl of stock restriction enzyme (~ 100–200 units) directly on to the surface of the gel lane, distributing it as uniformly as possible.
5. Seal the reservoir with plastic film, or set the reservoir in a glass baking dish with water in its bottom and then seal with plastic film. Incubate at 37 °C for 4 h or longer.
6. To stop the reaction, fill the reservoir with TE buffer (*Protocol 1*) and incubate with gentle agitation at room temperature for 30 min.
7. Proceed with the second dimension of electrophoresis (*Protocol 3*). Modify the conditions of electrophoresis, as appropriate, to account for the smaller size of the fragment of interest.

## 3.7 Hybridization with specific probes

Detection of the DNA molecules of interest on the blots of 2D gels requires hybridization with specific probes. Many procedures are available. Because replicating molecules are rare, whatever procedures are used should be optimized to provide the highest possible signal to noise ratio. If possible, use a PhosphorImager (Molecular Dynamics) or equivalent instrument, because it provides greater sensitivity than X-ray film, and it permits accurate signal quantitation over at least five orders of magnitude.

## 3.8 Additional information

Space limitations have prevented my including references to and discussion of all of the available literature on 2D gel replicon mapping. Valuable information on running and interpreting 2D gels can be obtained from further reading.

# Acknowledgements

The work performed in my laboratory was supported by grants from the National Science Foundation, the American Cancer Society, and the National Institutes of Health. I am grateful to Bonita Brewer for providing unpublished protocols and for helpful comments on part of the manuscript.

# References

1. Brewer, B. J. and Fangman, W. L. (1987). *Cell,* **51**, 463.
2. Huberman, J. A., Spotila, L. D., Nawotka, K. A., El-Assouli, S. M., and Davis, L. R. (1987). *Cell,* 51, 473.

3. Handeli, S., Klar, A., Meuth, M., and Cedar, H. (1989). *Cell*, **57**, 909.
4. Vassilev, L. and Johnson, E. M. (1989). *Nucl. Acids Res.*, **17**, 7693.
5. Burhans, W. C., Vassilev, L. T., Caddle, M. S., Heintz, N. H., and DePamphilis, M. L. (1990). *Cell*, **62**, 955.
6. Umek, R. M., Linskens, M. H. K., Kowalski, D., and Huberman, J. A. (1989). *Biochim. Biophys. Acta*, **1007**, 1.
7. Fangman, W. L. and Brewer, B. J. (1991). *Annu. Rev. Cell Biol.*, **7**, 375.
8. Campbell, J. L. and Newlon, C. S. (1991). In *The molecular biology of the yeast* Saccharomyces (ed. J. R. Broach, J. R. Pringle, and E. W. Jones), Vol. 1, p. 41. Cold Spring Harbor Laboratory Press, NY.
9. DePamphilis, M. L. (1993). *Annu. Rev. Biochem.*, **62**, 29.
10. Sundin, O. and Varshavsky, A. (1980). *Cell*, **21**, 103.
11. Bell, L. and Byers, B. (1983). *Analyt. Biochem.*, **130**, 527.
12. McDonell, M. W., Simon, M. N., and Studier, F. W. (1977). *J. Mol. Biol.* **110**, 119.
13. Nawotka, K. A. and Huberman, J. A. (1988). *Mol. Cell. Biol.*, **8**, 1408.
14. Brewer, B. J., Sena, E. P., and Fangman, W. L. (1988). In *Eukaryotic DNA replication* (ed. T. Kelly and B. Stillman), Vol. 6, p. 229. Cold Spring Harbor Laboratory Press, NY.
15. Vaughn, J. P., Dijkwel, P. A., and Hamlin, J. L. (1990). *Cell*, **61**, 1075.
16. Vaughn, J. P., Dijkwel, P. A., Mullenders, L. H. F., and Hamlin, J. L. (1990). *Nucl. Acids Res.*, **18**, 1965.
17. Shinomiya, T. and Ina, S. (1991). *Nucl. Acids Res.*, **19**, 3935.
18. Gamper, H., Lehman, N., Piette, J., and Hearst, J. (1985). *DNA*, **4**, 157.
19. Krysan, P. J. and Calos, M. P. (1991). *Mol. Cell. Biol*, **11**, 1464.
20. Hyrien, M. and Méchali, M. (1992). *Nucl. Acids Res.*, **20**, 1463.
21. Huberman, J. A., Zhu, J., Davis, L. R., and Newlon, C. S. (1988). *Nucleic Acids Res.*, **16**, 6373.
22. Zhu, J., Newlon, C. S., and Huberman, J. A. (1992). *Mol. Cell. Biol.*, **12**, 4733.
23. Linskens, M. H. K. and Huberman, J. A. (1990). *Nucl. Acids Res.*, **18**, 647.

# 11

# The isolation of functional mitotic organelles from tissue culture cells

LINDA WORDEMAN

## 1. Introduction

The repertoire of dynamic activities involved in the equipartitioning of chromosomes during cell division is complex. A complete and dramatic reorganization of the cell cytoplasm takes place within an interval of, at most, one-fifteenth of the total cell-cycle time. In tissue culture cells the centrosomes separate and proceed to opposite sides of the nucleus, the nuclear envelope breaks down and the condensed chromosomes, which are now free to interact with the microtubules of the mitotic spindle, travel back and forth from one spindle pole to the other, eventually reaching an equilibrium position at the spindle equator. When all the chromosomes have reached the spindle mid-zone, the sister chromatids separate and sweep towards their respective spindle poles, re-establishing the interphase state of the cell cycle.

Devising a mechanistic description of spindle function is a formidable task in view of all this dynamic intracellular motility. One obvious approach, however, is to dissect apart the component parts of the mitotic apparatus and reconstitute function in a controlled setting *in vitro*. Centrosomes, chromosomes, and even the mitotic spindle itself, have been isolated from marine eggs and tissue culture cells by a number of researchers (1–6) (see *Figure 1*). These isolated organelles are useful for identifying hitherto unknown components which are enriched in such isolates or for reconstituting function.

Isolated mitotic spindles have been successfully used to identify spindle-associated phosphoproteins and spindle-associated kinase activity (7, 8). In addition, ATP-dependent spindle elongation and microtubule polymerization have been reactivated in mitotic spindles isolated from diatoms (9, 10). Microtubule and tubulin-binding activities, microtubule motors, and ATP-dependent motor activities have been demonstrated to be associated with the kinetochores of isolated mitotic chromosomes (5, 11, 12). Finally, isolated centrosomes have been used to identify centrosome-specific proteins and to reconstitute cell-cycle dependent microtubule nucleation (13–15).

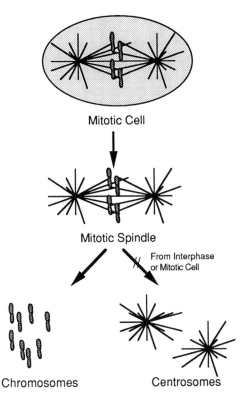

**Figure 1.** Diagram illustrating the isolation of mitotic spindle components.

In this chapter I have provided detailed protocols for the isolation of mitotic spindles, chromosomes, and centrosomes from tissue culture cells. These protocols are rapid, gentle, and make use of mild buffers and salt concentrations that promote protein–protein interactions. The goal of these protocols is to produce mitotic organelles in which biochemical activities are reactivatable or easily reconstituted *in vitro*. In many instances, the functional characteristics of the organelles isolated using these protocols have been partially elucidated. In other cases, less is known, and the astute protein chemist may see ways to improve the protocol in order to more effectively preserve function.

## 2. Mitotic arrest in tissue culture cells

### 2.1 Cell lines

Accumulating substantial quantitites of mitotic tissue culture cells, uncontaminated with interphase cells, is a critical prerequisite for the isolation of mitotic organelles. Procedures for synchronizing cultured cells are discussed in detail in Chapter 1. In brief, most methods for synchronizing tissue culture cells involve exposing the cells to a drug that arrests the cells at a particular stage of the cell cycle. The population of unsynchronized cells will progress normally through the cell cycle until they reach the arrest point where they will stop. This results in the slow accumulation of cells at a particular point in the cell cycle, whether it be S phase (thymidine block) or prometaphase (microtubule inhibitors). Deleterious effects associated with the exposure of cells to these drugs are reversible if the length of exposure time is kept to a minimum. However, the longer the cells are exposed to many of these drugs the greater the chance of irreversibly damaging them. Therefore, cells that are suitable for synchronization are cells that grow rapidly enough, under normal conditions, that a high percentage of cells will accumulate at the arrest point during the shortest period of time (usually a 16 hour overnight incubation). Hence, the optimal cells for synchronization are cells which have a doubling time equal to or less than 16 hours.

Several cell lines are available that fulfil these requirements. HeLa cells, both adherent and non-adherent variants, Chinese hamster ovary (CHO) cells, and HTC cells. If you are able to endure a precipitous loss in yield of mitotic cells then some of the slower-growing cell lines may also be employed. The synchronization protocols described here rely on arresting the cells in mitosis, where they arrest with a rounded morphology and can be easily harvested by mitotic shake-off which leaves interphase cells behind on the dish. When using suspension cells, however, it is advisable to use a cell line in which it is possible to achieve a very high mitotic index due to the difficulty in separating interphase from mitotic cells.

## 2.2 Methods

As described in Chapter 1, the simplest method for obtaining large numbers of mitotic cells is called mitotic 'shake-off' (16). This technique takes advantage of the fact that cells round up and detach from the substrate during mitosis. The cells will detach from the substrate when the flask is shaken, or banged against the benchtop, or if a stream of medium from a Pasteur pipette is directed at the cells. This technique works well for log-phase CHO and adherent HeLa cell lines. Subsequent manipulations must be accomplished quickly, however, because the cells will only remain in mitosis for a number of minutes.

High yields of cells arrested in mitosis can be achieved by culturing the cells in the presence of microtubule inhibitors such as Colcemid, nocodazole, or vinblastine. Depolymerization of the microtubules will cause the cells to arrest as they try to enter mitosis where they will remain for several hours. In some cases the destruction of the microtubule arrays is useful (such as in the purification of mitotic chromosomes). In such instances the drug-arrested cells may be used directly in the purification procedure. However, for the purification of mitotic spindles the microtubules must be intact and stabilized. Therefore, the microtubule inhibitors must be washed out and the microtubule arrays allowed to re-form prior to cell lysis. In this situation the choice of inhibitor, concentration, and arrest time is critical. The effects of Colcemid and nocodazole have been found to be more easily reversible, for example, than the effects of vinblastine, which induces the formation of tubulin paracrystals in the cell cytoplasm. The appropriate synchronization method has been incorporated into each isolation protocol in order to eliminate confusion.

Prior to arresting tissue culture cells in mitosis it is often useful, particularly in the case of slower growing cells, to 'presynchronize' the cells at another part of the cell cycle (17). This can be accomplished in a couple of different ways. DNA synthesis inhibition (S-phase arrest) using inhibitors such as hydroxyurea or thymidine (18), or nutritional deprivation (19) ($G_1$-phase arrest) can be simple and effective methods of achieving high synchrony for at least one cell cycle. In the case of nutritional deprivation, either serum or isoleucine can be removed from the media for a period of time. When restored, the cells will resume transmission through the cell cycle in a relatively synchronous manner. Synchronization procedures are included in the following protocols for completeness, and to emphasize that high synchrony is crucial to the success of many of these protocols. However, there is no doubt that other synchronization methods or variants (Chapter 1) would be equally, or perhaps more, successful.

## 3. The mitotic apparatus

### 3.1 Structure and components

The mitotic apparatus is responsible for the equipartitioning of chromosomes at mitosis. In the simplest structural sense it appears to be a bipolar array of microtubules with attached chromosomes. However, video analysis and fluorescence photobleaching have revealed the mitotic spindle to be an extraordinarily dynamic structure (20–22). Furthermore, in addition to microtubules and chromosomes, the mitotic spindle appears to be a repository for a huge spectrum of additional components such as vesicles, phosphatases, kinases, 'matrix' components of unknown function, centrosomal proteins, and a rapidly increasing array of motor molecules. Determining the role of these proteins in mitotic spindle function is an extremely challenging task and is complicated by the fact that many components that have been localized to the mitotic apparatus are also found in the cytoplasm. In an attempt to simplify the study of mitotic spindle function several researchers have developed methods to free the mitotic apparatus from the surrounding cytoplasm and study its composition *in vitro*.

## 3.2 Isolation of CHO mitotic spindles

The techniques used for the isolation of mitotic spindles from CHO cells are derived from those used for the isolation of mitotic spindles from marine eggs, synchronous populations of which can be obtained in large quantity (1). Mitotic spindle isolations simply involve detergent lysis of the cells into a microtubule and chromosome stabilizing buffer followed by collection of the mitotic spindles by a pelleting step (2, 23, 24). Biochemically, it is still an incredibly complex structure and still probably rife with contaminants. However, it represents a substantial enrichment for mitotic spindle-specific components.

---

**Protocol 1.** Isolating CHO mitotic apparatus (hexylene glycol method)

*Materials*
- MEM-alpha medium plus 10% fetal calf serum
- Thymidine (Sigma)
- Glycerol
- Nocodazole (Sigma)
- Hoechst 33258 (1 mg/ml stock in double-distilled $H_2O$)
- Hexylene glycol (Polysciences)
- Swelling buffer: 1 M hexylene glycol, 2 mM Pipes (pH 6.8), 0.5 mM $MgCl_2$, 0.5% aprotinin
- Lysis buffer: 1 M hexylene glycol, 2 mM Pipes (pH 6.8), 1 mM EGTA, 1% NP-40, 0.5% aprotinin
- Post-lysate buffer: 1 M hexylene glycol, 50 mM MES (pH 6.3), 0.5% aprotinin

*Method*

A. *Cell synchronization*

1. Plate cells (10–20 × 150 mm plates) in MEM-alpha medium plus 10% fetal calf serum.
2. Aspirate medium from subconfluent cultures (2 × $10^7$ cells per plate) and replace with medium containing 5 mM thymidine for 16 h.
3. Wash with fresh medium and incubate for 4 h in thymidine-free medium to allow the cells to traverse the $G_2$ phase of the cell cycle. Then add 0.035 µg/ml of nocodazole and incubate for another 3–4 h.
4. Shake off rounded cells which are blocked in prometaphase into a plastic conical tube and pellet at 800 *g* for 5 min at room temperature, in a clinical centrifuge. Resuspend in fresh medium and incubate for 10–15 minutes in a 37 °C water bath to allow the mitotic spindles to re-

**Protocol 1.** *Continued*

form and to allow the cells to progress as far as metaphase. Place a drop of cell-containing medium on a slide with a small amount of lysis buffer and 1 µl of Hoechst 33258 and examine under a fluorescence microscope.

5. When the majority of cells have reached metaphase immediately centrifuge the cells for 5 min at 800 $g$ at room temperature and aspirate the medium.
6. Gently resuspend the cells in 10 ml of swelling buffer (room temperature) and pellet for 3 min at 800 $g$.

B. *Lysis*

1. Add 10 ml of lysis buffer (at 37 °C) to the pellet and mix for 13 sec at setting 5 on a Scientific Products vortex mixer.
2. Add 4 ml of ice-cold postlysate buffer, gently mix, and place on ice for 10 min. During this incubation, gently invert the tube every 2–3 min.
3. Centrifuge the lysate at 150 $g$ for 5 min at 4 °C to remove large debris.

C. *Glycerol cushion*

1. Spin the supernatant from Step B3 through a cushion of 40% glycerol containing 50 mM MES (pH 6.3) at 700 $g$ for 10 min, at 4 °C.
2. Resuspend the mitotic apparatus-containing pellet in postlysate buffer.

The use of hexylene glycol to stabilize microtubules during spindle isolation is derived from protocols for isolating mitotic spindles from marine eggs (25). Other spindle-stabilizing agents such as glycerol, DMSO, and polyethylene glycol have also been successfully employed (26–28). In an attempt to replace these non-specific stabilizing agents with one specific for microtubules Kuriyama *et al.* (2) have devised a simple protocol for isolating taxol-stabilized mitotic spindles from CHO cells. Both of these protocols have been successfully employed by the author to isolate mitotic spindles from CHO cells.

**Protocol 2.** Isolating CHO mitotic apparatus (taxol method)

*Materials*

- MEM-alpha medium plus 10% fetal calf serum
- Thymidine (Sigma)
- Colcemid (Sigma)
- Taxol (2 mg/ml stock in DMSO), (National Cancer Institute)

- Isolation medium: 2 mM Pipes (pH 6.8), 0.25% Triton X-100, 20 µg/ml taxol

*Method*

A. *Cell synchronization*

1. Plate cells (10–20 × 150 mm plates) in MEM-alpha medium plus 10% fetal calf serum.
2. Aspirate medium from subconfluent cultures (2 × $10^7$ cells per plate) and replace with medium containing 5 mM thymidine for 16 h.
3. Wash with fresh medium and incubate for 4 h in thymidine-free media to allow the cells to traverse the $G_2$ phase of the cell cycle.
4. Replace medium with medium containing 0.1 µg/ml Colcemid for 3–4 h. Collect mitotic cells and pellet at 800 $g$ for 5 min at room temperature. Aspirate medium and wash once with fresh medium.
5. Resuspend the cells in fresh medium and allow the mitotic cells to recover for 15 min at 37 °C. Then add taxol to a final concentration of 5 µg/ml and allow cells to incubate for 5 min more.

B. *Lysis*

1. Pellet cells at 800 $g$ for 5 min at room temperature and wash the sides of the tube with distilled water while avoiding disturbing the pellet.
2. Add 1.0 ml of room temp. isolation buffer to the cell pellet and disrupt the cells by 10 seconds of vortex-mixing at setting 5 (Scientific Products adjustable speed vortex mixer).
3. Centrifuge the lysate at 300 $g$ for 15 min at room temp. or 4 °C to recover the spindles.

These protocols can be adapted for the isolation of HeLa cells grown as adherent monolayers. It is likely that either nocodazole (hexylene glycol method) or Colcemid (taxol method) can be used interchangeably in the cell-synchronization step.

## 3.3 Other systems for mitotic apparatus isolation

Mitotic spindle isolation requires the hyperstabilization of microtubules so that the spindle will remain intact during lysis and recovery. As a result, spindle functions that require dynamic microtubules will be necessarily inhibited. Nevertheless, it may still be possible to reactivate certain spindle functions and the interpretation of the results of such studies may even be simplified by the inhibition of the functional contribution of dynamic microtubules. Nevertheless, functional reactivation of mitotic spindles isolated from tissue culture cells is a field that is in its infancy. Mitotic spindles have been isolated from *Saccharomyces cerevisiae*, *Schizosaccharomyces pombe*, sea urchins, and diatoms (1, 3, 8–10, 25, 29, 30). At this time, the greatest strides have been made in reactivating spindle elongation (anaphase B) in spindles isolated from diatoms (8–10). In addition, information on the reactivation of chromosome separation (anaphase A) has been obtained from spindles assembled *in vitro* in *Xenopus* egg extracts (31). These model systems may provide useful modifications for the CHO cell system that will improve the functionality of the isolated CHO cell spindles.

## 4. Centrosomes

### 4.1 CHO cell centrosomes

Isolated centrosomes may be prepared from confluent monolayers of tissue culture cells. They inhibit tubulin-binding and microtubule nucleating activity (14, 32). Furthermore, both mitotic and interphase activity may be observed when centrosomes isolated from interphase cells are placed, respectively, in either a mitotic or interphase cell extract (13). Hence, for most studies, centrosomes isolated from interphase cells can be used, eliminating the need to produce synchronous populations of mitotic cells.

The success of the centrosome preparation relies on the preservation of an extremely low-salt environment. Low ionic-strength buffers preserve microtubule nucleation capacity. In addition, the centrosomes must be kept dilute or they will irreversibly clump together. Therefore, hard pelleting steps must be avoided.

---

**Protocol 3.** Isolating centrosomes from plated CHO cells

*Materials*

- MEM-alpha medium plus 10% fetal calf serum
- Nocodazole (Sigma), 10 mg/ml stock in DMSO
- Cytochalasin B (Sigma), 10 mg/ml stock in DMSO
- Phosphate-buffered saline (PBS)—ice cold
- 1/10 PBS, 8% sucrose—ice cold
- 8% sucrose—ice cold
- Lysis buffer: 1 mM Tris–HCl (pH 8.0) (from Tris base), 0.1% beta-mercaptoethanol, 0.5% NP-40
- 50 × PE: 500 mM Pipes (pH 7.2) 50 mM EDTA, 5% beta-mercaptoethanol

## 11: Isolation of mitotic organelles

- 20% Ficoll (w/w in PE + 0.1% NP-40)
- 40 micron Spectra/Mesh (Spectrapor)
- Tabletop clinical centrifuge
- Superspeed centrifuge (i.e. Sorvall RC-5)
- 15 ml glass Corex centrifuge tubes
- Ultracentrifuge
- Beckman SW-28 tubes and rotor
- Refractometer

*Method*

A. *Cells*

1. Plate cells on ten 150 mm plastic Petri dishes in MEM-alpha medium with 10% fetal calf serum and grow until cells are just confluent.
2. Aspirate medium. Add 15 ml of fresh media containing 20 µg/ml nocodazole and 10 µg/ml cytochalasin B.
3. Incubate in $CO_2$ incubator at 37 °C for 90 min.

B. *Washes*

1. Aspirate medium. Add 30 ml ice-cold PBS. Pour off and aspirate.
2. Add 30 ml 1/10 PBS, 8% sucrose. Pour off and aspirate.
3. Add 30 ml 8% sucrose.
4. Aspirate sucrose *completely*.

C. *Lysis*

1. Add 10 ml lysis buffer to each plate.
2. Incubate for 10 min in a slow rotary shaker.
3. Collect into two 50 ml plastic tubes. Add 1 ml 50 × PE.
4. Spin 1–2 min in a tabletop clinical centrifuge at top speed.
5. Filter supernatants through 40 micron Spectra/Mesh.

D. *Concentrating centrosomes*

1. Transfer lysates to four Corex tubes. Underlay with 2 ml 20% Ficoll.
2. Centrifuge 25 000 *g* (13 K r.p.m.), for 15 min at 0–2 °C (Sorvall HB-4 swinging bucket rotor).
3. Aspirate supernatant to 2 ml above the cushion.
4. Collect clear interface with a Pasteur pipette. Collect from just above the cushion, but collect most of the cushion itself.
5. Pool interfaces.

**Protocol 3.** *Continued*

E. *Sucrose gradient—further concentration and purification*
1. Make up, the evening before, a 20% (w/w) to 62.5% (w/w) continuous sucrose gradient in PE + 0.1% NP-40; total volume of 20 ml. Pour from above into a 38 ml SW-28 tube.
2. Layer Ficoll interface over gradient—*not* carefully, a little mixing may be good to diminish the interface and prevent the centrosomes from accumulating and sticking there.
3. Centrifuge at 120 000 $g$ for 1.5 h at 0–2 °C.
4. Collect 30-drop fractions from the bottom at 4 °C.
5. Measure the refractive index with 1–3 µl. Pool centrosomes between 50 and 58% sucrose. Microtubule nucleating activity may be stable for several days at 0 °C or several months aliquoted at −70 °C.

A high yield of interphase centrosomes can be theoretically obtained from N115 cells because they contain multiple centrosomes per cell. However, they sometimes prove more difficult to work with because they are more difficult to grow to confluence and their centrosomes clump together more readily during the course of the preparation.

**Protocol 4.** Isolating centrosomes from plated N115 cells

*Materials*
- DMEM (4.5 g glucose/litre) + 10% fetal calf serum
- Nocodazole (Sigma), 10 mg/ml stock in DMSO
- Cytochalasin B (Sigma), 10 mg/ml stock in DMSO
- PBS—ice cold
- 1/10 PBS, 8% sucrose—ice cold
- 8% sucrose—ice cold
- Lysis buffer (1 mM Tris–HCl (pH 8.0) (from Tris base), 0.1% beta-mercaptoethanol
- NP-40
- 50 × PE (500 mM Pipes (pH 7.2), 50 mM EDTA, 5% beta-mercaptoethanol
- 20% Ficoll (w/w in PE + 0.1% NP-40)
- 40 micron Spectra/Mesh (Spectrapor)

## 11: Isolation of mitotic organelles

- Tabletop clinical centrifuge
- Superspeed centrifuge (i.e. Sorvall RC-5)
- 15 ml glass Corex centrifuge tubes
- Ultracentrifuge
- Beckman SW-28 tubes and rotor
- Refractometer

*Method*

A. *Cells*

1. Plate cells on ten 150 mm plastic Petri dishes in DMEM medium with 10% fetal calf serum, and culture until cells are just confluent.
2. Aspirate medium. Add 15 ml of fresh media containing 20 µg/ml nocodozole and 10 µg/ml cytochalasin B.
3. Incubate in a $CO_2$ incubator at 37 °C for 90 min.

B. *Washes*

1. Bang dishes to detach cells and using a wide-bore 25 ml pipette suspend cells in drug-containing media.
2. Transfer cells to 50 ml conical tubes—$10^8$ cells/tube, approximately 3 plates of cells/tube. Spin at room temperature for 1 min in a tabletop clinical centrifuge at top speed. The cells are delicate so centrifuge briefly.
3. Aspirate medium. Add 50 ml ice-cold PBS, resuspend cells with a wide-bore pipette, centrifuge (as in Step B2) and aspirate.
4. Add 50 ml 1/10 PBS, 8% sucrose to the tube and resuspend, centrifuge (as in Step B2) and aspirate.
5. Add 50 ml 8% sucrose to tube and resuspend and centrifuge (as in Step B2).
6. Aspirate sucrose *completely*. Wipe the inside of the tube with a paper tissue.

C. *Lysis*

1. Add 13 ml of lysis buffer per tube and resuspend cells.
2. Add 13 ml of lysis buffer + 1% NP-40 per tube.
3. Centrifuge 3 min at room temperature in tabletop centrifuge at top speed and filter supernatant through 40 µm Spectra/Mesh.

D. *Concentrating centrosomes*

1. Continue the isolation procedure the same as for CHO centrosomes (*Protocol 3D*).

## 5. Metaphase chromosomes

### 5.1 CHO cell chromosomes

Isolated CHO chromosomes exhibit a wide variety of function properties *in vitro*, including tubulin binding and microtubule capture at the kinetochore, ATP-dependent and -independent directional motility along microtubules, and phosphorylation-dependent motile properties (5, 11, 12). Furthermore, many chemical and structural properties of kinetochores are more easily studied using isolated chromosomes, rather than chromosomes *in situ* (33). *Protocol 5* describes the isolation of metaphase chromosomes from CHO cells, the functional properties of which have been extensively studied *in vitro*.

---

**Protocol 5.** Isolating mitotic CHO chromosomes

*Materials*

- MEM-alpha medium plus 10% fetal calf serum
- Vinblastine sulfate (Sigma)
- Swelling buffer: 5 mM NaCl, 5 mM $MgCl_2$, 5 mM Pipes, 0.5 mM EDTA (pH 7.2 with KOH)
- Lysis buffer: 10 mM Pipes, 2 mM EDTA, 0.1% $\beta$-mercaptoethanol, 1 mM spermidine HCl, 0.5 mM spermine HCl (pH 7.2 with KOH), 0.1% digitonin (Sigma) (from 10% stock in DMSO) + 2 µg/ml $\alpha_2$-macroglobulin (Sigma)
- Lysis buffer without digitonin
- 15 ml Dounce homogenizer and tight pestle
- Hoechst 33258 (1 mg/ml stock in double-distilled $H_2O$)
- 15 ml glass Corex tubes

*Method*

A. *Cells*

1. Plate CHO cells on to ten 150 mm plastic dishes and use just prior to confluence.
2. Aspirate media and replace with fresh media containing 10 µg/ml vinblastine sulfate.
3. Incubate cells at 37 °C for between 10–12 h.

B. *Harvest*

1. Collect mitotic cells by squirting a stream of media over them with a Pasteur pipette. Pellet the cells (500 *g*, 2 min, room temperature).

## 11: Isolation of mitotic organelles

2. Resuspend cells in swelling buffer at 37 °C. Incubate for 2–10 min to allow swelling. Repellet the cells.

### C. Lysis

1. Add 10 ml ice-cold lysis buffer to the cell pellet, resuspend quickly, and immediately transfer to an ice-cold 15 ml Dounce homogenizer.
2. Quickly homogenize with 10–20 strokes using the tight pestle.
3. During the homogenization step, periodically observe the chromosomes in a small amount of lysate and Hoechst 33258 (final concentration 1 µg/ml) to determine the extent to which the chromosomes are broken up into single sister chromatid pairs.
4. When the majority of the cells have been disrupted, sediment the lysate at 250 $g$, for 1 min, at 4 °C to remove unbroken cells and chromosome clusters.

### D. Sucrose gradient

1. Layer the supernatant over two 9 ml sucrose gradients consisting of 20–60% (w/v) sucrose in lysis buffer minus the digitonin.
2. Centrifuge the gradient at 2500 $g$ for 15 min in a Sorval HB-4 or HS-4 swinging bucket rotor at 4 °C.
3. Collect the flocculent white mass of chromosomes from the side (they will be on the side) of the tube with a Pasteur pipette. Use immediately or aliquot and freeze in liquid nitrogen and store at −70 °C.

*Protocol 5* can be modified to produce chromosomes from suspension cell lines such as HeLa and HTC by the insertion of a low-speed spin (such as in Step C3, *Protocol 6*) to remove contaminating interphase nuclei. The step will not work well, however, if the mitotic index is below about 60%. *Protocol 5* can also be used successfully on other adherent cell lines, but the mitotic arrest step may have to be modified, or more cells grown to make up for the low synchrony in slowly growing cell lines. *Protocol 5* has been successfully used on HeLa, HTC, chick fibroblasts, and, with difficulty, muntjac cells.

Although chromosomes isolated using the above procedure are functionally robust, they are also biochemically complex. Lewis and Laemmli (4) along with W. C. Earnshaw (35) have devised a Percoll gradient purification procedure for HeLa chromosomes that exhibit high purity and extensive enrichment for chromosome specific proteins. While the functional characteristics of these chromosomes have not been well characterized, they are extremely useful for biochemical analysis and immunochemistry (36).

**Protocol 6.** Isolating mitotic HeLa chromosomes (Percoll gradient method of Lewis and Laemmli, modified by W. C. Earnshaw)

*Materials*

- RPMI 1640 plus 10% fetal calf serum
- Colcemid (Sigma), 100 µg/ml filter-sterilized stock in double-distilled $H_2O$
- 10% digitonin (Sigma) in DMSO
- 10% Ammonyx Lo (Continental Chemical Co.)
- Sorvall 50 ml centrifuge tube
- RNase A (Sigma): 5 mg/ml in 10 mM sodium acetate (pH 4.0), boil 7 min, store at 4 °C
- Tabletop centrifuge (e.g. Clay Adams Dyna)
- Beckman TJ6 and J21 centrifuges
- Two Wheaton Dounce homogenizers (one 15 ml plus pestle A, one 40 ml plus pestle B)
- Swelling buffer: 10 mM Tris–HCL (pH 7.4), 10 mM NaCl, 5 mM $MgCl_2$
- Buffer A: 15 mM Tris–HCl (pH 7.4), 80 mM KCl, 2 mM K–EDTA (pH 7.4), 0.75 mM spermidine, 0.3 mM spermine
- Aprotinin (Sigma)
- PMSF (Sigma)
- Hoechst 33258 (1 mg/ml stock in double-distilled $H_2O$)
- Percoll (Pharmacia)
- 20 × Percoll *buffer*: 100 mM Tris–HCl (pH 7.4), 40 mM KCl, 40 mM K–EDTA (pH 7.4), 7.5 mM spermidine
- Percoll *solution*: 2.5 ml 20 × Percoll buffer, 2 ml polyamines (solution consisting of 75 mm spermidine, 30 mM spermine), 0.5 ml 10% Ammynox Lo, add Percoll to 50 ml

*Method*

A. *Cell synchronization and preliminaries*

1. From a suspension culture of HeLa cells, take 125 ml cells plus 125 ml RPMI 1640 (10% fetal calf serum) and culture overnight in a 500 ml bottle with a gently rotating stir bar.
2. Dilute culture to 500 ml with culture media in the morning. At the end of the day add Colcemid to 0.1 µg/ml (0.5 ml of a 100 µg/ml stock).

## 11: Isolation of mitotic organelles

3. Prepare two 20 ml 15–60% sucrose gradients containing 0.1% digitonin, 0.1% Ammynox Lo, 1 µg/ml aprotinin, 0.1 mM PMSF underlayered with 10 ml of 80% sucrose plus additions. Use two Sorvall 50 ml tubes.

### B. Harvest

1. Collect cells (800 $g$, 5 min, at room temp. in a Beckman JA10 rotor).
2. Resuspend in 50 ml media and transfer to Corning 50 ml tube. Spin for 3 min at setting 50 in a Clay Adams Dynac centrifuge at room temp., or calibrate your own tabletop clinical centrifuge to just pellet the cells (approx. 800 $g$).
3. Gently resuspend cells in 10 ml of room temp. swelling buffer plus aprotinin/PMSF. Add another 45 ml of swelling buffer and let sit for 5 min.
4. Place 1 drop of swollen cells on a slide, add 1 µl of Hoechst 33258. Examine cells under phase/fluorescence to check that they are swollen (membrane raised well away from chromatin mass) and determine the mitotic index. The lower the percentage of mitotic cells, the poorer the preparation will be. Use cells in which the mitotic index is at least 50%.
5. Pellet cells as in Step 2.

### C. Lysis

1. Aspirate supernatant and add 100 µl of RNase A to each pellet.
2. *Rapidly resuspend* cells in 10 ml ice-cold buffer A plus 0.1% digitonin and *quickly* transfer the cells to the 15 ml Dounce homogenizer. Dounce for 10 strokes.

   *Important*: The yield of mitotic chromosomes will drop radically if more than 30 sec is allowed to elapse between the aspiration of the swelling buffer to the first dounce stroke. Once the cells are permeabilized the chromosomes will begin to clump irreversibly together, therefore the cells must be quickly disrupted and cellular contents dispersed by homogenization.

3. Spin homogenate at 800 $g$, 4 °C for 5 min in the Beckman TJ6 centrifuge to remove debris, unlysed cells, and chromosome clusters.

### D. Sucrose gradient

1. Layer supernatant from Step C3 over the sucrose gradients prepared in Step A3.
2. Spin at 1000 $g$ for 5 min. After 5 min and without stopping reset the speed to 2000 $g$ and spin for 30 min in the Beckman TJ6 centrifuge.

**Protocol 6.** *Continued*

3. Remove chromosome band (approx. 50% sucrose) from the bottom and the pellet of chromosomes adhering to the wall of the tube.
4. Bring the volume of the saved chromosomes up to 10 ml with 1 × Percoll buffer plus 0.1% Ammynox Lo.

E. *Percoll gradient*

1. Make up Percoll *solution*. Add Aprotinin and PMSF. Place on ice.
2. Add 0.5 ml polyamines (see Step E1) to each tube of 10 ml sucrose/chromosomes. Mix.
3. Add 10 ml Percoll *solution* to each tube. Homogenize each 20 ml aliquot 10 times in the 40 ml Dounce homogenizer.
4. Add 14 ml more Percoll *solution* to each tube and mix.
5. Spin at 30 000 $g$ in a Beckman J21 centrifuge (JA21 rotor) at 4 °C for 30 min. Chromosomes contaminated with cytoskeletal material will float on the top of the gradient. Purified chromosomes can be found in a band at the bottom of the gradient.
6. Dilute and pellet chromosomes in 1 × Percoll buffer to wash away the Percoll *solution*. Pellet the chromosomes at 2600 r.p.m. in the Beckman J21 centrifuge for 15 min.
7. Carefully resuspend pellet in 5 mM Tris–HCL (pH 7.4), 2 mM KCl, 0.4 mM spermidine plus aprotinin. Add glycerol to 50% to freeze the chromosomes if desired.

# Acknowledgements

I would like to acknowledge the helpful advice and encouragement, over the last several years, of R. Kuriyama and W. C. Earnshaw. This work was supported by the Bank of America–Giannini Foundation Fellowship.

# References

1. Mazia, D. and Dan, K. (1952). *Proc. Natl Acad. Sci. USA*, **38**, 826.
2. Kuriyama, R., Keryer, G., and Borisy, G. G. (1984). *J. Cell Science*, **66**, 265.
3. Kuriyama, R. (1986). In *Methods in enzymology*, Vol. 134 (ed. R. B. Vallee), pp. 190–199.
4. Lewis, C. D. and Laemmli, U. K. (1970). *Cell*, **29**, 171.
5. Mitchison, T. J. and Kirschner, M. W. (1985). *J. Cell Biol.*, **101**, 755.
6. Mitchison, T. J. and Kirschner, M. W. (1986). In *Methods in enzymology*, Vol. 134 (ed. R. B. Vallee), pp. 261–268.

7. Kuriyama, R. (1989). *Cell Motil. Cyt.*, **12**, 90.
8. Wordeman, L., McDonald, K. L., and Cande, W. Z. (1987). *Cell*, **50**, 535.
9. Masuda, H., McDonald, K. L., and Cande, W. Z. (1988). *J. Cell Biol.*, **107**, 623.
10. Hogan, C. and Cande, W. Z. (1900). *Cell Motil. Cyt.*, **16**, 99.
11. Mitchison, T. J. and Kischner, M. W. (1985). *J. Cell Biol.*, **101**, 766.
12. Hyman, A. A. and Mitchison, T. J. (1991). *Nature*, **351**, 206.
13. Belmont, L. D., Hyman, A. A., Sawin, K. E., and Mitchison, T. J. (1990). *Cell*, **62**, 579.
14. Stearns, T., Evans, L., and Kirschner, M. (1991). *Cell*, **65**, 825.
15. Moudjou, M., Paintrand, M., Vigues, B., and Bornens, M. (1991). *J. Cell Biol.*, **115**, 129.
16. Peterson, D. F., Anderson, E. C., and Tobey, R. A. (1968). In *Methods in cell physiology* (ed. D. M. Prescott), pp. 347–370. Academic Press, New York.
17. Kuriyama, R. and Borisy, G. G. (1981). *J. Cell Biol.*, **91**, 814.
18. Stubblefield, E. (1968). In *Methods in cell physiology* (ed. D. M. Prescott), pp. 25–43. Academic Press, New York.
19. Ley, K. D. and Tobey, R. A. (1970). *J. Cell Biol.*, **47**, 453.
20. Mitchison, T. J. (1989). *J. Cell Biol.*, **109**, 637.
21. Hayden, J. H., Bowser, S. S., and Rieder, C. L. (1990). *J. Cell Biol.*, **111**, 1039.
22. Sheldon, E. and Wadsworth, P. (1990). *J. Cell Sci.*, **97**, 273.
23. Brady, R. C., Schilbler, M. J., and Cabral, F. (1986). In *Methods in enzymology*, Vol. 134 (ed. R. B. Vallee), pp. 217–225.
24. Mullins, J. M. and McIntosh, J. R. (1982). *J. Cell Biol.*, **94**, 654.
25. Kane, R. E. (1965). *J. Cell Biol.*, **25**, 137.
26. Forer, A. and Zimmerman, A. M. (1974). *J. Cell Science*, **16**, 481.
27. Sakai, H. and Kuriyama, R. (1974). *Dev. Growth and Diff.*, **16**, 123.
28. Kuriyama, R. (1982). *Cell Struct. and Funct.*, **7**, 307.
29. Hyams, J. S. and King, S. M. (1986). In *Methods in enzymology*, Vol. 134, (ed. R. B. Vallee), p. 225.
30. Masuda, H., Hirano, T., Yanagida, M., and Cande, W. Z. (1990). *J. Cell Biol.*, **110**, 417.
31. Sawin, K. E. and Mitchison, T. J. (1991). *J. Cell Biol.*, **112**, 941.
32. Mitchison, T. J. and Kirschner, M. (1984). *Nature*, **312**, 232.
33. Wordeman, L., Steurer, E. R., Sheetz, M. P., and Mitchison, T. J. (1991). *J. Cell Biol.*, **114**, 285.
34. Earnshaw, W. C. and Laemmli, U. K. (1983). *J. Cell Biol.*, **96**, 84.
35. Gasser, S. M. and Laemmli, U. K. (1987). *Exp. Cell Res.*, **173**, 85.

# 12

# *Xenopus* egg extracts as a system for studying mitosis

MARIE-ANNE FÉLIX, PAUL R. CLARKE, JULIA COLEMAN,
FULVIA VERDE, and ERIC KARSENTI

## 1. Introduction

The ultimate goal (or dream) for a biochemist is to isolate the individual components of a complex system, for example all the enzymes involved in cell-cycle control, to mix them back together, and finally show that the system functions again as it did in the context of the whole cell. This is a very difficult task, however, and no highly complex process has been brought to this point yet. A simpler way to analyse complex systems is to use functional complementation assays. In organisms like *Drosophila* or yeast, this is feasible using a genetic approach. A single function is eliminated by mutation of a gene and then restored by complementation with the corresponding, fully functional gene. This approach has limitations however, as it is not always possible to locate exactly where and when, in a cascade of events, a gene product function is required. Complex *in vitro* systems that closely mimic the *in vivo* situation are required to fill in the gaps between genetic approaches and the reconstitution of functional processes from purified components *in vitro*. Frog egg extracts belong to this category of systems.

The philosophy behind the development of extracts from eggs or early embryos as tools to study the cell cycle has its roots in observations made *in vivo* at the beginning of this century. Cell biologists studying fertilization in frog and sea urchin eggs, noticed that multiple sperm heads introduced into maturing oocytes or into eggs would undergo cell-cycle changes governed by the cytoplasm (1–3). Later, it was found that somatic cell nuclei would respond similarly to the cytoplasmic state of these eggs (4, 5). More recently, these observations have been complemented by the demonstration that all the building blocks required for the assembly of nuclei and spindles around injected naked DNA are present and stored in the form of proteins in the cytoplasm of *Xenopus* eggs (6, 7). Finally, the striking demonstration that the cell cycle continues in eggs in the absence of DNA replication or spindle assembly (8), supported the idea that a relatively simple cytoplasmic clock could drive the early embryonic cell cycle in frogs. These observations also meant that the egg cytoplasm contained all the morphogenetic information required to assemble complex cellular structures like nuclei or spindles, the only missing information being the template provided by the DNA. Most interestingly, whether a nucleus or a spindle would assemble seemed to depend primarly on the 'state' of the cytoplasm.

In this context, it was not so unreasonable to try to reproduce such complicated events *in vitro* by making highly concentrated egg lysates containing all essential cytoplasmic components. Classical methods of cell disruption are harsh. They often cause lysosome disruption and as a consequence, massive release of proteases. Moreover, the cells are usually disrupted under rather dilute cytoplasmic conditions. This is a problem because this upsets the balance between the activity of many enzymes, in particular between key regulatory kinases and phosphatases. This results in discrepancies between *in vivo* and *in vitro* protein phosphorylation patterns and does not leave much hope of faithfully reproducing complex cellular processes like spindle assembly. The method first used successfully by Lokha and Masui (9) to produce egg extracts in which nuclear reconstitution and spindle assembly could take place, involved simple crushing of the eggs at low speed in a minimum volume of buffer. This method works for frog eggs because of their size and the presence of heavy yolk granules that disrupt the plasma membrane upon centrifugation. The method avoids dilution and destruction of vesicles including the lysosomes, and has been used subsequently by many groups with great success. When working with such extracts, it is often tempting to say that the extract is 'alive' when it 'works' or 'dead' when it does not work. This natural reaction of the experimenter means that the exquisite balance between the various components present in the egg which characterizes its living state is preserved, or not preserved, during the preparation of the extract. An extensive description of the methods used to prepare frog egg extracts for cell-cycle assays, chromatin assembly, and DNA replication has already been published and these papers include reviews on past work (10–13; see also Chapters 8 and 13). In this article we recall the essential features involved in the preparation of egg extracts with our own modifications. We report on the preparation of extracts arrested in different cell-cycle states and on their use in conjunction with bacterially expressed cell-cycle protein regulators. We also explain how to use such extracts to study microtubule dynamics during the cell cycle using a simple fluorescence method.

## 2. Preparation of frog egg extracts

First we will present the different types of extracts that can be prepared (Section 2.1), before describing how to obtain the eggs (Section 2.2), and how to prepare the extracts (Section 2.3). A simple assay for the cell-cycle state of the extracts, the measure of cdc2 histone H1 kinase activity, will then be described (Section 2.4).

## 2.1 The different types of extracts

Extracts can differ in:

- The time when the eggs are collected; hence the cell-cycle state of the extracts and their dynamic properties upon incubation at room temperature.
- The speed of the centrifugation used to prepare the extracts; this can affect both the events that are reproduced in the extracts and the regulation of cdc2 kinase activity.

*Xenopus* eggs are laid arrested in second meiotic metaphase. Under natural conditions, they are fertilized immediately after being laid. Penetration of the sperm induces a calcium wave that inactivates a cytostatic factor thereby releasing the metaphase block. After resumption of the second meiotic division, the eggs enter interphase after 10 minutes, the first mitotic metaphase occurs at 70 minutes and cleavage ensues at 90 minutes (at 20 °C, *Figure 1 a*). This is followed by a series of rapid cell cycles alternating between S and M phases every 30 minutes. Large amounts of eggs arrested in second meiotic metaphase can be obtained by stimulating the frogs to lay by hormone injections. 'Mitotic extracts' can be prepared from such eggs. It is not as easy to obtain large amounts of eggs at different times in the first cell cycle. Synchronous fertilization is not easy to achieve on a large scale. Instead, one can 'activate' the eggs by an electric shock which induces a calcium wave similar to that produced by fertilization. In this case, the eggs enter the first embryonic cell cycle, surface contraction waves can be observed as well as periodic activation of MPF (maturation promotion factor) or histone H1 kinase, but cleavages do not occur because the sperm centriole is required for this latter event (*Figure 1 b, c*). Low speed extracts (centrifuged at 10 000 g), prepared from activated eggs will undergo several cell cycles upon incubation at room temperature (13). As *in vivo*, protein synthesis should occur in these extracts to obtain cell-cycle progression, since it is the continuous synthesis of cyclin protein and its periodic degradation that drives the early embryonic cell cycle (14). If activated eggs are pre-incubated in cycloheximide and kept in cycloheximide during the electric shock, protein synthesis is inhibited, cyclin is degraded during the activation process, the eggs exit from metaphase and arrest in a stable interphase state. Extracts prepared from such eggs are called *interphase extracts*. They are devoid of endogenous cyclins and have a low cdc2 kinase activity.

It is also possible to prepare extracts from activated eggs without blocking the cell cycle with cycloheximide and which represent different phases of the first cell cycle. In this case, the extracts are centrifuged at a higher speed

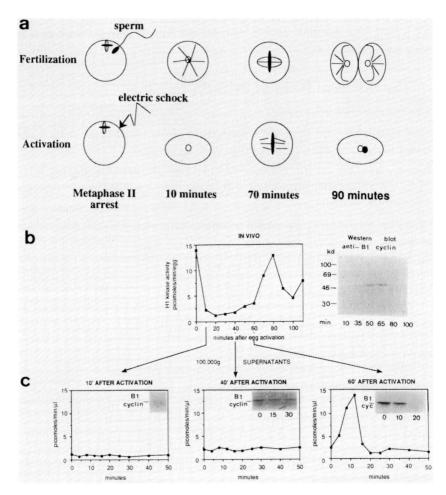

**Figure 1.** The first embryonic *Xenopus* egg cell cycle and when to make extracts. (a) DNA replication occurs between 10 and 20 min following fertilization (interphase period with a pronucleus and a microtubule aster). First mitotic metaphase occurs at 70 min and first cleavage at 90 min. In eggs activated by an electric shock, a similar situation is found, but the spindle is irregular and cleavage does not take place because the sperm centriole is absent. However, surface contraction waves occur in phase with the normal cell cycle (symbolized here by a change in egg height. Similar waves occur in fertilized eggs but this is not shown for simplicity). (b) The graph shows the level of histone H1 kinase activity measured in single eggs sampled at different times during the first cell cycle after activation by an electric shock. At zero time, the eggs are arrested in the second metaphase of meiosis and, therefore, have high histone H1 levels. At the time of first mitotic metaphase, 70–80 min the kinase level is high again. The level of cyclin B1 during this period is shown on the immunoblot made from a gel loaded with one egg equivalent at corresponding times during the first cell cycle. (c) This shows the kinetics of histone H1 kinase activity in 100 000 *g* extracts prepared from eggs sampled at the indicated times during the first cell cycle (according to *Protocols 2–4* and *Figure 2*) and

(100 000 g). The supernatant is devoid of polyribosomes and therefore protein synthesis does not occur in such extracts. They contain the amount of cyclins present in the eggs at the time they are sampled for extract preparation (see *Figure 1 c*). A threshold amount of cyclins is required for cyclin B–cdc2 kinase activation (15, 16), and a threshold level of cdc2-kinase activity is also required to activate the cyclin degradation pathway and the subsequent inactivation of cdc2 kinase (17). In both cases, once the threshold is reached, there is lag of 10–15 minutes before the initiation of cdc2 kinase activation or cyclin degradation, respectively. This system of thresholds and time-lags forms the basis for the cell-cycle period in these early embryos.

*In vivo*, the threshold level of cyclin B required for cdc2 kinase activation is reached 50 minutes after fertilization or artificial activation. Extracts prepared before this time (for instance at 40 minutes) do contain some A and B-type cyclins (*Figure 1 c*). There is some cyclin A-associated cdc2 kinase activity, but the cyclin B–cdc2 complex does not activate in 100 000 g extracts upon incubation at room temperature because there is no protein synthesis and the threshold level of cyclin is never reached. This state is stable because there is not enough cdc2 kinase activity to induce cyclin degradation. We call such stable extracts $G_2$-*prophase extracts* because they are prepared from eggs which have passed the period of DNA replication, but not yet entered mitosis, and contain stably inactive cyclin B–cdc2 complexes. In extracts prepared from eggs sampled 60 minutes after activation, the cyclin level is above the threshold required for cdc2 kinase activation, but because of the lag, cdc2 kinase is not yet active (*Figure 1 c*). However, upon incubation at room temperature, the cdc2 kinase activates spontaneously in 10 minutes (*Figure 1 c*). Since the second threshold of cdc2 kinase activity is automatically reached in this case, cyclin degradation follows the activation of cdc2 kinase 10 minutes later and cdc2 kinase is inactivated. We call such extracts *one-cycle extracts*. They are easier to prepare in a reproducible fashion than genuine cycling extracts and they can be frozen.

An alternative way of controlling the cell-cycle state of the extracts is to prepare interphase extracts devoid of cyclins and to add exogenous bacterially-produced cyclins to them. These cyclins can activate the endogenous cdc2 (see Section 3).

Finally one may wish to obtain '*soluble*' *extracts* to examine the role of membranes or particulate material in structural events, such as nuclear and spindle assembly or in the regulation of cdc2 kinase. Because the extracts are extremely concentrated (40–50 mg/ml of proteins), they are very viscous.

---

incubated at room temperature. In extracts prepared 10 min after activation, no cyclin is detected and the kinase never activates in the extract. In 40 min extracts, some cyclin can be detected, but the H1 kinase never activates. In 60 min extracts maximum cyclin level is reached, the H1 kinase activity cycles once and the endogenous cyclin is degraded while the kinase becomes inactivated.

Centrifugation at 100 000 $g$ is insufficient to sediment all membranes. A further centrifugation in an airfuge or a tabletop ultracentrifuge is required. Note that this very high-speed centrifugation results in the loss of mitotic cdc2 kinase activity, unless a special solubilizing buffer is used.

## 2.2 The eggs

*X. laevis* female frogs are induced to lay by consecutive injections of two gonadotropic hormones. The first one, pregnant mare serum gonadotropin (PMSG), accelerates the growth of oocytes. The second one, human chorionic gonadotropin (HCG) provokes oocyte maturation and egg laying.

We routinely inject PMSG on Fridays for the following week, and HCG at the end of the afternoon for the following day. There should be at least 3 days between PMSG and HCG injections. The frogs which have been used are then put together in a tank labelled with the month of injection. We keep them at least 3 months before injecting them again. Frog maintenance is described in (ref. 18).

---

**Protocol 1.** Hormonal stimulation of *Xenopus*

*Materials*

- Pregnant mare serum gonadotropin (PMSG, Intervet)
- Human chorionic gonadotropin (HCG, Sigma cat. no. CG-10)
- 1 ml syringes, 27-gauge needles
- 0.1 M NaCl (4 M stock)

*Method*

A. *PMSG injection*

1. Dilute the PMSG to 200 U/ml in sterile water. It can be frozen and stored at −20 °C.
2. Inject 0.5 ml (100 U) per frog. The frogs are injected subcutaneously in the back near the border of the leg. If the needle has punctured the rubber of the hormone flask it is best to discard it and use another one for the injection (it is likely to be clogged by some rubber or else dulled).
3. Keep the frogs for 3–10 days. Do not feed them (this will keep the water in which they lay clean).

B. *HCG injection*

1. Dilute the HCG to 2000 U/ml in sterile water. It can be stored at 4 °C for up to one week.

## 12: Xenopus egg extracts for studying mitosis

2. Inject 0.5 ml (1000 U) per frog as above.
3. Keep the frogs in 0.1 M NaCl overnight (the eggs activate spontaneously in pure water). The temperature should not exceed 25 °C, 18–20 °C is best for the frogs and for the eggs. Egg laying is slower at lower temperatures.

Each frog can lay several thousand eggs. Batch quality is judged from the aspect of the eggs; they should all show two well-distinguishable poles, a dark animal pole and a clear vegetal one, separated by a thin white zone; they should activate upon an electric shock without bursting or becoming white upon incubation. Batches of eggs containing more than a few per cent of 'bad-looking' or spontaneously activating eggs are discarded. In any case, such eggs are removed from the batch; this yields extracts of much better quality. Avoid mixing eggs from different frogs since you may contaminate a good batch with a less good one.

It is the experience of many laboratories that summer is a difficult period to prepare egg extracts: the frogs lay few eggs of generally poor quality. Prepare extracts before June if possible and store them for the summer period!

### 2.3 Preparation of the extracts

The preparation of extracts from the eggs is quite straightforward: it involves removing the jelly which surrounds the eggs after they are laid and crushing them by low-speed centrifugation in a minimal amount of buffer. The jelly is held together by disulfide bonds which can be broken by reducing agents such as cysteine.

The eggs are packed in a minimal amount of buffer to obtain extracts as concentrated as possible (40–50 mg of proteins per ml of extract). The buffer represents about 25% of the final volume of the extract. It is possible to remove more buffer by replacing it by mineral oil (Versilube F-50, General Electric) layered above the extract during the 200 $g$ spin: the buffer being less dense than the oil, it floats over it. The oil then floats over the extract during the 10 000 $g$ crushing/centrifugation step (13). We have never found it necessary to use this oil, although it may yield extracts with a higher protein synthesis rate when needed. The acetate buffer contains cytochalasin D to avoid clumping of actin during incubation of the extract. This is an obvious restriction to the use of these extracts to study actin-dependent processes. To prepare mitotic extracts from non-activated eggs, it is necessary to add EGTA to the buffer to chelate the $Ca^{2+}$ ions released from intracellular stores during extract preparation. Free calcium ions induce inactivation of the cytostatic factor and as a consequence loss of the histone H1 kinase activity in the extract. We use EGTA in the preparation of interphase extracts as well as when we need to compare them with mitotic extracts. Interphase extracts can be prepared in the absence of EGTA. We have never found any difference between extracts containing or not containing EGTA concerning the regulation of microtubule dynamics or of cdc2 kinase.

### 2.3.1 Mitotic extracts

Mitotic extracts are prepared from eggs naturally arrested in metaphase of the second meiotic division. Therefore, the state of the cytoplasm corresponds to a 'meiotic' state rather than to a true mitotic state. It is possible to release the metaphase II arrest in extracts by adding calcium. If one wishes to do this, it is preferable to add as little EGTA as possible to the acetate buffer in order to avoid extensive precipitation of the chelation complexes upon calcium addition. The minimal amount of EGTA and calcium required should be determined experimentally (15).

---

**Protocol 2.** Preparation of mitotic egg extracts

*Materials*

- MMR/4: 25 mM NaCl, 0.4 mM KCl, 0.25 mM $MgSO_4$, 0.5 mM $CaCl_2$, 1.25 mM Hepes, 25 μM EDTA (pH 7.2) (500 ml). A 10 × MMR stock is kept at room temperature and diluted 40 times

- L-cysteine (Sigma cat. no. C 7880): 200 ml of 2.5% cysteine in water, (pH 7.8 neutralized with 6 M NaOH). It should be prepared less than one hour before use

- 50 ml plastic tubes

- Plastic Pasteur pipette or glass Pasteur pipette with a cut tip (about 3 mm diameter) that has been fire polished

- Acetate buffer: 100 mM potassium acetate, 2.5 mM magnesium acetate, 60 mM EGTA, 1 mM DTT, 50 μg/ml cytochalasin D (Sigma) pH 7.2. The buffer without DTT and cytochalasin D can be stored at 4 °C for some days or in 50 ml aliquots at −20 °C. Dilute the cytochalasin D stock to 1 mg/ml in DMSO; the DTT stock to 100 mM in water; store in aliquots at −20 °C and add to the acetate buffer before use. Fill the centrifugation tubes used in step B with ice-cold acetate buffer before starting to handle the eggs. Keep some more acetate buffer on ice for washing

- Centrifugation tubes and centrifuges: we use routinely SW-50.1 (5 ml tubes) or SW-60 (4 ml tubes) rotors of Beckman ultracentrifuges. Any swinging-out rotor corresponding to the desired volume would do, for instance a TLS-55 for a tabletop ultracentrifuge (Beckman) or a HB-4 rotor with appropriate adaptors for a Sorvall centrifuge (for the low-speed centrifugation only). Precool the rotors.

- 5 ml syringes and needles that can pierce the centrifugation tubes

## 12: Xenopus egg extracts for studying mitosis

- Stock solutions for the ATP regenerating system: 50 mM ATP (pH 7), 500 mM creatine phosphate (Boehringer–Mannheim, cat. no. 621714) in water, 4 mg/ml creatine kinase (Boehringer–Mannheim, cat. no. 126969) in 50% glycerol in water. Store aliquots at −20 °C, 1 µl of each will be added to 50 µl of extract
- Liquid nitrogen

*Method*

A. *Dejellying the eggs*

1. Wash the eggs twice in MMR/4 in a 50 ml plastic tube. Allow the eggs to settle to the bottom of the tube between each wash. There should be no more than 30–35 ml of eggs per tube otherwise the action of cysteine will be very slow. The yield is about 8–10 ml of packed eggs after the jelly is removed. Small numbers of eggs can be washed and dejellied in Petri dishes (avoid them sticking to the dish during the dejellying by gentle swirling).

2. Rinse twice with cysteine. Fill up the tube completely the second time and close it. Avoid air bubbles inside the tube. Contact of the eggs with air causes activation.

3. Incubate in cysteine for a few minutes. Keep the tube horizontal in your hands and turn it continuously (about one half turn every second). Stop the incubation when you see that the eggs tend to cluster: this is the sign that the jelly is removed. It takes about 2–3 min. From now on you should work quickly until the centrifugation step to avoid egg activation.

4. Wash the eggs three times with MMR/4. The eggs now pack rapidly and tightly at the bottom of the tube, their heavier vegetal poles point downwards.

B. *Egg crushing and centrifugation*

1. Wash the eggs with one volume ice-cold acetate buffer.

2. Pick up the eggs with a plastic Pasteur pipette (or a glass Pasteur pipette with a cut tip) and transfer them to a centrifuge tube filled with ice-cold acetate buffer. Be quick but do not lyse the eggs. Allow the eggs to settle and remove as much buffer as possible with a 1 ml Gilson pipette. *Keep everything at 4 °C from now on.*

3. Centrifuge at 10 000 *g* for 10 min at 4 °C (9000 r.p.m. in the SW-50.1 rotor, 8000 r.p.m. in the SW-60 rotor). Use maximal acceleration rate to crush the eggs rapidly. The yolk sediments (about 40% of the volume). The dark pigments are found within and immediately above the yolk. If all pigment is sedimented, the supernatant is yellow, otherwise it is

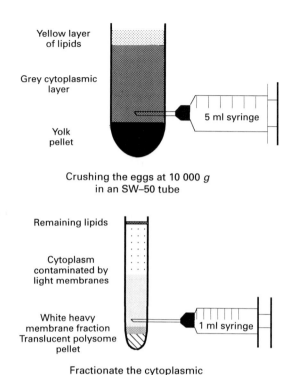

**Figure 2.** Preparation of concentrated egg extracts. Eggs are packed in an SW-50 tube and centrifuged at 10 000 $g$ for 10 min according to *Protocols 2–4*. The crushed eggs give rise to 3 fractions (shown in the top tube). The grey layer is removed through the wall of the tube with a syringe and spun again at 100 000 $g$ in a SW-50.1 tube (bottom tube). This second spin gives rise to separation of the cytoplasm in several layers. The shaded and dotted cytoplasm contaminated with membranes are aspirated through the tube wall as shown.

**Protocol 2.** *Continued*

more or less grey (*Figure 2*). The residual pigments can be eliminated by a second spin if desired (this is not necessary if the extracts are then centrifuged at a higher speed). Finally a yellow layer of lipids is found at the top.

4. Pierce the tube with a large needle above the pigment and yolk layers. Remove the viscous and turbid supernatant (without the bright yellow lipids at the top). Avoid making bubbles in the extract. If the needle is clogged after it has been used to pierce the tube, rapidly change it before taking out the extract.

*12: Xenopus egg extracts for studying mitosis*

5. Add the ATP regenerating system: 1 mM ATP, 10 mM creatine phosphate, 80 µg/ml creatine kinase (1/50th volume of each stock solution for one volume of extract).
6. If desired, centrifuge at 100 000 $g$ for 60 min at 4 °C (30 000 r.p.m. in an SW-50.1 rotor, 28 000 r.p.m. in an SW-60 rotor). A golden translucent pellet of polyribosomes and glycogen is found at the bottom of the tube. Fast-sedimenting membranous compartments sediment above (most likely Golgi stacks and lysosomes). The supernatant is still yellowish and contains membranes and other particulate material such as monosomes. Pierce the tube with a needle above the heavy membranes and take off the supernatant above this fraction but not any yellow lipids (*Figure 2*).
7. Freeze the extract in 50–100 µl aliquots in liquid nitrogen. The extracts are stable for months in liquid nitrogen. Keeping them at −80 °C alters their properties for microtubule dynamics studies. Do not freeze them a second time. Freezing of low-speed extracts is more problematic than freezing of 100 000 $g$ supernatants, probably because the latter contain less lysosomes. Moreover the protein synthesis rate is much lower after freezing. The extracts can be kept for a few hours on ice.

### 2.3.2 Interphase extracts

To prepare interphase extracts the eggs must be activated. To obtain extracts which are truly arrested in interphase and do not contain any endogenous cyclins, the eggs are incubated in cycloheximide before and during activation.

**Protocol 3.** Preparation of interphase egg extracts

*Materials*
- Same material as for mitotic extracts (*Protocol 2*)
- Activation chamber: we use a homemade activation chamber. For standard extract preparation, a 10 cm × 10 cm square chamber, 6 cm high is convenient. Larger ones can be manufactured to scale-up the extract preparation. The bottom of the chamber and the lid (which is a 9.8 cm square that can enter the chamber) form the two electrodes. They are made of stainless steel, the chamber of Plexiglass. The bottom of the chamber is covered with 1 cm of 1% agarose in MMR/4. The eggs rest on the agarose. The lid lays above the eggs on 0.5 cm high plastic feet fixed at the four corners below it. There should be enough MMR/4 buffer to cover the eggs and the lid (about 200 ml)

**Protocol 3.** *Continued*

- Power supply: able to deliver 12 V AC to the chamber (not every power supply can do it)
- Agarose 1% in MMR/4 (100 ml): the agarose is made fresh and the chamber cleaned before each use
- Cycloheximide (Sigma cat. no. C 6255): 200 µg/ml in MMR/4 (200 ml)

*Method*

A. *Dejellying the eggs* (see above *Protocol 2, A*).

B. *Egg activation by an electric shock*

1. Gently pour the eggs from the 50 ml tube into the activation chamber coated with agarose and filled with MMR/4 and cycloheximide. Close the lid of the chamber. The eggs should stay below the lid. Make sure that the buffer covers the lid.
2. Activate the eggs with a 2 sec 12 V pulse.
3. Remove the lid and observe the eggs for 4–6 min. If activation is successful the pigment at the animal pole should contract: the black part then covers less than half of the egg surface. The extent of contraction upon a good activation is variable from batch to batch and may be easier to see under a dissecting microscope. Moreover, consecutive to the elevation of the fertilization membrane, the egg becomes free to rotate according to gravity: all eggs turn their dark animal pole up. Remove damaged eggs or eggs which have turned totally white (or grey). If there are too many of them (more than 5%), it is best to discard the whole batch.
4. Incubate the eggs for 60–90 min at 20 °C in MMR/4 plus cycloheximide. Discard eggs which have turned bad during this time.

C. *Egg crushing/centrifugation* (see above, *Protocol 2, B*).

### 2.3.3 Extracts prepared during the first embryonic cell cycle

The characteristics of these extracts have been presented above in Section 2.1. Basically, low-speed extracts from activated eggs (in the absence of cycloheximide) can display several cycles of cdc2 kinase activation and inactivation *in vitro* (2–3 cycles). We find this most reproducible when these low-speed extracts are prepared 60 minutes after egg activation, probably because some event occurring during the first long interphase is difficult to reproduce *in vitro*.

## 12: Xenopus egg extracts for studying mitosis

Supernatants (100 000 g) prepared at different times after egg activation contain different amounts of cyclins. In extracts prepared 60 minutes after egg activation, one cycle of cdc2 kinase activation and inactivation is very reproducibly observed during the incubation of the extract. These extracts can be easily frozen.

---

**Protocol 4.** Preparation of egg extracts during the first embryonic cell cycle

*Materials*

• Same material as for interphase extracts (*Protocol 3*)

*Method*

A. *Dejellying the eggs* (see above *Protocol 2, A*).

B. *Egg activation by an electric shock*

See above, *Protocol 3, B* with the following modifications for the incubation:

(a) Buffer: the eggs should be incubated in MMR/4 without cycloheximide. It is very important to carefully control the incubation temperature (20 °C), because the cell-cycle time is much faster at higher temperatures.

(b) Time: the incubation time can be chosen to yield different types of extracts (10, 40, 60, 80 minutes, *Figure 1*).

C. *Egg crushing/centrifugation* (see above, *Protocol 2, B*).

---

An example of such an experiment is shown in *Figure 1 c*. The level of cyclin B1 has been monitored during the experiment by immunoblotting. There is no cyclin in the extract prepared at 10 min and a subthreshold level of cyclin in the extract prepared at 40 min: the H1 kinase never activates in these extracts. In the extract prepared at 60 min, the threshold has been reached: the H1 kinase activates and the cyclin is degraded after H1 kinase activation. We have used prophase extracts (40 minutes) for studying the role of type 2A phosphatase in cyclin activation (19) and the regulation of microtubule nucleation by centrosomes (20), for example. They can also be used to study the regulation of cdc2 kinase by different cyclins at different times in the first cell cycle or to study cyclin degradation (17).

### 2.3.4 Soluble extracts

Because the extracts are so viscous, it is difficult to sediment small vesicles and particles. To obtain clear extracts devoid of membranes or particulate material, a further centrifugation must be performed after the 100 000 g centrifugation. This final centrifugation step is made in a small volume in an airfuge or a tabletop ultracentrifuge. After this centrifugation the supernatant is no longer yellow. An analysis of the supernatant by electron microscopy is useful for determining whether all vesicles or particulate components have been removed. The total protein concentration goes down to about 20 mg/ml during this step.

It is not possible to prepare soluble extracts with a high mitotic cdc2 kinase activity from metaphase II eggs in the acetate buffer described here. Metaphase II extracts lose their cdc2 H1 kinase activity upon high-speed centrifugation in the airfuge or the TL100 tabletop centrifuge unless they are prepared in the MPF extraction buffer that contains high concentrations of $\beta$-glycerophosphate (21). The cyclin–cdc2 complexes themselves are associated to subcellular structures that sediment during high-speed centrifugation. They can be solubilized by relatively mild salt treatments (22). Care should, therefore, be taken as to the localization and level of cdc2 kinase activity during high-speed centrifugation steps.

---

**Protocol 5.** Preparation of soluble extracts

*Materials*

- Airfuge or tabletop ultracentrifuge with a TL-100 rotor (Beckman) with the corresponding tubes (maximum 200 μl per tube)

*Method*

1. Prepare a 100 000 g supernatant by centrifuging the 10 000 g lysate (*Protocol 2 B*, 3) at 30 000 r.p.m. (100 000 g) in an SW-50.1 rotor and collect it as described in *Protocol 2 B*, 4–6.
2. Centrifuge the 100 000 g supernatants at 70 000 r.p.m. (200 000 g) for 30 min in the TL-100 rotor, or at 30 p.s.i. for 30 min in the airfuge, at 4 °C.
3. Take the clear supernatant and freeze in aliquots in liquid nitrogen. The pellet can be resuspended in the acetate buffer, washed, resuspended again, and frozen.

## 2.4 Monitoring cdc2 H1 kinase activity as a cell-cycle marker in egg extracts

### 2.4.1 Rationale for the choice of the assay

It is recommended that each extract is tested before use and its state monitored during the experiments. A universal way of monitoring the interphase versus mitotic state of a cell or of an extract is to measure its cdc2 kinase activity. This assay can be used on other types of extracts or on cultured cells. The mitotic cdc2 kinase activity is assayed on histone H1. It can be shown that the mitotic level of H1 kinase activity is due to the activity of cdc2 by checking that the activity binds to a specific protein, p13$^{suc1}$. Since there are different types of cyclins acting at different times in the cell cycle, especially during mitosis (23), it can be useful to complement the H1 kinase assay by a specific immunoprecipitation or a Western blot for the various A and B-type cyclins.

An alternative cell-cycle test is to add demembranated frog sperm to the extract. In an interphase extract the chromatin decondenses and a nucleus is formed, in a mitotic extract the chromatin is reorganized in condensed chromosomes and spindles form (24, 25). Although this assay is quicker we prefer the cdc2 H1 kinase assay for several reasons:

(a) It assays a well-defined enzymatic activity, not a very complex process such as nuclear formation with its many structures and enzymes. For instance, nuclear formation cannot occur in the 100 000 $g$ supernatants, not because the extracts are in a mitotic state, but because some membrane components are lacking.

(b) The assay is quantitative and allows the measurement of intermediary states at precise timepoints.

(c) Cdc2 kinase is not a secondary enzyme in the cell cycle, but plays a key role in inducing cell-cycle events.

### 2.4.2 Principle of the enzymatic assay

The enzymatic activity assayed is the initial rate of the reaction when the substrates are not limiting. The number of moles of reaction product formed after a certain time (number of moles of phosphate transferred to H1 histones) is a measure of the initial reaction rate if the reaction is still linear at this time. The assay described below fulfils these conditions: neither ATP nor H1 histones are limiting, and the reaction is still linear after 15 minutes (as long as the activity is not very much higher than the normal activity in a mitotic extract: at least until 50 picomoles/min/µl of extract).

In practice, the extract is first diluted into the H1 kinase assay buffer which contains $\beta$-glycerophosphate and stabilizes cdc2 H1 kinase activity. The 20-fold dilution is also required to lower the enzyme activity in the assay (at the same time it dilutes the amount of ATP contributed by the extract). An aliquot of the extract can be taken at any time during an incubation, diluted in H1 kinase assay buffer (1 μl extract in 20 μl assay buffer), and kept frozen at −20 °C until used for the kinase assay.

The reaction is then performed by mixing this diluted extract with a mix containing exogenous calf thymus H1 histones, [γ-$^{32}$P]ATP, and unlabelled ATP. Unlabelled ATP is required so that the ATP substrate for the enzyme is not limiting. The concentration of radioactive ATP is negligible compared to the added unlabelled ATP. The radioactive terminal phosphate of the ATP is transferred to the histones during the reaction. The histones are separated from the ATP by making use of their affinity for phosphocellulose. Alternatively, the reaction product can be run on SDS–PAGE to check that the only labelled polypeptides are the histones. The assay described in *Protocol 6* is very convenient and we routinely use it. It allows one to assay very large numbers of samples, rapidly and accurately.

**Protocol 6.** Assay of histone H1 kinase activity

*Materials*

- γ[$^{32}$P]ATP (0.5–1 × 10$^6$ c.p.m./nmol); take all appropriate precautions when using the $^{32}$P-labelled radioactive material
- 4 mM ATP solution in water: dilute from a 100 mM stock, adjust pH to 7.0 using NaOH; store at −20 °C
- 20 mg/ml Histone H1 (Sigma) solution in water, store at −20 °C
- Extraction buffer (EB): 80 mM β-glycerophosphate (pH 7.3), 20 mM EGTA, 15 mM MgCl$_2$, store in 50 ml aliquots at −20 °C; just prior to use, add 1 mM dithiothreitol and protease inhibitors: 25 μg/ml aprotinin (Boehringer–Mannheim), 25 μg/ml leupeptin (Boehringer–Mannheim), 1 mM benzamidine (Sigma), 0.5 mM PMSF (Serva)
- 96 micro-well plate (Nunclon)
- Phosphocellulose P81 paper (Whatman)
- 0–20 μl pipette and tips
- Forceps and scissors
- Beaker with wash solution (150 mM H$_3$PO$_4$)
- Technical grade ethanol
- Scintillation counter and vials

*Method*

A. *Kinase reaction*

1. Prepare the assay mix of γ[$^{32}$P]ATP-histones at 4 °C (sufficient for 60

assays): 3 μl of γ[$^{32}$P]ATP (0.5–1 × 10$^6$ c.p.m./nmol), 45 μl of 4 mM ATP, 222 μl of EB, 30 μl of 20 mg/ml histone H1.

2. Pipette 5 μl of each sample to be assayed into labelled micro-well plate, including two blanks of EB at 4 °C.

3. Warm the micro-well plate to 20 °C. Start each reaction by adding 5 μl of assay mix to each sample, mix well by pipetting up and down several times. Start and stop each reaction at 15 sec intervals. Let each reaction continue for 15 min at 20 °C.

4. Stop the reaction by transferring 6 μl of the reaction mixture to phosphocellulose P81 paper precut into 2 cm strips with 5 squares for timepoints. Paper filters should be identified with pencil (no ink!) so they can be washed in the same beaker.

5. Allow mixture to adsorb for a few seconds then dip into a beaker containing (per 10 strips) 300 ml of 150 mM $H_3PO_4$.

6. Wash 3 × 15 min. in 150 mM $H_3PO_4$ at 20 °C.

7. Rinse briefly in ethanol and air-dry on paper towels.

8. Cut strips and transfer squares into scintillation vials. Add scintillation fluid and count.

9. Count two 5 μl samples of the assay mix adsorbed on to phosphocellulose paper to determine the ATP specific radioactivity.

B. *Calculations*

1. Average the c.p.m. values for the controls to get the background. Subtract this from the c.p.m. for each sample.

2. Subtract the control value from the c.p.m. values for the 5 μl mix and calculate the average.

   *NB* The ATP concentration in the assay is that due to 5 μl of assay mix (3000 pmol) plus that due to 5 μl of diluted extract (250 pmol, assuming 1 mM ATP in undiluted extract)

3. Kinase activity in pmol/min/μl of original extract:

$$= \text{c.p.m.}^* \times \frac{\text{pmol ATP/assay}}{\text{c.p.m.}^{**} \text{ in ATP}} \times \frac{\text{sample vol.}}{\text{counted vol.}} \times \frac{1}{\text{vol. orig. extr/assay}} \times \frac{1}{\text{assay time}}$$

$$= \text{c.p.m.}^* \times \frac{3250}{\text{c.p.m.}^{**} \text{ in ATP}} \times \frac{10 \text{ μl}}{6 \text{ μl}} \times \frac{1}{0.25 \text{ μl}} \times \frac{1}{15 \text{ min}}$$

$$= \text{c.p.m.}^* \times \frac{1444}{\text{c.p.m.}^{**} \text{ in ATP}}$$

c.p.m.* = counts in experimental samples. c.p.m.** = total c.p.m. as determined in *Step A* 9.

## 3. Assay of cyclin activity in egg extracts

All the protein components, other than the mitotic cyclins, that are required for the cell cycle are apparently present in *Xenopus* eggs. This means that egg extracts devoid of endogenous cyclins provide a useful system in which the cell cycle can be induced at will by adding exogenous cyclins. This allows changes in cellular processes, such as microtubule dynamics to be studied, as described in Section 5. These extracts are also useful for studying the cyclins themselves, the regulation of cyclin-dependent protein kinases and the molecular mechanism of the induction of mitosis. By combining molecular biology techniques with the use of egg extracts, the functions of cyclins which have never been purified by biochemical techniques can be studied (16, 26). Once a clone encoding a cyclin has been identified, the protein can be produced using a variety of expression systems. Mutant proteins can also be expressed to study the cyclin molecules themselves, for instance their susceptibility to proteolysis and the role of cyclin degradation in the cell cycle (15, 27, 28). We use the interphase extracts described above prepared as a post-ribosomal supernatant with the addition of the protein synthesis inhibitor cycloheximide to prevent the accumulation of cyclins (*Protocol 3*). We have used A and B type cyclins expressed in *Escherichia coli*, although cyclins translated *in vitro* in reticulocyte lysates or produced in the baculovirus insect cell-expression system have also been used. A-type cyclins, such as those from human cells, have been successfully produced as soluble and functional full-length proteins (29). A functional truncated sea urchin cyclin B, designated Δ90 because it lacks the N-terminal 90 residues of the full-length protein, has been successfully renatured from inclusion bodies (27). We have used these two cyclins to study the effects of A- and B-type cyclins on cellular processes, such as microtubule dynamics and nucleation by centrosomes (20, 23, 30). We have also studied the mechanisms of kinase activation by the two cyclins in egg extracts (26). However, problems have been encountered with the expression of some other cyclins in *E. coli*. Full-length B-type cyclins in particular are often expressed as insoluble components of inclusion bodies.

In Section 3.2, we describe how to produce and purify a B-type cyclin as a soluble fusion protein with the maltose binding protein (MBP), and how to assay the activity of this cyclin in interphase extracts.

### 3.1 Production of cyclins as fusion proteins

To aid solubility and purification, proteins can be expressed as fusions with bacterial proteins that act as tags, such as protein A, glutathione-S-transferase (GST), maltose-binding protein (MBP), or a poly-histidine sequence. The bacterial strains, vectors, and methodological information are available as kits from a number of commercial suppliers (e.g. Pharmacia for GST, New England Biolabs for MBP systems). These fusion proteins have a number of advantages. First, they promote soluble expression, presumably because the bacterially-derived part of the fusion protein helps correct folding during protein synthesis. Secondly, the proteins used as tags have a strong affinity for an easily available molecule that can be used as an affinity matrix for their purification, such as glutathione beads for GST-fusions and amylose beads for MBP-fusions. The bound protein can then be eluted using free ligand (glutathione or maltose) which can subsequently be removed by dialysis. Commonly, fusion proteins can be purified to homogeneity in a single step from bacterial lysates. Thirdly, the fusion proteins can often be retrieved from extracts using the same affinity matrix to analyse post-translational modifications or the binding of other proteins that occur in the extract. If necessary, the fusion protein can be cleaved from the protein of interest using a specific protease that recognizes a sequence added as a linker between the protein of interest and the expression system protein. Frequently, however, the expression system protein does not seem to interfere with function, and the complete fusion protein can be used.

We have used the maltose-binding protein fusion system (New England Biolabs) to express full length *Xenopus* cyclins A, B1, and B2. The original clones (generous gifts of J. Minshull and T. Hunt) were ligated into the pMAL$^{TM}$ vector and used to transform *E. coli* TB1. A purification strategy was then developed based on the manufacturer's instructions which may also be applicable to other proteins expressed in this vector. The fusion protein can be cleaved by making use of a factor Xa protease site, although it is difficult to subsequently inhibit the factor Xa and care must be taken that it does not interfere with the function of the expressed protein (it does not seem to affect kinase activation in egg extracts). We have found that the maltose-binding fusion protein is difficult to retrieve from extracts, probably because of competing molecules, hence a different tag may be more useful for this purpose.

---

**Protocol 7.** Production of MPB-Xenopus cyclins

*Materials*

- Protein fusion and purification kit (New England Biolabs) which contains *E. coli* strain TB1, pMAL$^{TM}$ vector, amylose resin, and instructions for LB medium
- *E. coli* TB1 strain (New England Biolabs) transformed with cyclin clone in pMAL$^{TM}$ vector (any cyclin clone can be introduced in this vector according to the kit protocol)

**Protocol 7.** *Continued*

- LB medium
- Ampicillin
- IPTG (isopropyl-thio-$\beta$-D-galactoside) solution in water (0.1 M)
- Leupeptin (Boehringer–Mannheim), aprotinin (Boehringer–Mannheim), soybean trypsin inhibitor (Sigma), pepstatin, benzamidine (Sigma)
- Lysozyme (Sigma)
- Lysis buffer: 10 mM phosphate, 30 mM NaCl, 10 mM EDTA, 10 mM EGTA, 10 mM $\beta$-mercaptoethanol (pH 7.0)
- Binding buffer (BB): 10 mM Na–phosphate, 0.5 M NaCl, 1 mM EGTA, 10 mM $\beta$-mercaptoethanol, 0.25% Tween-20 (pH 7.0)
- 0.5 M sodium phosphate buffer (pH 7.2) *Solution A*: 69 g $NaH_2PO_4 \cdot H_2O$/litre = 0.5 M; *Solution B*: 134 g $Na_2HPO_4 \cdot 7H_2O$/litre **or** 89 g $Na_2HPO_4 \cdot 2H_2O$/litre = 0.5 M
  Mix 117 ml of A with 383 ml of B. (The pH should be 7.2.)
- Elution buffer A (EBA): 50 mM Tris–HCl, (pH 8), 100 mM KCl, 5 mM $MgCl_2$, 5 mM DTT

*Method*

A. *Growth of bacteria and induction with IPTG*

1. Inoculate 20 ml of LB plus 1% glucose plus 100 µg/ml ampicillin with the transformed bacteria and incubate with shaking overnight at 37 °C.
2. Inoculate 1 litre of LB/glucose/ampicillin with 10 ml of the overnight culture and incubate with shaking at 37 °C until $OD_{600}$ = 0.4.
3. Induce by adding IPTG to 0.4 mM and incubate with shaking for 2 h at 37 °C.
4. Pellet the cells by centrifugation at 4000 $g$ for 20 min in a GSA rotor at 4 °C. Pour off the supernatant and freeze pellet at −20 °C.

B. *Lysis*

1. Resuspend the pellet in 50 ml lysis buffer/litre of culture plus protease inhibitors (10 µg/ml leupeptin, 10 µg/ml aprotinin, 10 µg/ml soybean trypsin inhibitor, 10 µg/ml pepstatin, 100 µM benzamidine).
2. Add 1 mg/ml lysozyme and incubate with rocking on ice for 30 min.
3. Sonicate 2 × 1 min. at speed 6 with microtip. The suspension becomes viscous.
4. Centrifuge at 10 000 $g$ for 30 min in SS-34 rotor at 4 °C. Pour off supernatant and discard pellet.

C. *Purification on amylose resin*

1. Measure the concentration of the crude extract by the Bio-Rad Protein Assay (should be approx. 7–9 mg/ml). Dilute to 50% with BB to give a protein concentration of approx. 3.5–4.5 mg/ml.

2. *Column size*: NB 1 litre of culture yields approximately 10 mg of unpurified fusion protein (rough estimation from the intensity of the Coomassie stained band on the gel). One millilitre of amylose resin binds approx. 1 mg of protein. Therefore, use at least 10 ml of resin per litre of culture.

3. Wash the resin with 5 column volumes of BB. Pipette the slurry of resin into the column and allow to settle.

4. Allow the buffer to run out so that just the top of the packed resin is wet, then gently pour the diluted crude extract on to the column (with as little disturbance as possible). Fill up all the excess column volume with extract.

5. *Flow rate* = 10 × (column diameter in $cm^2$) ml/h. Therefore, use a short column with a large diameter to reduce the time of the fractionation.

6. Wash with:

    (a) 3 column volumes BB

    (b) 5 column volumes of EBA

7. Elute fusion protein with EBA plus 10 mM maltose. Collect 0.5 ml fractions and detect protein either with a UV detector or measure $OD_{280}$ of the fractions using quartz cuvettes. The fusion protein elutes as a sharp peak soon after the void volume (approx. fractions 2–6).

8. Pool the positive fractions. Quickly distribute into 10 or 20 µl aliquots and freeze in liquid nitrogen. Store at −70 °C.

D. *Analysis*

1. Measure protein concentration of purified fusion protein by Bio-Rad Protein Assay.

2. Compare total protein from non-induced, induced bacteria, and the protein eluted from the amylose resin on a 10% polyacrylamide gel (see *Figure 3*).

## 3.2 Analysis of cyclin activity in egg extracts

In egg extracts, the cdc2 protein kinase requires the association of cyclins for activation, but is itself also regulated by phosphorylation/dephosphorylation (31). In fact, when cyclin binds to cdc2, three sites on cdc2 can become

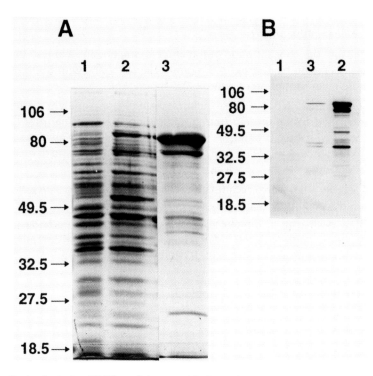

**Figure 3.** Analysis by PAGE and immunobloting of *Xenopus* MBP–cyclin B1 fusion protein expressed in bacteria. (A) Uninduced (lane 1) and IPTG-induced bacteria (lane 2) were dissolved in hot gel sample buffer and the lysate run on a 10% polyacrylamide Bio-rad minigel. Lane 3 shows the fusion protein obtained from induced bacteria lysed in lysis buffer (*Protocol 7B*) and purified on amylose resin (*Protocol 7C*). Coomassie blue staining. (B) Same as in A, but the proteins were transferred from the gel to nitrocellulose and probed with a monoclonal antibody raised against *Xenopus* cyclin B1. Lanes 1–3 correspond to the Coomassie-stained lanes shown in A.

phosphorylated. One is required for kinase activity, but the other two are inhibitory. Final activation of the kinase requires that these two inhibitory sites are dephosphorylated. Following addition of cyclins to an extract, the phosphorylation/dephosphorylation reactions may not always occur in the same way, depending on the nature of the added cyclin.

In order to assess the ability of cyclins to activate cdc2 and related kinases, cyclins are added to cycloheximide-treated interphase extracts (prepared as described in *Protocol 3*) in a range of concentrations in a buffer lacking protein kinase or phosphatase inhibitory compounds. The concentration of cdc2 kinase in the extracts is about 0.5 μM, so a similar concentration of cyclin should produce maximal kinase activity. It is important, however, to investigate both the concentration dependence and the time course of kinase activation. We use cyclins in the range of 10 to 1000 nM in the extract and

## 12: Xenopus egg extracts for studying mitosis

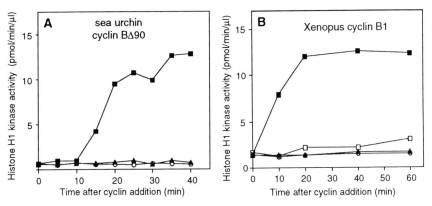

**Figure 4.** Activation of cdc2 protein kinase in *Xenopus* egg extracts by bacterially expressed cyclins. Control incubations were carried out without cyclin (open circles). In (A), sea urchin cyclin BΔ90 was added at 20 nM (closed triangles) or 100 nM (closed squares). In (B) *Xenopus* cyclin B1-MBP was added at 1 µM in the absence (closed triangles) or presence of 1 µM microcystin-LR (closed squares). The effect of microcystin-LR alone is also shown (open squares). The lack of activation of cdc2 kinase in the absence of microcystin-LR is not due to the presence of MBP associated with cyclin because the same result was obtained after removal of the MBP by proteolytic cleavage with factor Xa (there is a cleavage site between MBP and cyclin in the construct).

follow kinase activity for up to 120 min (see *Figure 4*). Samples are removed at 5–15 minute intervals into extraction buffer (EB) which stabilizes the activity of the cyclin/kinase complex. Samples diluted in this way can then be assayed immediately or stored frozen at −20 °C before use. Kinase activity is stable for at least several months when frozen, and assays can be repeated if necessary. Incubations are usually performed in microtubes and samples diluted into EB in wells on a microtitre plate. This allows multiple samples to be taken over a time course from several different incubation conditions. The volumes used below can be adapted to suit individual experiments, but dilution of the extracts changes the kinetics of kinase activation and should not be more than 50%. Kinase assays are also carried out using microtitre plates. Routinely, we transfer samples from each well to a corresponding well on a fresh plate for the kinase assay, ensuring that the lanes and columns on the plate are clearly marked and taking care not to transfer to the wrong well. The use of these plates allows up to 96 kinase assays per plate to be performed together, without having many individual assay tubes to contend with.

*Figure 4* shows two examples of the activation of cdc2 kinase in egg extracts by cyclins. In *Figure 4 A*, a sea urchin cyclin which has been produced as a truncated form lacking the first 90 amino acids from its amino-terminal, called cyclin BΔ90, is used. This cyclin is not degradable in the extracts (27) and provides a useful tool for studying the mechanism of kinase activation (16, 26). When a low concentration of cyclin (20 nM) is added, no kinase activity is stimulated. In fact a threshold concentration of the cyclin (40 nM) needs to be exceeded. Even above this threshold (100 nM cyclin), activation occurs only after a lag of about 10 minutes. This shows that a time course of activation using different concentrations of cyclin should be carried out to investigate the ability to activate cdc2 kinase. It can happen that a given cyclin never activates cdc2 kinase in egg extracts even at high concentrations, suggesting that the bacterially produced protein added is inactive. This occurs with the *Xenopus* cyclin B1 produced as a fusion with the maltose-binding protein as described in *Protocol 7*. Even high concentrations (1 μM) of the cyclin do not produce kinase activation. The cyclin is functional since it binds to and induces full phosphorylation of cdc2. Moreover, it activates cdc2 kinase very efficiently in extracts to which an inhibitor of serine/threonine phosphatases 1 and 2A, such as microcystin-LR (Calbiochem) or okadaic acid (Gibco), has been added (*Figure 4 B*). What happens with this cyclin is that dephosphorylation of the inhibitory sites on cdc2 never occurs under normal conditions. Inhibition of a type 2A phosphatase in the extract is required to allow this reaction to occur. Therefore, it is recommended that incubations with microcystin-LR or okadaic acid are performed when testing the activity of new cyclins in egg extracts.

---

**Protocol 8.** Cyclin-dependent protein kinase activation in egg extracts using exogenous cyclins

*Materials*

- Cycloheximide-treated interphasic egg extract, one 100 μl aliquot stored in liquid nitrogen
- Cyclin, diluted in Tris buffer: 50 mM Tris–HCl (pH 8.0), 100 mM KCl, 5 mM $MgCl_2$, 5 mM dithiothreitol
- Extraction buffer (EB): (see *Protocol 6*)
- 96 micro-well plates (Nunclon)
- 0.5 ml microtubes (Eppendorf or similar)
- 0–20 μl pipette and tips

*Method*

1. Aliquot 19 μl of EB into each well on a micro-well plate. Label each row with a letter corresponding to an incubation condition (A, B, C, . . .) and each column with the time of the sample to be taken from each incubation (0, 10, 20, . . .). Keep plate on ice.
2. Aliquot 10 μl of egg extract into 0.5 ml microtubes on ice labelled A, B, C, . . .

3. Place tubes in rack at 20 °C, or at room temperature, and remove 1 μl into corresponding wells (A0, B0, C0, . . .) on the micro-well plate on ice. Mix well by pipetting up and down.

4. Add 1 μl of cyclin dilution (0–10 μM) to start the reaction. Stagger addition to each incubation tube by 15 sec. Control incubations without cyclin should also be used.

5. Remove 1 μl samples from each tube every 10 min into corresponding wells (A10, B10, C10, . . .). Stagger sampling from each incubation tube by 15 sec.

6. When completed, either transfer 5 μl from each well to a well on a fresh plate, and carry out kinase assays (*Protocol 6*), **or** freeze plate at −20 °C until use.

## 4. Assay of other cell-cycle regulatory molecules in egg extracts

Egg extracts provide a useful cell-free system to study the mechanism of cdc2 kinase activation, since proteins and chemicals can be easily added. Although it seems that all the proteins required for cdc2 kinase activation and induction of the mitotic state are present except cyclins, the activity state of some of the other regulatory proteins may be modulated during the cell cycle. This can be studied by adding these molecules to the extracts and following their effect on cdc2 kinase activation (this would correspond to an overexpression), or by modulation of their phosphorylation state by adding chemical inhibitors such as protein kinase inhibitors or protein phosphatase inhibitors. One can also deplete regulatory proteins, for instance by immunoprecipitation. The relationship between cdc2 kinase activation and other cellular processes, such as the completion of DNA replication can also be studied. Although the extracts normally have no nuclei present, chromatin can be added to them (32). Complete functional nuclei are formed, and the coupling between kinase activation and nuclear processes examined. Finally, these extracts provide a potentially good system for assaying the effect of drugs on the mechanism of the cell cycle and to screen for mitotic inhibitors.

## 5. Regulation of microtubule dynamics in egg extracts

During the interphase to metaphase transition, microtubule dynamics change dramatically. Overall, the microtubule turnover increases more than 20-fold and microtubules are reorganised from a radial array into a bipolar spindle (33). Since spindles can assemble in egg extracts, it is reasonable to assume that microtubule dynamics are comparable to the situation *in vivo*. The tubulin concentration in egg extracts is between 10–20 µM, close to the levels found *in vivo*. Using both fixed timepoints (34) and real-time video microscopy (35), it has been possible to study microtubule dynamics in interphase and mitotic extracts. This is achieved by adding isolated human centrosomes to the extract of nucleate microtubules. The parameters of microtubule dynamics are very similar to those found in intact tissue culture cells. This system has allowed the demonstration of the role of cdc2 kinase and different cyclins in regulating microtubule dynamics (30). One interesting aspect of these extracts is that the number of centrosomes added is relatively small compared to the volume of the extract. So, the polymer mass of microtubules does not significantly reduce the free tubulin concentration. Therefore, the effect of cell-cycle regulators on microtubule dynamics can be studied in the absence of changes in free tubulin concentration which interfere with microtubule dynamics. Our goal in using these extracts to examine microtubule dynamics is to analyse the respective contribution of specific components of the extract to the overall dynamics. This could ultimately be achieved by comparing the dynamics of microtubules assembled from pure centrosomes tubulin with the dynamics of microtubules in these extracts, and then isolating factors from the extract that could be responsible for the specific dynamics observed in the total extract. Here, we describe a simply assay which allows the study of the steady-state length of microtubules in interphase and mitotic extracts. The assay involves fixation of the extracts after different incubation times at room temperature and analysis by immunofluorescence microscopy. The analysis of microtubule dynamics by video microscopy using the method first developed by T. Mitchison's group (35) is not described here as there are many aspects that would go beyond the scope of this article. This video microscopy technique and the method for analysis of microtubule dynamics is described in refs 35 and 30, respectively. The method for labelling tubulin with rhodamine is described in ref. 36.

## 5.1 Analysis of microtubule dynamics using fixed timepoints and immunofluorescence microscopy

In *Protocol 9*, we describe the methods for the fixation and immunofluorescence of dynamic microtubules nucleated by centrosomes in interphase extracts. The method is the same both in mitotic or cyclin-treated extracts and can also be used to fix and observe more complex structures, such as nuclei or spindles. *Figure 5* shows how interphase and mitotic microtubules look in these extracts. It is important to measure the H1 kinase activity in aliquots of the same samples that are used to examine microtubules, in order to make sure that the extract was indeed mitotic or interphasic during the experiment.

## 12: Xenopus egg extracts for studying mitosis

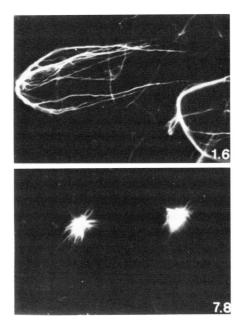

**Figure 5.** Immunofluorescence staining of centrosome-nucleated microtubules in egg extracts. In the upper picture, an interphase extract was incubated for 15 min at room temperature in the presence of centrosomes, fixed and processed as described in *Protocol 9*. The lower picture shows microtubules in the same extract to which purified starfish cdc2 kinase was added before the addition of centrosomes. In the upper panel, microtubules are longer than 20 μm, whereas in the lower panel their size does not exceed 5–6 μm. The numbers in the lower right corners indicate the H1 kinase activity measured in the same extracts on aliquots taken in EB (*Protocols 9* and *6*).

---

**Protocol 9.** How to fix and observe microtubules in egg extracts

*Materials*

- Cycloheximide-treated interphase egg extract, one 100 μl aliquot stored in liquid nitrogen (*Protocol 3*)
- Centrosomes prepared according to refs 37, 38. The preparation should contain at least $10^7$ centrosomes/ml, store at $-70\,°C$ in 50–55% sucrose
- RG1 (ReGrowth buffer 1): 80 mM Pipes, 1 mM EGTA, 1 mM $MgCl_2$, 1 mM GTP (pH 6.8 with KOH). This buffer can be prepared in advance and kept frozen at $-20\,°C$
- RG2 (RG1 without GTP)
- GTP: 0.5 M stock solution in water, stored in 100 μl aliquots at $-20\,°C$
- Electron microscopy grade glutaraldehyde (2.5%); just before use, prepare 10 ml of 0.25% glutaraldehyde in RG1

**Protocol 9.** *Continued*

- 25% (v/v) glycerol in RG2 (5 ml/sample to analyse). We use extra pure anhydrous glycerol from Merck (cat. no. 4093)
- Anti-tubulin antibody. We routinely use the anti-α and/or β tubulin monoclonal from Amersham (cat. no. 356), but any good polyclonal can be used
- Secondary anti-mouse or rabbit polyclonal antibody, affinity purified and preferably labelled with rhodamine. We use antibodies produced in goat by Dianova (cat. no. 715 075 137)
- Phosphate-buffered saline (PBS): 100 mM NaCl, 20 mM Na–phosphate (pH 7.2)
- 0.1% sodium borohydride (Sigma, S 9125) in PBS; prepare just prior to use by dissolving the powder in PBS. Hydrogen-gas bubbles should be visible
- PBS containing 3% BSA
- 1 ml Eppendorf tubes
- Round glass coverslips (12 mm in diameter) stored in ethanol
- 15 ml Corex tubes and Plexiglass adaptors as described in ref. 39
- Sorvall centrifuge and HB-4 rotor with adaptors for 15 ml Corex tubes
- −20 °C methanol
- Mowiol (cat. no. 4–88 Hoechst): to 6 g of glycerol add 2.4 g of Mowiol 4–88 in a 50 ml disposable conical tube and stir thoroughly. Add 6 ml distilled water and leave for 2 h at room temperature. Add 12 ml of 0.2 M Tris (pH 8.5) and incubate at 53 °C until the Mowiol is dissolved (*a long time*); stir occasionally. Clarify by centrifugation at 5000 *g* for 20 min and store in 1 ml aliquots at −20 °C. Once defrosted, it is stable at room temperature for one month

*Method*

1. Aliquot 19 μl of EB into each well on a micro-well plate. Label each well with the time of the sample to be taken (i.e. 0, 2, 5, 10, 20, 30 min.). Keep plate on ice.
2. Aliquot 15 μl of interphase extract into 1.5 ml Eppendorf tubes on ice, labelled with the time as in Step 1.
3. Still on ice, add 1 μl of centrosomes (about $10^4$ centrosomes) to the interphase extract and mix well.
4. Transfer 1 μl of extract to the zero time point in the micro-well plate on ice, and fix the rest of the sample with glutaraldehyde as described in Step 7 (below).

## 12: Xenopus egg extracts for studying mitosis

5. Bring the tubes to 20 °C in a dry heating-block adjusted to this temperature.
6. At each timepoint, transfer 1 µl of extract to the corresponding micro-well on the plate, and fix the rest of the sample as described in Step 7 (below).
7. To fix each sample, add 1 ml of 0.25% glutaraldehyde in PBS *directly* to the extract in the Eppendorf tube. Add the fixative slowly along the side of the tube so that the extract becomes well mixed with the fixative but without too much disturbance so that microtubules are not destroyed. Incubate for 6 min at room temperature.
8. Prepare 15 ml Corex tubes with the Plexiglass adaptors at the bottom (39). Add to each tube a glass coverslip taken out from ethanol with forceps and wiped dry with clean paper tissue. Fill up each tube with 5 ml of 25% glycerol in RG2.
9. Layer the fixed extracts over the glycerol cushion using a 1 ml Eppendorf pipette with a blue tip, cut at the extremity. This should be done carefully in order not to break the microtubules. Do not pipette the microtubules down the side of the tube. Rather pipette slowly at the surface in the centre of the glycerol cushion.
10. Centrifuge at 12 000 r.p.m. (20 000 $g$) in the Sorvall centrifuge at room temperature.
11. Recover the coverslip from the Corex tube, remove excess glycerol buffer by putting the edge of the coverslip on to a paper tissue. Post-fix by immersion for 5 min in −20 °C methanol. Before the methanol fixation step it is important to remove as much buffer as possible from the surface of the coverslip without complete drying to avoid damaging the microtubules.
12. Remove the coverslip from the ethanol and incubate in sodium borohydride for 10 min at room temperature.
13. Incubate with 10 µl of the primary antibody, at the recommended dilution, in PBS containing 3% BSA in a wet atmosphere for 10 min at room temperature. Wash 3 times in PBS, 5 min each time. Incubate with the secondary antibody for 10 min, wash as before.
14. Mount in Mowiol by putting 4 µl of Mowiol on the microscope slide (*not* the coverslip). Place the coverslip gently on the Mowiol drop and let it dry for 1 h without pressing on the coverslip.
15. Observe using a 40 ×, 63 × or 100 × neofluar lens using the microscope in the appropriate fluorescence mode.
16. Measure H1 kinase activity from the samples taken in the micro-wells according to *Protocol 6*.

## Acknowledgements

We thank M. Glotzer and T. Hunt for their generous gifts of cyclin molecules, clones, and many reagents, and S. Reinsch for critical reading of the manuscript. M.-A. Félix was supported by the CNRS (France), P. R. Clarke by the Wellcome Trust (UK), and part of the work described here by an HFSP grant to E. Karsenti.

## References

1. Bataillon, E. (1928). *C. R. Hebd. Seances Acad. Sci.*, **187**, 520.
2. Brachet, A. (1922). *Arch. Biol.*, **32**, 205.
3. Diberardino, M. A. (1979). In *Nuclear transplantation*, Vol. 9 (suppl.) (ed. J. F. Danielli and M. A. DiBerardino), p. 129. Academic Press, New York, San Francisco, London.
4. Gurdon, J. B. (1968). *J. Embryol. Exp. Morphol.*, **20**, 401.
5. Ziegler, D. and Masui, Y. (1973). *Dev. Biol.*, **35**, 283.
6. Forbes, D. J., Kirschner, M. W., and Newport, J. W. (1983). *Cell*, **34**, 13.
7. Karsenti, E., Newport, J., and Kirschner, M. (1984). *J. Cell Biol.*, **98**, 1730.
8. Hara, K., Tydeman, P., and Kirschner, M. (1980). *Proc. Natl Acad. Sci. USA*, **77**, 462.
9. Lohka, M. J. and Masui, Y. (1983). *Science*, **220**, 719.
10. Leno, G. H. (1991). In Xenopus laevis: *practical uses in cell and molecular biology*, Vol. 36 (ed. B. K. Kay and H. B. Peng), p. 561. Academic Press.
11. Wolffe, A. P. and Schild, C. (1991). In Xenopus laevis: *practical uses in cell and molecular biology*, Vol. 36 (ed. B. K. Kay and H. B. Peng), p. 541. Academic Press.
12. Newmeyer, D. D. and Wilson, K. L. (1991). In Xenopus laevis: *practical uses in cell and molecular biology*, Vol. 36 (ed. B. K. Kay and H. B. Peng), p. 607. Academic Press.
13. Murray, A. (1991). In Xenopus laevis: *practical uses in cell and molecular biology*, Vol. 36 (ed. B. K. Kay and H. B. Peng, p. 581. Academic Press.
14. Murray, A. W. and Kirschner, M. W. (1989). *Nature*, **339**, 275.
15. Murray, A. W., Solomon, M. J., and Kirschner, M. W. (1989). *Nature*, **339**, 280.
16. Solomon, M. J., Glotzer, M., Lee, T. H., Philippe, M., and Kirschner, M. W. (1990). *Cell*, **63**, 1013.
17. Félix, M. A., Labbé, J. C., Dorée, M., Hunt, T., and Karsenti, E. (1990). *Nature*, **346**, 379.
18. Wu, M. and Gerhart, J. (1991). In Xenopus laevis: *practical uses in cell and molecular biology*, Vol. 36 (ed. B. K. Kay and H. B. Peng), p. 3. Academic Press.
19. Félix, M. A., Cohen, P., and Karsenti, E. (1990). *EMBO J.*, **9**, 675.
20. Buendia, B., Draetta, G., and Karsenti, E. (1992). *J. Cell Biol.*, **116**, 1431.
21. Félix, M.-A., Pines, J., Hunt, T., and Karsenti, E. (1989). *EMBO J.*, **8**, 3059.
22. Leiss, D., Félix, M.-A., and Karsenti, E. (1992). *J. Cell Sci.*, **102**, 285.
23. Buendia, B., Clarke, P. R., Félix, M. A., Karsenti, E., Leiss, D., and Verde, F. (1991). *CSH Quant. Symp. Biol.*, **56**, 523.

24. Lohka, M. and Maller, J. (1985). *J. Cell Biol.*, **101**, 518.
25. Sawin, K. E. and Mitchison, T. J. (1991). *J. Cell Biol.*, *112*, 925.
26. Clarke, P. R., Leiss, D., Pagano, M., and Karsenti, E. (1992). *EMBO J.*, **11**, 1751.
27. Glotzer, M., Murray, A. W., and Kirschner, M. W. (1991). *Nature*, **349**, 132.
28. Luca, F. C., Shibuya, E. K., Dohrmann, C. E., and Ruderman, J. V. (1991). *EMBO J.*, **10**, 4311.
29. Pagano, M., Pepperkok, R., Verde, F., Ansorgue, W., and Draetta, G. (1992). *EMBO J.*, **11**, 961.
30. Verde, F., Dogterom, M., Stelzer, E., Karsenti, E., and Leibler, S. (1992). *J. Cell Biol*, **118**. (In press.)
31. Clarke, P. R. and Karsenti, E. (1991). *J. Cell Sci.*, **100**, 409.
32. Dasso, M. and Newport, J. W. (1990). *Cell*, **61**, 811.
33. Karsenti, E. (1991). *Seminars in Cell Biology*, **2**, 251.
34. Verde, F., Labbé, J. C., Dorée, M., and Karsenti, E. (1990). *Nature*, **343**, 233.
35. Belmont, L. D., Hyman, A. A., Sawin, K. E., and Mitchison, T. J. (1990). *Cell*, **62**, 579.
36. Hyman, A., Drechsel, D., Kellogg, D., Salser, S., Sawin, K., Steffen, P., Wordeman, L., and Mitchison, T. (1991). In *Methods in enzymology*, Vol. 196 (ed. R. B. Vallee), p. 478. Academic Press, San Diego.
37. Mitchison, T. and Kirschner, M. (1984). *Nature*, **312**, 232.
38. Bornens, M., Paintrand, M., Berges, J., Marty, M. C., and Karsenti, E. (1987). *Cell Motil. Cytoskel.*, **8**, 238.
39. Evans, L., Mitchison, T. J., and Kirschner, M. W. (1985). *J. Cell Biol.*, **100**, 1185.

# 13

# Maturation-promoting factor and cyclin-dependent protein kinases

MARCEL DORÉE, THIERRY LORCA, and ANDRÉ PICARD

## 1. Introduction

The cell cycle has two main phases: interphase and mitosis. In interphase the cell nucleus remains intact and chromosomes are replicated. In mitosis the cytoplasmic network of microtubules breaks down and is reassembled into the mitotic spindle, whilst the chromosomes condense and attach to it (metaphase). Then the spindle segregates the chromosomes (anaphase), ensuring that each daughter nucleus will have one copy of every chromosome. At the end of mitosis the cell divides to form two complete daughter cells that are genetically identical to the parent.

For many years it has been suspected that mitotic events are triggered by some dominant factor present in the cytoplasm of mitotic cells. Indeed fusion of mitotic and interphase mammalian cells rapidly causes the interphase nucleus to prematurely condense chromosomes, regardless of its position in the cell cycle (1). These pioneer experiments have been supplemented by investigations on oocyte maturation which has led to the characterization of this dominant factor. In most animals, oocytes which are still in the ovary have a large nucleus (the germinal vesicle, GV) containing 4n chromosomes. In terms of the cell cycle, these oocytes are arrested at prophase of the first meiotic division. At the time of ovulation, they are released from cell-cycle arrest, enter metaphase, twice discard one-half of the chromosomes as polar bodies and then develop into mature, fertilizable ova with a single set of chromosomes. This process of oocyte maturation is controlled by specific hormones in many vertebrates and invertebrates (see ref. 2 for a review). Because of the experimental advantage they provide, amphibians amongst vertebrates and starfishes amongst invertebrates have been extensively used for investigating the hormonal control of oocyte maturation *in vitro*. The hormones which trigger the onset of oocyte maturation were identified as progesterone and 1-methyladenine (1-MeAde) in amphibians and starfishes, respectively. It was rapidly shown that these hormones do not need to enter the oocytes and are inefficient inducers of maturation when microinjected into oocytes. It was, therefore, anticipated that some intracellular factor which triggers entry into metaphase is produced under the influence of the maturation hormones.

Most likely this factor was cytoplasmic, since nuclear transplantation in amphibian eggs and protozoa had shown that the cytoplasm controls nuclear activity during the cell cycle. In agreement with this view, it was shown that cytoplasm taken from a maturing oocyte after treatment with the specific hormone induced germinal vesicle breakdown (GVBD) and subsequent meiotic maturation upon injection into an intact immature oocyte in both amphibians and starfish (3–5). The dominant factor produced under the influence of the hormone and responsible for oocyte maturation was called maturation-promoting factor (MPF). Subsequently, similar MPF activity was shown to be present in the cytoplasm of maturing oocytes of many species (6). MPF taken from donors in a given species can induce maturation when microinjected into recipient oocytes from other zoological groups that do not respond to the same maturation-inducing hormone as the donors. Moreover, MPF activity was found not only in maturing oocytes, but also in mitotically dividing cells. In contrast, interphase cells have no MPF activity. Thus MPF acts in a non-species specific manner across the different phyla and promotes M phase in all eukaryotic cells regardless of whether they are undergoing meiosis or mitosis.

The state of phosphorylation of numerous proteins cycles in phase with MPF activity during meiotic maturation, increasing when cells enter metaphase and decreasing when cells exit from metaphase (7, 8). This supported the view that MPF activity was correlated with a high activity of protein phosphorylation. Indeed it was found that a major histone H1 kinase that does not require $Ca^{2+}$, cyclic nucleotides, or diacylglycerol (thus different from the main histone kinases known at that time) greatly increases in activity when oocytes from starfish or amphibians are released from prophase arrest by hormonal stimulation. The activity of this mitotic protein kinase then oscillates during meiosis and early embryogenesis in parallel with MPF (9–12). This kinase activity becomes maximal when cells enter M phase, begins to decrease at anaphase, and is low during interphase. It had been known for many years that inhibiting protein synthesis blocks entry into mitosis and cleavage in marine invertebrate embryos (13), despite the fact that most components are already present in the oocyte, and no growth occurs during early embryogenesis. The mitotic H1 kinase activity did not reappear after its drop at anaphase when protein synthesis was suppressed during the meiotic and the early mitotic cell cycles. Moreover, the drop in H1 kinase and MPF activities could be blocked in starfish oocytes by microinjecting high amounts of protease inhibitors, suggesting that some component required for the mitotic kinase activity was proteolysed at the end of each cell cycle. In starfish, the only newly synthesized protein whose proteolysis was detected at anaphase was cyclin (14), and antipain microinjection prevented cyclin proteolysis. The cyclins (A and B) had been originally discovered in fertilized clam and sea urchin eggs as proteins whose abundance oscillated dramatically during the cell cycle (15, 16). Taken together these results suggested that MPF could be identical to the M-phase specific H1 kinase, and that cyclin could be an activator of this kinase.

Only some of the events required for M phase to take place are important for its timing. These events can be distinguished; speeding up the rate at which they are completed will significantly advance a cell into M phase, whilst speeding up other events will have little effect on the timing of M phase. This reasoning was the basis for the isolation, in fission yeast, of mutants in regulatory genes that influence the cell-cycle timing of M phase (17). Only a very limited number of such genes were found to regulate the coordination of the $G_2$ to M phase transition. They form a regulatory gene network controlling entry into mitosis, in which $cdc2^+$ acts downstream of the other genes. The central gene encodes a 34 kDa protein kinase. One pathway of the mitotic regulatory network is inhibitory and includes the $nim1^+$ and $wee1^+$ gene products, whereas the other pathway is activatory and includes the $cdc25^+$ gene product. The balance of these two pathways regulates function of the protein kinase $p34^{cdc2}$ and advances or delays the onset of mitosis (18, 19). In 1987, an homologue of the yeast $cdc2^+$ gene was found in a human cDNA library using a complementation strategy (20), suggesting that the mitotic regulatory network which operates in fission yeast could be conserved in higher eukaryotes.

In 1988, extensive biochemical purification of MPF and the M-phase specific H1 kinase demonstrated their identity to the product of $cdc2$ homologues in both starfish and amphibians (21–24). One year later it was shown that MPF is a heterodimer comprising one molecule of $p34^{cdc2}$ and one molecule of cyclin B (25).

## 2. Detection of MPF activity by direct cytoplasm transfer

### 2.1 In *Xenopus* oocytes

#### 2.1.1 Handling *Xenopus* oocytes and eggs

The cells most widely used for the detection of MPF activity are full-grown *Xenopus laevis* oocytes, due to their enormous size, about 500 000 times greater than a somatic cell, which facilitates the use of semiquantitative microinjection, and their availability throughout the year. Donor oocytes can be prepared (*Protocol 1*) from ovarian oocytes free of follicle cells and induced to mature *in vitro* with progesterone, which releases them from arrest at first meiotic prophase. Alternatively eggs, which are laid after undergoing physiological maturation including jelly coat formation, can be used as donor cells (*Protocol 2*). In either case, MPF activity of donor cells is assayed by microinjecting cytoplasm (*Protocol 3*, *Figure 1*) with a micropipette into fully-grown immature oocytes, which then undergo germinal vesicle breakdown (GVBD) in the case of successful transfer. Although *Xenopus* oocytes are opaque, due to the accumulation of large amounts of yolk platelets during oogenesis, GVBD is easy to detect because it is correlated with the appearance of a white spot, due to pigmentation changes in the animal hemisphere of the oocyte.

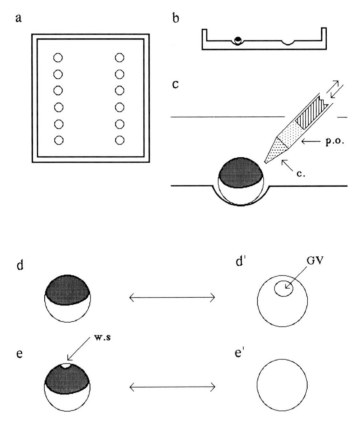

**Figure 1.** Microinjection into *Xenopus* oocytes. (**a**) Frontal view and (**b**) Cross-section of the vessel used to immobilize oocytes. (**c**) High magnification of a recipient oocyte positioned in a hole of the vessel (**c**) cytoplasm; p.o., paraffin oil). (**d**) and (**e**) External view of an oocyte before and after germinal vesicle breakdown, respectively (w.s: white spot). (**d'**) and (**e'**) Cross-sections corresponding to (**d**) and (**e**), respectively.

## Protocol 1. Preparation of *Xenopus* oocytes[a]

*Materials*

- Barth medium: 88 mM NaCl, 1 mM KCl, 0.82 mM $MgSO_4$, 0.33 mM $Ca(NO_3)_2$, 0.41 mM $CaCl_2$, 2.4 mM $NaHCO_3$, 1.5 mM Tris (pH 7.8)
- Modified Ringer medium: MMR: 100 mM NaCl, 2 mM KCl, 1 mM $MgCl_2$, 2 mM $CaCl_2$, 0.1 mM EDTA, 5 mM Hepes (pH 7.8)

1. Inject frogs 4–7 days before use with 100 units of pregnant mare serum gonadotropin (PMSG, Intervet)
2. Anaesthetize frogs by immersion in 1 g/litre tricaine (MS 222, Sandoz) at pH 7.
3. Remove ovarian tissue through a slit in the lateral body wall.
4. Select fully-grown oocytes (1.2–1.4 mm) with a relatively unpigmented equatorial band 0.2 mm wide (stage VI oocytes) as donor and recipient oocytes.
5. Prepare oocytes free of follicle cells by manual dissection with watchmaker's forceps, and transfer them in modified Barth medium or in modified Ringer medium.
6. Induce meiotic maturation in donor oocytes by adding progesterone (Sigma) to a final concentration of 1 μM from a 2 mM stock solution in methanol. Donor oocytes are washed free of external progesterone with Barth or MMR medium before transfer in the microinjection vessel.

[a] *Xenopus* eggs and oocytes are handled at room temperature (23–25 °C).

## Protocol 2. Preparation of unfertilized, fertilized, or activated *Xenopus* eggs[a]

**A. Eggs**

1. For egg laying, inject PMSG-treated females with 1000 units human chorionic gonadotropin (HCG, Sigma).
2. Leave frogs overnight in 100 mM NaCl to prevent activation of the eggs after laying. The time to the beginning of egg laying is variable, ranging from 7–14 h, with ovulation continuing for at least 6 h.
3. Recover eggs to be fertilized directly from a female into a dry Petri dish by gently pressing its abdomen during the period of egg laying.

**Protocol 2.** *Continued*

B. *Sperm*

1. To obtain sperm take a testis from a male killed by severing the spinal cord.
2. Lacerate the testis with a blade.

C. *Fertilization*

1. Induce fertilization by letting the lacerated testis gently wander on the eggs in the Petri dish.
2. Three to five minutes later add 5-fold diluted MMR (*Protocol 1*) and observe eggs under a dissecting microscope. A few minutes after fertilization, cortex contraction is observed, then the eggs rotate within the elevated vitelline membrane to a position with the vegetal end (the white area) downward.

D. *Removal of jelly coat*

The jelly coat must be removed after fertilization for microinjection experiments.

1. Remove excess saline and transfer eggs in 50 ml conical Falcon tubes containing 2% cysteine hydrochloride adjusted to pH 8–8.5 with NaOH.
2. Rotate the tubes manually and carefully and change the cysteine solution twice after about 3 min each time. The eggs will pellet spontaneously, and apparent volume will decrease by about 2–3 fold, due to solubilization of the jelly coat.
3. Discard the cysteine solution and wash the eggs several times, first with 100 mM NaCl, then with modified Barth medium, or MMR (*Protocol 1*).

E. *Parthenogenetic activation*

Parthenogenetic activation can be induced simply by pricking unfertilized and jelly coat-free eggs with a microinjection needle, or by applying an electric shock (12 V, 1 sec, 3 cm separation of electrodes) to a cell suspension.

[a] *Xenopus* eggs and oocytes are handled at room temperature (23–25 °C).

## 13: MPF and cyclin-dependent protein kinases

**Protocol 3.** Microinjection into *Xenopus* oocytes[a]

1. Prepare injection needles: pull 10 μl microdispenser tubes (Drummond Scientific) using a vertical pipette puller (Model 720, David Kopf Instrument).

2. Break off the sealed needles under a dissecting microscope against the sealed tip of a Pasteur pipette to form a tip 0.01 mm in diameter, and fill it with paraffin oil (used as hydraulic fluid).

3. Fit the injection needle to the end of a microlitre Drummond digital microdispensor clamped to a Prior micromanipulator.

4. Position both donor and recipient oocytes in hemispherical holes (1.5 mm in diameter) made in a square (50 × 50 mm) flat (10 mm depth) plastic vessel containing Barth medium (*Figure 1, a, b*). When eggs are used as donor, it may be helpful to carefully hold them with watchmaker's forceps to facilitate impalement.

5. Impale donor oocytes or eggs with the needle at 45° near the equatorial ring.

6. Suck up cytoplasm with the help of a negative pressure.

7. Inject recipient oocytes with 30–50 nl of cytoplasm near the equatorial ring (*Figure 1c*).

[a] All operations are carried out at room temperature (23–25 °C).

GVBD is detected in recipient oocytes by a white spot appearing at the animal pole within 2 h after the injection (*Figure 1e*). Almost without exception, oocytes which fail to mature within 2 h after injection also fail to mature by 24 h. Oocytes never undergo maturation if the amount of microinjected cytoplasm is too small. GVBD can be confirmed by boiling oocytes for 3 min in water, cutting them with a blade through the symmetry plane and comparing the section (absence of a nuclear envelope-limited area) with that of control non-injected oocytes (*Figure 1, d'–e'*). Successfully injected oocytes develop to the metaphase of the second meiotic division when they again arrest, as do progesterone-stimulated oocytes. They can in turn be used as donor oocytes. In such serial transfers, there is no appreciable decrease in the frequency with which the recipient oocytes undergo GVBD. Since at the most 10% of the cytoplasm is transferred at each injection, the original cytoplasm taken from the hormone-stimulated oocyte is diluted 1 × $10^{10}$ fold in the tenth serial recipient, indicating an 'autocatalytic amplification' of MPF during each successive passage. Successful transfer of cytoplasm occurs even if recipient oocytes are treated with 100 μg/ml cycloheximide, which blocks protein synthesis.

### 2.1.2 Time course of MPF activation in progesterone-stimulated *Xenopus* oocytes and fertilized or activated eggs

Progesterone-stimulated oocytes usually undergo GVBD after 4–8 hours, depending on the batch of oocytes. Synchrony is not excellent in this species. If the time when 50% of the stimulated oocytes which will ultimately undergo GVBD in a given experiment is taken as $GVBD_{50}$, MPF activity is first detected (10% of maturation following transfer) when donor oocytes have reached 0.65 $GVBD_{50}$. This also corresponds to the time when donor oocytes become independent of protein synthesis for progesterone to induce GVBD (26). MPF activity drops at the time of first polar body emission (about 1 h after GVBD), increases again, and reaches a plateau which is maintained for many hours, as long as oocytes remain arrested at metaphase II.

MPF activity in progesterone-matured oocytes is similar to that of unfertilized eggs. After parthenogenetic activation, MPF activity drops to undetectable levels within 8 min. It probably also does so in fertilized eggs, although the exact timing of MPF inactivation is difficult to determine due to the relatively time-consuming procedure for dejellying eggs after fertilization. MPF activity reappears at approximately two-thirds of the time required for first mitotic cleavage in fertilized eggs, and approximately at the same time in parthenogenetically-activated eggs. Protein synthesis is required for MPF activity to reappear after fertilization or parthenogenetic activation (27).

Arrest of vertebrate oocytes at metaphase II is due to the presence of a factor, called cytostatic factor (CSF) that prevents inactivation of MPF. This was demonstrated by showing that microinjection of cytoplasm taken from an unfertilized egg (metaphase II-arrested) into a blastomere of a two-cell embryo arrested the blastomere at metaphase, whilst the non-injected blastomere continued dividing (3). Purified MPF has no CSF activity. Although CSF has not been purified, the $p39^{c-mos}$ proto-oncogene, which is expressed during but not after meiotic maturation, appears to be an essential component of this activity. Indeed microinjection of *c-mos* mRNA into one of the blastomeres of a two-cell embryo arrests the recipient cell at metaphase. Moreover, extracts prepared from unfertilized eggs can be immunodepleted of their CSF activity with antibodies directed against $p39^{c-mos}$ (28).

## 2.2 Detection of MPF activity by direct cytoplasm transfer in other species

In most animals, the volume of fully-grown oocytes is at least $10^2$–$10^3$ times smaller than that of *Xenopus*. A variety of methods have been described for microinjection in small oocytes, but we recommend the volume-controlled pressure microinjection system of Hiramoto (*Protocol 4*, *Figure 2*), which is completely suitable for starfish and many other oocytes. Only a brief account

# 13: MPF and cyclin-dependent protein kinases

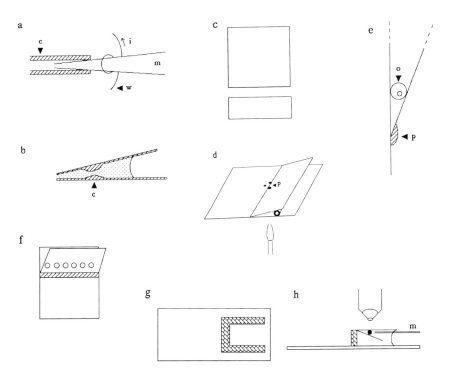

**Figure 2.** Microinjection into small oocytes. (**a**) Procedure for making a constriction near the tip of a micropipette. The micropipette (m) is inserted into the loop of a platinum wire (w) and its tip into a glass capillary (c), to avoid bending of the pipette during heating, when an electric current (i) is applied to the wire. (**b**) Cross-section of the tip of a micropipette filled wtih silicone oil and showing the constriction (c). (**c**) The coverslip and the coverslip fragment used to make the egg holder. (**d**) Preparing the holder. A capillary tube is used to support the coverslip fragment. Pieces of paraffin wax (p) are placed on the coverslip along the fragment. When the wax is heated with a match, it melts and spreads by capillary action along the coverslip fragment. The capillary is removed after the paraffin has hardened. (**e**) and (**f**) An oocyte suspension is introduced into the holder and the oocytes allowed to settle by gravity and are immobilized in a line near the paraffin. (**g**) The chamber supporting the oocyte holder: a U-shaped acrylic plate has been attached to a glass slide. (**h**) Cross-section of the microinjection system. The micropipette (m) is fitted to a needle holder connected to a micromanipulator. The space between the oocyte holder and the chamber is filled with sea water (in the case of starfish oocytes). (**a**), (**g**) and (**h**) were adapted from Kishimoto (29).

of this method is given in *Protocol 5*. The reader may refer to the excellent review by Kishimoto for a thorough and detailed description (29). The main feature of this method is the use of a constriction for controlling the volume of material delivered by the micropipette (*Figure 2, a, b*).

**Protocol 4.** Preparation of the microinjection needle

1. Pull a glass micropipette with a vertical pipette puller (Model 720, David Kopf Instruments).
2. Break its sealed tip under a microscope to about 5 μm diameter, against the edge of a coverslip.
3. Make a constriction (internal diameter: about 10 μm) approximately 5 mm from the tip by inserting the tip of the micropipette into the loop of a platinum wire which is heated by applying an electric current. To prevent the micropipette from bending when the wire is heated, a glass capillary is used to support the tip.
4. Fill the tip of the micropipette, including the constriction, with paraffin oil, using a long hypodermic needle (0.60 × 90 mm).
5. Fit the micropipette to a needle holder (Leitz), whose rear end is connected to thick plastic tubing (2.5 mm diameter) leading to a 2 ml syringe. The entire space from the tip of the needle holder to the syringe plunger is filled with silicone oil. Changes in pressure caused by the injection syringe are delivered to the micropipette through the pressure-tight system and the constriction near the tip, filled with silicone oil, attenuates the speed of flow in and out of the micropipette.

**Protocol 5.** Cytoplasmic transfer of starfish oocytes[a]

1. Make a holder by cementing a small fragment of an 18 × 18 mm coverslip with paraffin on to a 20 × 20 mm coverslip.
2. Keep the holder vertical to align and wedge (by gravity) the recipient and donor oocytes (washed to eliminate the substance used to induce maturation) (*Figure 2 c–f*).
3. Place the holder (oocytes down) on a supporting glass slide on which a U-shaped acrylic plate has been attached (*Figure 2 g-h*). The chamber between the holder, the U-shaped plate, and the glass slide is filled with a convenient medium (sea water in the case of starfish oocytes).
4. Adjust the tip of the micropipette and the donor cell to the same focal plane and the same microscope field.
5. Impale the donor oocyte on the micropipette by moving the mechanical microscope stage.
6. Suck cytoplasm out by reducing the pressure within the syringe[b].
7. Remove the donor and impale a recipient oocyte on the micropipette by moving the microscope stage. During this time the micropipette is not moved and remains in focus.

8. Inject cytoplasm into the recipient by increasing pressure within the syringe[b].

[a] All operations are carried out at room temperature (23–25 °C).
[b] The amount of microinjected cytoplasm can be easily calculated

$$v = \frac{\pi h}{12} (D_0^2 + D_0 D + D^2)$$

after measuring displacement of the meniscus ($h$) and its diameter before ($D_0$) and after microinjection ($D$).

Alternatively, the exact volume to be injected can be sucked up and microinjected together with a silicone oil droplet from the micropipette, in order to allow microinjected oocytes to be recognized. For successful induction of GVBD into a fully-grown prophase-arrested oocyte (*Figure 3*), it is usually necessary to microinject at least 5% of the volume of a donor oocyte at metaphase.

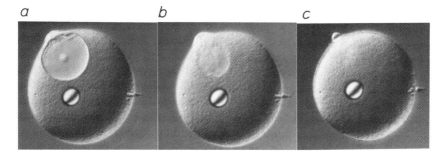

**Figure 3.** Induction of meiotic maturation in an oocyte of *Holothuria forskali* by transfer of cytoplasm taken from an oocyte of the starfish *Marthasterias glacialis* at first meiotic metaphase. (a) Micrograph taken just after microinjection of cytoplasm (followed by an oil droplet: the nuclear envelope of the large germinal vesicle is clearly visible. (b) Micrograph taken 40 min after microinjection: the nuclear envelope has disappeared. (c) Micrograph taken 75 min after microinjection: first polar body has been emitted.

In the case of the starfish *Marthasterias glacialis*, MPF is detected in the cytoplasm shortly before GVBD, which occurs about 20 min after adding 1MeAde in sea water. It persists for a period of about 30–40 min and drops at the time of first polar body emission. It reappears at second meiotic metaphase and drops again at the time of second meiotic cleavage. It takes about 20 min for recipient oocytes to undergo GVBD upon cytoplasm transfer. Less than 5 min after hormone addition, thus about 10 min before the appearance of cytoplasmic MPF, a factor appears in the germinal vesicle which triggers GVBD when transferred into the cytoplasm of a recipient oocyte (30). This nuclear MPF has not been identified. It is not cyclin B-cdc2 kinase because no cyclin B is detected in the germinal vesicle. In some species (for example the starfish *Astropecten aranciacus*) microinjection of GV material readily induces GVBD in recipient oocytes even in the absence of hormonal stimulation (31). Finally, MPF amplification does not occur in enucleated starfish oocytes (32). It is still unknown how the germinal vesicle material controls MPF amplification,

Nicotinamide (10 mM in sea water) and microinjected $Li^+$ at a final intracellular concentration higher than 12 mM are potent inhibitors of MPF amplification in starfish, and most likely in other oocytes (33–35). For this reason appropriate controls must be performed when microinjecting thiophosphorylated nucleotides as adenosine-5'-$O$-(3-thiotriphosphate) (ATP-γS) or guanosine-5'-$O$-(3-thiotriphosphate) (GTP-γS), which are usually purified and purchased as lithium salts.

## 3. Detection of MPF activity in cell-free extracts

### 3.1 Extracts and assays

Extracts containing detectable MPF activities have been prepared from a variety of mitotic cells, including metaphase II-arrested *Xenopus* and fish eggs, maturing starfish oocytes at first meiotic metaphase, HeLa cells and yeast (21, 36–39). For any MPF activity to be directly detectable by microinjection, dilution of extracts by extraction buffer must be as limited as possible. Therefore, the pelleted cells are first washed with a convenient isotonic extraction buffer, excess buffer is removed, and cells are crushed in a minimal amount of the same buffer. Inclusion of both EGTA and phosphorylated small molecules such as ATP and Na–β-glycerophosphate favours detection of MPF activities in crude extracts, as does the addition of ATP-γS, which converts phosphoproteins to thiophosphorylated forms resitant to protein phosphatases (40–42). To assay MPF activity in such cell-free extracts, drops of samples are transferred on to the surface of a piece of Parafilm under a dissecting microscope, sucked up in the microinjection pipette and microinjected as described above for cytoplasm transfer. Heterologous transfers can, in principle, be used because MPF appears to act in a general and non-specific manner (6), although there may be some exceptions (see below). We describe below standard methods used to prepare extracts from unfertilized *Xenopus* eggs (*Protocol 6*) and maturing starfish oocytes (*Protocol 7*) using modifications of a procedure initially described by Wu and Gerhart (43). These procedures can be used with slight modifications to prepare extracts from other cells.

---

**Protocol 6.** Preparing cell free extracts of *Xenopus* eggs

*Materials*

- Extraction buffer (EB): 80 mM Na β-glycerophosphate, 20 mM EGTA, 15 mM $MgCl_2$ (pH 7.3)

- Protease inhibitors: 20 mM TPCK, 30 mM BAME, 80 mM PMSF, 0.1 mg/ml SBTI, 0.01 mg/ml leupeptin
- ATP-γS

*Method*

1. Remove the jelly as described above (*Protocol 2*).
2. Pack dejellied eggs in ice-cold extraction buffer (EB).
3. Remove excess buffer immediately.
4. Add 1 vol. of EB buffer containing both a cocktail of protease inhibitors and 2 mM ATP-γS.
5. Crush eggs by pipetting up and down with an air displacement pipette (pipette tip < 1 mm).
6. Centrifuge the homogenate at 160 000 $g$ for 30 min at 4 °C.
7. Recover the material between the pellet and lipid cap. Its protein concentration is about 10 mg/ml. This extract can be frozen in liquid nitrogen for storage.

**Protocol 7.** Preparing cell-free extracts of starfish (*Marthasterias*) oocytes

*Materials*

- $Ca^{2+}$-free artificial sea water, pH 5.5
- Natural sea water at 20 °C
- 1-methyladenine
- Buffer A: isotonic buffer containing 144 mM Na β-glycerophosphate, 34 mM EGTA, 27 mM $MgCl_2$, 1.8 mM DTT, 200 mM sucrose, 200 mM KCl (pH 7.8)
- Potter–Elvehjem homogenizer with Teflon pestle

*Method*

1. Release the oocytes from ovaries minced with scissors in $Ca^{2+}$-free artificial sea water adjusted to pH 5.5. The absence of $Ca^{2+}$ releases the oocytes from the follicle cells, and the acidic pH solubilizes the jelly coat that surrounds the oocytes.
2. Wash the oocytes several times in pH 5.5 sea water.
3. Transfer the washed oocytes into natural sea water containing 1 μM 1-methyladenine, which releases them from arrest at first meiotic prophase.

> **Protocol 7.** *Continued*
>
> 4. Forty minutes later (when they are at first meiotic metaphase) wash the oocytes in ice-cold buffer A.
> 5. Pellet the oocytes immediately by low-speed centrifugation (about 600 $g$ for 1 min).
> 6. Add 1 vol. of 2-fold diluted buffer A.
> 7. Crush oocytes in a Potter–Elvehjem homogenizer with a Teflon pestle.
> 8. Centrifuge at 140 000 $g$ for 40 min, and freeze the supernatant in liquid nitrogen. Store it at $-70\,°C$ until use.

## 3.2 Nature of MPF activities in cell-free extracts

Only a single component with MPF activity has been highly purified, so far, from metaphase extracts prepared from *Xenopus*, fish, or starfish oocytes (21, 25, 36, 37). It corresponds to cyclin B–cdc2 kinase, the major M-phase specific H1 histone kinase in a variety of cells. However, cyclin B–cdc2 kinase is not sensitive to $Ca^{2+}$ at 0.3 mM, and its activity is not decreased by treatment with broad-spectrum phosphatases such as alkaline or acid phosphatase. Since EGTA and phosphorylated small molecules such as ATP and $\beta$-glycerophosphate greatly favour detection of MPF in crude extracts, this suggests that other components may have MPF activity. In starfish and *Xenopus* at least, immature oocytes arrested at first meiotic prophase contain the cyclin B–cdc2 complex in an inactive form (44, 45), due to phosphorylation of the cdc2 subunit on inhibitory residues (Threonine 14 and Tyrosine 15). Microinjection of active cdc2 kinase into *Xenopus* oocytes is sufficient to convert the inactive into the active cyclin B–cdc2 complex, even in the absence of protein synthesis, a process known as MPF amplification. Therefore, at least some components of the MPF amplification loop would be expected to gain M phase-promoting activity once activated by cdc2 kinase, and, conversely, to lose this activity by the action of phosphoprotein phosphatases.

In agreement with this view, factors distinct from, but sharing MPF activity in common with cdc2 kinase, can be detected in homogenates treated with ATP-$\gamma$S (46, 47). Moreover, MPF activity is higher in cell-free extracts containing NaF and phosphorylated small molecules, although these molecules do not directly increase the activity of cdc2 kinase *in vitro*. The relative contribution of components stabilized or activated by phosphatase inhibitors to the total MPF activity of crude extracts may exceed that of cyclin B–cdc2 kinase, especially when crude extracts are treated with ATP-$\gamma$S, which further increases MPF activity. In agreement with this view, monoclonal antibodies specific for thiophosphorylated proteins were reported to immunoprecipitate

## 13: MPF and cyclin-dependent protein kinases

MPF activity in *Xenopus* extracts treated with ATP-γS (48), although cyclin B–cdc2 kinase was not precipitated. In the absence of ATP-γS, however, a single peak of MPF activity coincident with cyclin B–cdc2 kinase is detected throughout its purification to final homogeneity (36, 37).

Components of the MPF amplification loop activated by thiophosphorylation have not yet been identified, but candidates include the tyrosine/threonine phosphatase cdc25 (or activators of this phosphatase) as well as putative inhibitors of the tyrosine/threonine kinase(s) that phosphorylate(s) cdc2 on inhibitory sites. Components unrelated to MPF can also induce GVBD when microinjected into *Xenopus* oocytes. This includes proteins such as the regulatory subunits and the thermostable inhibitor of cAMP-dependent protein kinases (49, 50). In contrast to MPF these agents do not cause GVBD in the presence of protein synthesis inhibitors.

Microinjection of active cyclin B–cdc2 kinase alone may not be sufficient to induce the $G_2$ to M-phase transition in recipient oocytes of some species, as observed in the starfish *Astropecten aranciacus*. Indeed, in this species, the injected kinase undergoes rapid inactivation in recipient oocytes instead of activating the MPF amplification loop (31). This indicates that other components may be required in conjunction with cdc2 kinase for the microinjected kinase to exert its M-phase promoting activity in recipient oocytes. These components have not been identified, but their effects are mimicked by okadaic acid, when it is used at a low concentration to partially depress type 2A phosphatase activity. When used at a higher concentration, okadaic acid alone induces GVBD in recipient oocytes. It has been suggested that the germinal vesicle component which is required, in addition to cyclin B–cdc2 kinase, for MPF amplification in starfish oocytes might be an inhibitor of type 2A phosphatase.

In most instances activity of microinjected MPF is not sufficient to induce GVBD by itself in recipient oocytes, and amplification of endogeneous MPF is first required. If the amount of microinjected MPF is too small, it undergoes inactivation in the cytoplasm of prophase-arrested recipient oocytes, due to phosphorylation of cdc2 on inhibitory residues, and the MPF amplification process cannot be triggered. This explains why the response to microinjected MPF seems to obey an 'all-or-none' law (43).

## 4. Purification of cyclin B–cdc2 kinase (MPF)

The main step in the purification procedure is affinity chromatography on the yeast $p13^{suc1}$ protein, which binds to the yeast cdc2 protein and its homologues in higher eukaryotes. Any kinase complex containing a cdc2 homologue should be purified using only minor modifications of this procedure, provided the catalytic subunit is free to interact physically with the recombinant p13$^{suc1}$ protein. In mature oocytes, which are highly synchronized at metaphase, cyclin B is by far the principal component associated with cdc2. Therefore, it is easy to purify the cyclin B–cdc2 kinase to apparent homogeneity. The same procedure can be used for other cells, but a variety of cdc2 (or cdc2-like) complexes are expected to co-purify.

A detailed procedure for the purification of cyclin B–cdc2 kinase to apparent homogeneity from starfish or *Xenopus* oocytes has been described recently (51).

In outline, the material eluted at 200 mM NaCl from a DEAE ion-exchange column is applied to a column of Sepharose covalently bound to recombinant p13$^{suc1}$ protein, eluted with excess p13$^{suc1}$, and passed through a Mono-S column of a fast-performance liquid chromatography (FPLC) system. p13$^{suc1}$ does not bind to the matrix and is recovered in the flow-through. In contrast cyclin B–cdc2 kinase binds and is recovered as a sharp peak eluting at ~ 330 mM NaCl. Complete removal of p13$^{suc1}$ is important because this protein severely inhibits MPF amplification, although it has no effect by itself on the catalytic activity of cdc2 kinase. This procedure allows the purification of cyclin B–cdc2 kinase from starfish oocytes at first meiotic metaphase within ~ 8 h and with a yield of > 50%. One milligram of the final preparation catalyses the transfer of about 5 μmol of phosphate from ATP to histone H1 in 1 min under standard conditions.

An essentially identical procedure can be used to prepare cyclin B–cdc2 kinase in its inactive form (also called pre-MPF) from non-hormone stimulated starfish or *Xenopus* oocytes. Pre-MPF undergoes a slow, but spontaneous, activation in cell-free extracts after removal of low molecular weight molecules including ATP, due to basal cdc25 activity and the impossibility of rephosphorylating cdc2 on inhibitory residues in the absence of ATP (12, 24). Complete activation of the pre-MPF would be observed after the ion-exchange step unless precautions were taken to avoid this process. This can be achieved by adding ATP and vanadate (both 1 mM) in extracts and buffers until elution from the p13$^{suc1}$–Sepharose column (45).

## 5. Cyclin-dependent kinases

Cyclin B–cdc2 kinase is the first among a variety of cyclin-dependent kinase complexes to have been purified. In such complexes cdc2 is the catalytically active subunit with cyclin playing a regulatory role. In the active complex, cdc2 is phosphorylated on threonine 161 (threonine 167 in yeast). This phosphorylation is an absolute requirement for catalytic activity (52, 53). Under physiological conditions, cdc2 undergoes complete dephosphorylation and loses its catalytic activity in the absence of an associated cyclin subunit.

Besides cyclin B, the 34 kDa catalytic subunit can associate with a variety of other regulatory subunits, collectively called cyclins, although it is not always demonstrated that they follow periodic changes in abundancy during the cell cycle. Other proteins are included in the cyclin family on the basis of their amino acid sequence, although it has not been formally demonstrated that they can form heterodimers with cdc2. Even within the central 'cyclin box', there is only about 30% identity between cyclins of a given type, and only a small number of dispersed residues are perfectly conserved among the various members of the cyclin family (for review see ref. 54).

Besides cdc2 homologues, cdc2-like proteins have been described in multicellular eukaryotes. These 'cdc2-like' proteins are defined by their inability to complement cdc2 mutants at the $G_2$ to M-phase transition in fission yeast, at variance with 'cdc2-homologues' (55). There is biochemical evidence that some of these cdc2-like proteins form active protein kinase complexes when associated with members of the cyclin family. These cdc2-like proteins are referred to as cdk proteins, the first discovered member being cdk2 (cdc2 homologues would correspond to cdk1). It has been demonstrated that at least some of these cyclin-dependent kinases (cdk) play a regulatory role in cell cycle, although this may not be true for all members of this class of protein kinases. In budding yeast for example, $CLN_{1-3}$ cyclins definitively control the timing of 'Start' in $G_1$ before commitment to DNA replication (reviewed in ref. 56).

Besides cyclin B, another member of the cyclin family, cyclin A, is believed to play a role in the regulation of mitosis, at least in early embryos. Both cyclins form protein kinase complexes with cdc2 homologues that peak in activity at M phase and decline abruptly as M phase is completed. In agreement with this view, microinjection of cyclin A or B readily induces GVBD in $G_2$-arrested oocytes (57–59). In association with cdk2, cyclin A is also involved in the control of S phase in somatic cells (60–62).

Inactivation of cyclin A– and cyclin B–cdc2 kinase at the end of M phase requires the destruction of the cyclin subunit. This process involves conjugation of the cyclin subunit with ubiquitin and is triggered by the activity of cyclin B–cdc2 kinase itself (63–65). Once the cyclin subunit has been destroyed, the cdc2 catalytic subunit undergoes dephosphorylation on threonine 161 and the kinase is inactivated (53). As mitotic cyclins accumulate during the next interphase, they combine with cdc2. Although this complex undergoes phosphorylation on threonine 161, it is not immediately active, at least in the case of cyclin B–cdc2, because it also undergoes inhibitory phosphorylation on threonine 14 and tyrosine 15 (only tyrosine 15 in yeast). Activation of the cyclin B–cdc2 complex depends on the removal of inhibitory phosphates, which is catalysed by a specific phosphatase with dual specificity for tyrosine and threonine at the end of $G_2$. This phosphatase has been identified as homologues of the product of the fission yeast gene $cdc25^+$, which advances cells in mitosis (45, 66, 67).

## 5.1 Assay and properties of mitotic cdc2 kinases

Cyclin A and cyclin B are localized in different parts of the cell (68, 69). Moreover cyclin A– and cyclin B–cdc2 kinases have clearly different effects in modulating microtubule dynamics or in inducing cyclin degradation in cell-free extracts (70, 72). Nonetheless, no difference of substrate specificity has been detected *in vitro* between purified cyclin A– or cyclin B–cdc2 complexes. This suggests that the cyclin subunit controls the cellular localization of cdc2 kinases rather than their intrinsic protein kinase activity. *Protocol 9* describes an assay for mitotic cdc2 kinases. Note that although histone H1 is one of the best substrates for mitotic cdc2 kinases, and cyclin B–cdc2 kinases contribute the major H1 kinase activity at M phase in both oocytes and yeast (9, 12, 73), eukaryotic cells contain many other protein kinases capable of phosphorylating histone H1. Moreover, the activity of some of these H1 phosphorylating kinases increases at M phase. Nonetheless, it is possible to greatly improve the specificity of the cdc2 kinase assay in crude or partially-purified extracts by making use of their affinity for the fission yeast p13$^{suc1}$ (or its homologues) as described in *Protocol 9*.

---

**Protocol 8.** Assay of mitotic cdc2 kinases

1. Prepare the reacton mixture containing 20 mM HEPES-NaOH pH 7.4, 10 mM MgCl$_2$, 0.2 mM ($\gamma$-$^{32}$P) ATP (50 cpm/pmol), 1 mg/ml calf thymus histone H1 (Boehringer–Mannheim, Germany).
2. Initiate the reaction by the addition of enzyme.
3. Incubate reaction mix at 30 °C for 5 min.
4. Terminate the reaction by pipetting aliquots of the mixture onto 1 × 1 cm pieces of P81 ion-exchange paper (Whatman).
5. Immerse the papers in tap water (1 l for 50 samples).
6. Change water four times (every 5 min) and wash once in ethanol.
7. Dry papers and count in liquid scintillation solution.

---

**Protocol 9.** Assay of mitotic cdc2 kinases after purification on Sepharose-bound p13$^{suc1}$

1. Deplete extracts (high-speed supernatant prepared from them) from cdc2 complexes by affinity chromatography on p13$^{suc1}$ covalently bound to Sepharose (see *Section 4*).

2. Wash the p13$^{suc1}$–Sepharose beads several times at pH 7–7.5 with a buffer containing 200 mM NaCl and 0.5% Nonidet P-40 (or Triton X-100).
3. Assay H1 kinase activity directly by adding the reaction mixture described in *Protocol 8*.

The p13$^{suc1}$ affinity matrix does not discriminate between the various complexes containing cdc2 homologues or cdc2-like proteins. Nonetheless, it is possible to specifically immunoprecipitate cdc2 or cdc2-like complexes with antibodies raised against the C-terminus of the catalytic subunit, which do not interfere with protein kinase activities. Antibodies directed against the N-terminus or the conserved PSTAIR region do not immunoprecipitate active cdc2 or cdc2-like proteins in association with cyclin (53). Alternatively, immunoprecipitation can be performed using antibodies directed against the cyclin subunit.

## 5.2 Properties of cdc2 kinases

Cyclin A– and cyclin B–cdc2 kinases can use either ATP or GTP as a phosphate donor. The kinases are inhibited neither by heparin (3 μg/ml) nor by $Ca^{2+}$ (0.3 mM), but pyrophosphate ($IC_{50}$ 2 mM) and $Zn^{2+}$ ($IC_{50}$ 1 mM) are potent inhibitors.

The consensus target sequence for a cdc2 kinase phosphoacceptor is:

K/R S/T P X K/R.

Nonetheless, other kinases including MAP (microtubule-associated protein) kinases may display substrate site specificities overlapping with that of cdc2 kinases (myelin basic protein is an excellent substrate for both categories of kinase). Moreover, there are variations to the canonical cdc2 site, and several proteins are phosphorylated, at least *in vitro* on sites containing simply S/T P. Finally, the cyclin subunit readily undergoes phosphorylation in purified cdc2 kinases on sites which do not even contain this motif.

Although histone H1 is one of the best substrates both *in vitro* and *in vivo*, many other proteins have been described as potential substrates of cdc2 kinases. They include lamins, vimentin, caldesmon, myosin light chain (which lacks the canonical S/T P motif) many chromatin-associated proteins, and small G-proteins. The number of these substrates continues to grow (reviewed in ref. 74).

## 5.3 Production of mitotic cdc2 kinases *in vitro*

Since mitotic cyclins are destroyed at the end of each cell cycle during early embryogenesis in most animals, including marine invertebrates and *Xenopus*, whilst cdc2 is stable, extracts can be prepared from cells arrested at interphase by treating maturing oocytes or embryos with protein synthesis inhibitors. If a recombinant cyclin is added to such extracts, it will combine with endogenous cdc2 and form either an active or an inactive kinase complex, depending on conditions prevailing in the extract and on the nature of the cyclin. Cyclin A forms active kinase complexes without a lag phase with cdc2, whilst cyclin B forms either inactive complexes or complexes which become active only after a lag phase. The lag phase is suppressed if type 2A phosphatase activity is blocked by okadaic acid, or if active cyclin A–cdc2 kinase is also added in the extract (75). In *Protocol 10* we describe a procedure to prepare a chimeric cyclin A–cdc2 kinase from starfish extracts. In this experiment, recombinant cyclin A was produced in a soluble form by standard methods in *Escherichia coli* (strain BL 21) transfected with a human cyclin A cDNA, and purified to better than 80% homogeneity on a Mono Q column.

---

**Protocol 10.** Production of a chimeric cyclin A–cdc2 kinase

*Materials*

- Dialysis buffer: 25 mM Tris–HCl (pH 7.0), 5 mM β-glycerophosphate, 1.5 mM $MgCl_2$, 1 mM EGTA, 1 mM DTT, 1 mM ATP
- Emetine

*Method*

1. Stimulate starfish oocytes with 1-MeAde (*Protocol 7*).
2. Treat oocytes with 0.1 mM emetine. In *Marthasterias glacialis* the protein synthesis inhibitor is added 25 min after hormonal stimulation. Cyclin degradation occurs 60–70 min after hormonal stimulation. In the absence of protein synthesis oocytes arrest after the first polar body is emitted.
3. Between 100–120 min after hormonal stimulation, prepare extracts and high-speed supernatants as described for the preparation of cyclin B–cdc2 kinase (Section 4). Keep them frozen at −70 °C until use.
4. Centrifuge the thawed high-speed supernatant again (50 ml; 1 g of protein) for 40 min at 100 000 *g* at 4 °C.
5. Add recombinant human cyclin A to a final concentration of about 1 μM to the supernatant.
6. Dialyse the mixture against the dialysis buffer first at room temperature, then at 4 °C for 2 h.

7. Purify the cyclin A–cdc2 kinase from the dialysed material as described for cyclin B–cdc2 kinase (Section 4). The major components in the final preparation are starfish p34$^{cdc2}$ and human cyclin A. The preparation also contains a minor 32 kDa protein distinct from cdc2 but recognized by antibodies against the PSTAIR region of cdc2, most likely a cdc2-like protein.

## 5.4 Assay of cyclin degradation *in vitro*

Mitotic cyclins are stable at interphase and undergo ubiquitin-dependent degradation at the end of mitosis (64). They do not require association with cdc2 to undergo degradation. In fact the cyclin box can be deleted, and a chimeric protein made by fusing protein A with the N-terminal domain of cyclin B has been reported to undergo cell-cycle dependent degradation in *Xenopus* extracts. Degradation of *in vitro* translated cyclins can be triggered in interphase extracts prepared from *Xenopus* eggs by adding purified cyclin B–cdc2 kinase (65) or bacterially-produced cyclin B, which combines with free cdc2. Alternatively, cyclin degradation can be triggered in extracts prepared from unfertilized eggs arrested at metaphase II by first inactivating CSF with $Ca^{2+}$ (63, 76). *Protocol 11* describes the preparation of interphase extracts from *Xenopus* for cyclin degradation studies. Metaphase extracts are prepared in the same way, except that eggs are not parthenogenetically activated.

**Protocol 11.** Preparation of interphase extracts from *Xenopus* eggs

*Materials*
- Buffer: 100 mM KCl, 1 mM MgCl$_2$, 0–1 mM CaCl$_2$, 10 mM K–Hepes (pH 7.7), 50 mM sucrose, 5 mM EGTA, 10 µg/ml cytochalasin B, 10 µg/ml cycloheximide
- ATP regenerating solution: 10 mM creatine phosphate, 80 µg/ml creatine kinase, 1 mM ATP.

*Method*
1. Prepare interphase *Xenopus* extracts 15 min after parthenogenetic activation of dejellied unfertilized eggs (see *Protocol 2*).
2. Transfer eggs to centrifuge tubes containing buffer at 4 °C.
3. Remove excess buffer prior to centrifugation at 4 °C for 10 min at 15 000 *g* (63).
4. Collect the cytoplasmic layer from the crushed eggs.

> **Protocol 11.** *Continued*
> 5. Add the ATP-regenerating solution system.
> 6. Spin again for 15 min at 15 000 g.
> 7. Collect the supernatant (between the pellet and the lipid cap) and store it at −70 °C until use.

Full-length cyclin cDNAs are transcribed by T3, T7, or SP6 RNA polymerase, including m$^7$GpppG (p'-5'-(7-methyl)-guanosine-p$^3$-5'-guanosine triphosphate in the transcription reaction. Then [$^{35}$S] methionine-labelled cyclins are prepared in reticulocyte lysate programmed with the corresponding mRNAs. Besides the full-length proteins, truncated cyclins lacking the amino terminus are often generated in the reticulocyte system, owing to internal initiation of translation. Such truncated cyclins may excape cell-cycle dependent degradation if they lack the 'destruction box', a 9-amino acid region, RXALGX'IXN, which contains two invariant residues (R and L), four highly conserved residues, and one residue that serves to differentiate A and B-type cyclins (X', which is always V in A-type cyclins).

Full-length $^{35}$S-labelled cyclins (either B- or A-type) are stable in interphase extracts, but they undergo rapid degradation when cyclin B–cdc2 kinase is added, starting about 15 min after kinase addition. To assay degradation, samples are taken as a function of time and processed for determination of residual cyclins by fluorography after separating proteins by SDS–PAGE (*Figure 4*). Cyclin A–cdc2 kinase is inefficient in inducing cyclin

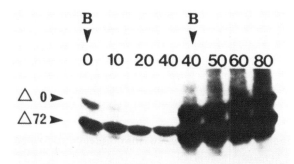

**Figure 4.** Cyclin degradation in *Xenopus* extracts. Cyclin B–cdc2 kinase, purified from starfish oocytes, was added at zero time to an interphase extract to give an activity of 15 pmol of phosphate transferred to histone H1/min/μl of extracts, simultaneously with [$^{35}$S]methionine-labelled starfish cyclin B, translated in the reticulocyte system. Samples were taken at the indicated times (min) and processed by SDS–PAGE and fluorography to monitor cyclin degradation. *In vitro* translation of starfish cyclin B produces both the full-length protein (Δ0) and a truncated protein lacking 72 amino acids from the N-terminus (Δ72), which is resistant to proteolysis. At 40 min, [$^{35}$S]methionine-labelled cyclin was added for a second time: Δ0 was not degraded because cyclin B–cdc2 kinase was inactivated during the first round.

degradation (71, 72). The threshold activity of cyclin B–cdc2 kinase required to trigger degradation is about 10 pmol of phosphate transferred to histone H1 min/µl of extract. A single round of cyclin degradation is observed, and the cyclin degradation pathway closes once cyclin B–cdc2 kinase has been inactivated, unless okadaic acid is simultaneously added to the extracts in order to block type 2A phosphatase (77).

Increasing free [$Ca^{2+}$] only transiently to the micromolar range is sufficient to induce cyclin degradation in extracts prepared from metaphase II-arrested *Xenopus* eggs. For this, $CaCl_2$ is added to a final concentration of 0.6 mM in extracts containing 5 mM EGTA. Although p39$^{mos}$ does not undergo proteolysis, it is probably inactivated by a $Ca^{2+}$-calmodulin dependent process, and this allows cyclin degradation by a ubiquitin-dependent pathway. In contrast, if free [$Ca^{2+}$] is increased to the millimolar range by adding $CaCl_2$ in excess (final concentration 6 mm), calpain is activated, and the $Ca^{2+}$-dependent protease destroys both p39$^{mos}$ and full-length cyclins. Unexpectedly, calpain does not proteolyse truncated cyclins lacking the destruction box (74).

# References

1. Johnson, R. T. and Rao, P. N. (1970). *Nature*, **226**, 717.
2. Masui, Y. and Clarke, H. J. (1979). *Int. Rev. Cytol.*, **57**, 185.
3. Masui, Y., and Markert, C. L. (1979). *J. Exp. Zool.*, **177**, 129.
4. Smith, L. D. and Ecker, R. E. (1979). *Dev. Biol.*, **25**, 232.
5. Kishimoto, T. and Kanatani, H. (1976). *Nature*, **260**, 321.
6. Kishimoto, T., Kuriyama, R., Kondo, H., and Kanatani, H. (1982). *Exp. Cell Res.*, **137**, 121.
7. Maller, J., Wu, M., and Gerhart, J. C. (1977). *Dev. Biol.*, **58**, 295.
8. Dorée, M., Peaucellier, G., and Picard, A. (1983). *Dev. Biol.*, **99**, 489.
9. Picard, A., Peaucellier, G., Le Bouffant, F., Le Peuch, C., and Dorée, M. (1985). *Dev. Biol.*, **109**, 311.
10. Picard, A., Labbé, J. C., Peaucellier, G., Le Bouffant, F., and Dorée, M. (1987). *Dev. Growth Diff.*, **29**, 93.
11. Meijer, L., and Pondaven, P. (1988). *Exp. Cell Res.*, **174**, 116.
12. Labbé, J. C., Picard, A., Karsenti, E., and Dorée, M. (1988). *Dev. Biol.*, **127**, 157.
13. Wagenaar, E. B. (1983). *Exp. Cell Res.*, **144**, 393.
14. Standart, N., Minshull, J., Pines, J., and Hunt, T. (1987). *Dev. Biol.*, **124**, 248.
15. Evans, T., Rosenthal, E. T., Yougblom, J., Distel, D., and Hunt, T., (1983). *Cell*, **33**, 389.
16. Minshull, J., Pines, J., Golsteyn, R., Standart, N., Mackie, S., Colman, A., Blow, J., Ruderman, J. V., Wu, M., and Hunt, T. (1989). *J. Cell Sci. (Suppl.)*, **12**, 77.
17. Nurse, P. (1990). *Nature*, **344**, 503.
18. Russell, P. and Nurse, P. (1986). *Cell*, **45**, 145.

19. Russell, P. and Nurse, P. (1987). *Cell*, **49**, 559.
20. Lee, M. G. and Nurse, P. (1987). *Nature*, **327**, 31.
21. Lohka, M., Hayes, M. K., and Maller, J. L. (1988). *Proc. Natl Acad. Sci. USA* **85**, 3009.
22. Gautier, J., Norbury, C., Lohka, M., Nurse, P., and Maller, J. (1988). *Cell*, **54**, 433.
23. Arion, D., Meijer, L., Brizuela, L., and Beach, D. (1988). *Cell*, **55**, 371.
24. Labbé, J. C., Lee, M. G., Nurse, P., Picard, A., and Dorée, M. (1988). *Nature*, **335**, 251.
25. Labbé, J. C., Capony, J. P., Caput, D., Cavadore, J. C., Derancourt, J., Kaghad, M., Lelias, M., Picard, A., and Dorée, M. (1989) *EMBO J.*, **8**, 3053.
26. Wasserman, P. and Masui, Y. (1975). *Exp. Cell Res.*, **91**, 381.
27. Gerhart, J., Wu, M., and Kirschner, M. (1984). *J. Cell Biol.*, **98**, 1247.
28. Sagata, N., Watanabe, N., Vande Woude, G. F., and Ikawa, Y. (1989). *Nature*, **342**, 512.
29. Kishimoto, T. (1986). In *Methods in cell biology*, Vol. 27 (ed. T. E. Schoeder), p. 379. Academic Press, New York.
30. Picard, A. and Dorée, M. (1984). *Dev. Biol.*, **104**, 357.
31. Picard, A., Labbé, J. C., Barakat, H., Cavadore, J. C., and Dorée, M. (1991). *J. Cell Biol.*, **115**, 337.
32. Kishimito, T., Hirai, S., and Kanatani, H. (1981). *Dev. Biol.*, **81**, 177.
33. Sano, K., Hishimoto, T., Koide, S. L., Kanatani, H., and Koide, S. S. (1979). *Dev. Growth Diff.*, **21**, 457.
34. Picard, A. and Dorée, M. (1983). *Exp. Cell Res.*, **147**, 41.
35. Dorée, M., Sano, K., and Kanatani, H. (1982). *Dev. Biol.*, **90**, 13.
36. Yamashita, M., Fukada, S., Yoshikumi, M., Bulet, P., Hirai, T., Yamaguchi, A., Lou, Y. H., Zhao, Z., and Nagahama, Y. (1992). *Dev. Biol.*, **149**, 8.
37. Labbé, J. C., Picard, A., Peaucellier, G., Cavadore, J. C., Nurse, P., and Dorée, M. (1989) *Cell.* **57**, 253.
38. Adlakha, R. C., Wright, D. A., Sahasrabuddhe, C. G., Davis, F. M., Prashad, N., Bigo, H., and Rao, P. N. (1985). *Exp. Cell Res.*, **160**, 471.
39. Tachibana, K., Yanagishima, N., and Kishimoto, T. (1987). *J. Cell Sci.*, **88**, 273.
40. Wasserman, W. J. and Masui, Y. (1976). *Science*, **191**, 1266.
41. Drury, K. (1978). *Differentiation*, **10**, 181.
42. Hermann, J., Bellé, R., Tso, J., and Ozon, R. (1983). *Cell Diff.*, **13**, 143.
43. Wu, M. and Gerhart, J. C. (1980). *Dev. Biol.*, **79**, 465.
44. Gautier, J. and Maller, J. L. (1991). *EMBO. J.*, **10**, 177.
45. Strausfeld, U., Labbé, J. C., Fesquet, D., Cavadore, J. C., Picard, A., Sadhu, K., Russell, P., and Dorée, M. (1991). *Nature*, **351**, 242.
46. Dorée, M., Labbé, J. C., and Picard, A. (1989). *J. Cell Sci. (Suppl.)*, **12**, 39.
47. Kuang, J., Penkala, J. E., Ashorn, C. L., Wright, D. A., Saunders, G. F., and Rao, P. N. (1991). *Proc. Natl Acad. Sci. USA*, **88**, 11 530.
48. Cyert, M., Scherson, T., and Kirschner, M. W. (1988). *Dev. Biol.*, **129**, 209.
49. Maller, J. L. and Krebs, E. G. (1977). *J. Biol. Chem.*, **252**, 1712.
50. Dorée, M., Kishimoto, T., Le Peuch, C., Demaille, J. G., and Kanatani, H. (1981). *Exp. Cell Res.*, **135**, 237.
51. Labbé, J. C., Cavadore, J. C., and Dorée, M. (1991). In *Methods in enzymology*, Vol. 200 (ed. T. Hunter and B. M. Sefton), p. 291. Academic Press, New York.

52. Solomon, M., Lee, T., and Kirschner, M. W. (1992). *Mol. Biol. Cell*, **3**, 13.
53. Lorca, T., Labbé, J. C., Devault, A., Fesquet, D., Capony, J. P., Cavadore, J. C., Le Bouffant, F., and Dorée, M. (1992). *EMBO J.*, **11**, 2381–90.
54. Xiong, Y. and Beach, D. (1991). *Current Biol.*, **1**, 362.
55. Paris, J., Le Guellec, R., Couturier, A., Le Guellec, K., Omilli, F., Camomis, J., McNeill, S., and Philippe, M. (1991). *Proc. Natl Acad. Sci., USA*, **88**, 1039.
56. Lew, D. J. and Reed, S. I. (1992). *Trends Cell Biol.*, **2**, 77.
57. Swenson, K. I., Farrell, K. M., and Ruderman, J. V. (1986). *Cell*, **47**, 861.
58. Tachibana, K., Ishivra, M., Uchida, T., and Kishimoto, T. (1990). *Dev. Biol.*, **140**, 241.
59. Roy, L. M., Swenson, K. I., Walker, D. H., Gabrielli, B. G., Li, R. S., Piwnica-Worms, H., and Maller, J. L. (1991). *J. Cell Biol.*, **113**, 507.
60. Girard, F., Strausfeld, U., Fernandez, A., and Lamb, N. J. C. (1991). *Cell*, **67**, 1.
61. Pagano, M., Pepperhok, R., Verde, F., Ansorge, W., and Draetta, G. (1992). *EMBO J.*, **11**, 961.
62. Zindy, F., Lanma, E., Chenivesse, X., Sobejak, J., Wang, J. Fesquet, D., Henglein, B., and Brechot, C. (1992). *Biochem. Biophys. Res. Commun.*, **182**, 1144.
63. Murray, A. W., Solomon, M. J., and Kirschner, M. W. (1989). *Nature*, **339**, 280.
64. Glotzer, M., Murray, A. W., and Kirschner, M. W. (1991). *Nature*, **349**, 132.
65. Félix, M. A., Labbé, J. C., Dorée, M., Hunt, T., and Karsenti, E. (1990). *Nature*, **346**, 379.
66. Dunphy, W. G. and Kumagai, A., (1991). *Cell*, **67**, 189.
67. Gautier, J., Solomon, M. J., Booher, R. N., Bazan, J. F., and Kirschner, M. W. (1991). *Cell*, **67**, 197.
68. Pines, J. and Hunter, T. (1991). *J. Cell Biol.*, **115**, 1.
69. Gallay, P. and Nigg, E. A. (1992). *J. Cell Biol.*, **117**, 213.
70. Buendia, B., Draetta, G., and Karsenti, E., (1992). *J. Cell Biol.*, **116**, 1431.
71. Luca, F. C., Shibuya, E. K., Dohrman, C. E., and Ruderman, J. V., (1991). *EMBO J.*, **10**, 4311.
72. Lorca, T., Labbe, J. C., Devault, A., Fesquet, D., Strausfeld, U., Nilsson, J., Nygren, P. A., Uhlen, M., Cavadore, J. C., and Dorée, M. (1992). *J. Cell Sci.*, **102**, 55.
73. Moreno, S. and Nurse, P. (1990). *Cell*, **61**, 549.
74. Nigg, E. A. (1991). *Seminars in Cell Biol.*, **2**, 261.
75. Devault, A., Cavadore, J. C., Fesquet, D., Labbé, J. C., Lorca, T., Picard, A., Strausfeld, U., and Dorée, M. (1991). *Cold Spring Harbor Symp. Quant. Biol.* Vol. 56. p. 503. Cold Spring Harbor Laboratory Press, Cold Spring Harbor, NY.
76. Lorca, T., Galas, S., Fesquet, D., Devault, A., Cavadore, J. C., and Dorée, M. (1991). *EMBO J.*, **10**, 2087.
77. Lorca, T., Fesquet, D., Zindy, F., Le Bouffant, F., Cerruti, M., Brechot, C., Devauchelle, G., and Dorée, M. (1991). *Mol. Cell. Biol.*, **11**, 1171.

# A1
# Addresses of suppliers

**Accurate Chemical and Scientific Corporation**, 300 Shames Drive, Westbury, NY 11590, USA.

**Aldrich Chemical Co. Ltd**, The Old Brickyard, New Road, Gillingham, Dorset SP8 4JL UK; Milwaukee, Wisconsin, USA.

**Amersham International plc**, Northern Europe Region, Lincoln Place, Green End, Aylesbury, Bucks HP20 2TP, UK; 2636 South Clearbrook Drive, Arlington Heights, IL 60005, USA.

**Atochem UK Ltd**, Colthrop Way, Thatcham, Newbury, Berks RG13 4LW, UK.

**BDH**: see Merck

**Beckman Instruments (UK) Ltd**, Progress Road, Sands Industrial Estate, High Wycombe, Bucks HP12 4JL, UK; 2500 Harbor Boulevard, Fullerton, CA 92634–3100, USA.

**Becton Dickinson (UK) Ltd**, Between Towns Road, Cowley, Oxford OX4 3LY, UK; 1 Becton Drive, Franklin Lakes, NJ 07417–1886, USA.

**Berkeley Antibody Company (BAbCo)**, 4131 Lakeside Drive, Richmond, CA 94806, USA.

**Bio 101 Inc.**, PO Box 2284, La Jolla, CA 92038, USA; Stratech Scientific Ltd, 61/63 Dudley Street, Luton, Beds LU2 0NP, UK.

**Bio-Rad Laboratories**, 3300 Regatta Boulevard, Richmond, CA 94804, USA; Bio-Rad House, Maylands Avenue, Hemel Hempstead, Herts HP2 7TD, UK.

**Boehringer–Mannheim GmbH**, Sandhofer Strasse 116, D–6800 Mannheim 31, Germany; Bell Lane, Lewes, E. Sussex BN7 1LG, UK.

**BRL**, see GIBCO BRL.

**Calbiochem**, PO Box 12087, San Diego, CA 92112–4180, USA; Calbiochem Novabiochem (UK) Ltd, 3 Heathcoat Building, Highfields Science Park, University Boulevard, Nottingham NG7 2QJ, UK.

## Addresses of suppliers

**Citifluor Ltd**, Connaught Building, City University, Northampton Square, London EC1V 0HB, UK.

**Cole–Palmer International**, 7425 Northoak Park Avenue, Chicago, IL 60648, USA; CP Instrument Company Ltd, PO Box 22, Bishop's Stortford, Herts CM23 3DX, UK.

**Continental Chemical Company**, 270 Clifton Boulevard, Clifton, NJ 67011, USA.

**Corning Science Products**, PO Box 5000, Corning, NY 14830, USA; M. J. Patterson (Scientific) Ltd, Unit 21 Apex Business Centre, Avenue One, Letchworth, Herts SG6 2BB, UK.

**Corning Tissue Culture Products**, Bibby Sterilin Ltd., Tilling Drive, Stone, Staffordshire ST15 0SA, UK.

**Costar**, One Alewife Center, Cambridge, MA 02140, USA; Victoria House, 28–38 Desborough Street, High Wycombe, Bucks HP11 2NF, UK.

**Coulter Corporation**, 590 West Twentieth Street, Hialeah, FL 33010, USA; Coulter Electronics Ltd, Northwell Drive, Luton, Beds LU3 3RH, UK.

**David Kopf Instruments**, 7324 Elmo Street, Tujunga, CA 91042, USA.

**Dianova**, Raboisen 5, D–2000 Hamburg 1, Germany.

**Difco**, Difco Laboratories, Detroit, Michigan, USA.

**Dow-Corning**, see Corning.

**Drummond Scientific Company**, 500 Parkway, Broomall, PA 19008, USA.

**DuPont (UK) Ltd**, Biotechnology Systems Division, NEN Research Products, Wedgwood Way, Stevenage, Herts SG1 4QN, UK.

**Dynal (UK) Ltd**, Station House, 26 Grove Street, New Ferry, Wirral, Merseyside L62 2AB, UK; Dynal, Inc., 45 North Station Plaza, Great Neck, NY 11020 USA.

**Electron Microscopy Sciences**, 321 Morris Road, Box 251, Fort Washington, PA 19034, USA.

**Enzo Diagnostics, Inc.**, 325 Hudson Street, New York, NY 10013, USA; Cambridge BioScience, 25 Signet Court, New Market Road, Cambridge CB5 8LA, UK; Universal Biologicals (UK agent), 30 Merton Road, London SW18 1QY, UK.

**Falcon**: see Becton Dickinson

**Fisher Scientific Company**, 711 Forbes Avenue, Pittsburgh, PA 15219, USA;

## Addresses of suppliers

**A. R. Horwell Ltd**, 73 Maygrove Road, West Hampstead, London NW6 2BP, UK.

**Fluka Chemicals Ltd**, The Old Brickyard, New Road, Gillingham, Dorset SP8 4JL, UK.

**Fungal Genetics Stock Center**, University of Kansas Medical Center, Department of Microbiology, Kansas City, Kansas 66103, USA.

**Gelman Sciences**, 600 S. Wagner Road, Ann Arbor, MI 48106–1448, USA; Brackmills Business Park, Caswell Road, Northampton NN4 0E7, UK.

**Genetics Society of America**, Bethesda, Maryland.

**GIBCO BRL Life Technologies**, 3175 Staley Road, Grand Island, NY 14072, USA; 3 Fountain Drive, Inchinnan Business Park, Inchinnan, PA4 9RF, Scotland.

**Hoechst (UK) Ltd**, Hoechst House, 50 Salisbury Road, Hounslow, Middlesex TW4 6JH, UK; Hoechst Celanese Corporation, Specialty Chemicals Group, Building 5200, 77 Center Drive, PO Box 1026, Charlotte, NC 28201–1026, USA.

**Hoefer Scientific Instruments**, 654 Minnesota Street, Box 77387, San Francisco, CA 94107, USA; Hoefer Scientific Instruments, PO Box 351, Newcastle, Staffs ST5 0TT, UK.

**IBI Ltd**, 36 Clifton Road, Cambridge CB1 4ZR, UK.

**ICN Biomedicals Ltd (ICN)**, Eagle House, Peregrine Business Park, Gomm Road, High Wycombe, Bucks HP13 7DL, UK.

**InterSpex Products, Inc.**, 11558 Chess Drive, Suite 114, Foster City, CA 94404, USA.

**Intervet Inc.**, PO Box 318, 405 State Street, Millsboro, DE 19966, USA; Science Park, Milton Road, Cambridge CB4 4BH, UK.

**Jackson Immuno Research Laboratories Inc.**, 872, West Baltimore Pike, PO Box 9, West Grove, PA 19390, USA; UK supplier see Stratech.

**Leitz**, Wild Leitz USA Inc., 24 Link Drive, Rockleigh, NJ 07647, USA.

**Life Science Laboratories**, Sarum Road, Luton, Bedfordshire LU3 2RA, UK.

**Maravac Ltd**, 5821 Russell Street, Halifax, Nova Scotia, Canada B3K 1X5.

**Merck**, Broom Road, Poole, Dorset BH12 4NN, UK.

**Molecular Dynamics**, 880 East Arques Avenue, Sunnyvale, CA 94086, USA; 4 Chaucer Business Park, Kemsing, Sevenoaks, Kent TN15 6PL, UK.

Addresses of suppliers

**Molecular Probes Inc.**, PO Box 22010, Eugene, Oregon, OR 97042–9144, USA.

**MSE (Measuring Scientific Equipment)**, Manor Royal, Crawley, Sussex RH10 2QQ, UK.

**Murex Diagnostics Inc.**, 4401 Research Common, 79 T. W. Alexander Drive, PO Box 13451, Research Triangle Park, NC 27709–3451, USA; Temple Hill, Dartford, Kent, DA1 5AH, UK.

**New England Biolabs, Inc.**, 32 Tozer Road, Beverly, MA 01915–5599, USA; 6231 Schwalbach/Taunus, Germany.

**Novo BioLabs Ltd.**, St John's Innovation Centre, Cowley Road, Cambridge CB4 4WS, UK.

**Nunc:** see GIBCO BRL

**Oncogene Sciences**, 106 Charles Lindbergh Boulevard, Uniondale, NY 11553–3649, USA.

**Pharmacia LKB Biotechnology Inc.**, 800 Centennial Avenue, PO Box 1327, Piscataway, NJ 08855, USA; Davy Avenue, Knowlhill, Milton Keynes, MK5 8PH, UK; Björkgatan 30, S–751 82 Uppsala, Sweden.

**Polysciences Ltd.**, 24 Low Farm Place, Moulton Park, Northampton NN3 1HY, UK; 400 Valley Road, Warrington, PA 18976, USA.

**Prior Scientific Instruments**, Unit 4, Wilbraham Road, Fulbourn, Cambridge CB1 5ET, UK.

**Sandoz**, CH–4002, Basel, Switzerland.

**Saxon Biochemical GmbH**, D–3000 Hannover, Germany.

**Scientific Products**, 1430 Waukegan Road, McGaw Park, Illinois 60085, USA.

**SCM Specialty Chemicals**, PO Box 1466, Gainesville, FL 32602 USA.

**Sera Labs Ltd.**, Crawley Down, Sussex RH10 4FF, UK.

**Serva Biochemicals**, 200 Shames Drive, Westbury, NY 11590, USA; Postfach 105260, Carl Benz-Strasse 7, D–6900 Heidelberg, Germany.

**Sigma Chemical Company**, PO Box 14508, St Louis, MO 63178, USA; Fancy Road, Poole, Dorset BH17 7NH, UK.

**Snake Farm, South Africa**, Woods Snake Park, Fish Hoek, PO Box 6, Cape Province, South Africa.

**Spectrapor**—tradename of: **Spectrumedical**, 8430 Santa Monica Boulevard,

Los Angeles, CA 90069, USA; Medicell International Ltd., 239 Liverpool Road, London N1 1LX, UK.

**Stratech Scientific Ltd**, 61–63 Dudley Street, Luton, Beds LU2 0NP, UK.

**Vector Laboratories**, 16 Walfire Square, Bretton, Peterborough PE3 8RF, UK.

**VWR Scientific**, PO Box 13645, Philadelphia, PA 19101, USA.

**Whatman International Ltd**, St. Leonard's Road, Maidstone, Kent ME16 0LS, UK.

# Index

In the list of protocols, the procedures are listed first, to assist their independent location. Within the general index, most of the procedures appear under the names or types of organism.

$\alpha_1$-mating factor 39
  arrest, *S. cerevisiae* 73
  *MATAa* strain, *Saccharomyces cerevisiae* 39
acetate buffer 260
acridine orange stain (AO) 54
F-actin, rhodamine-phalloidin staining 108
actin structures, staining 108
activation chamber 263
addresses, equipment 311–15
adenosine-5'-O-(3-thiotriphosphate)
  (ATP–γS) 268–9, 296, 298
affinity chromatography, purification of
  MPF 299–300
antibody screens, cloning 121–2
aphidicolin, arrest of replication 6
arrest of mitosis, cautionary note 19–20
*Aspergillus nidulans*
  cell cycle analysis 127–42
    generation time 127
    scoring 128–30
    using conditional mutations 140–1
  cell cycle synchrony methods 140–1
    hydroxyurea block-release 140
  life cycle *(illus.)* 128–9
  *nim*T23 mutation 141
  number of nuclei 127
  transformation 135–9
    conditional mutations, *alc*A promoter 136–7
    heterokaryon gene disruption, phenotype analysis *(illus.)* 137–9
    *pyr*G gene 135
*Astropecten see* starfish oocytes
ATP–γS *see* adenosine-5'-O-(3-thiotri-phosphate) (ATP–γS)
automatic synchronization apparatus *(illus.)* 17
autoradiographic methods
  mammalian cell cycle 46–50
    [³H]TdR labelling index 46–7
    fraction of labelled mitoses (FLM) analysis 47–9
    tumour growth fraction estimation 49–50

β-glycerophosphate 266, 267, 296
*bar*1 mutant cells and Bar1 protease 73
Barth X solution 183

benomyl
  obtaining 141
  and related compounds, *S. pombe* 111
biotin-11-dUTP labelling, *Xenopus* cell-free
  egg extracts 190–1
BND-cellulose (benzoylated naphthoylated
  diethylaminoethyl cellulose) 220
bromodeoxyuridine (BrdU)
  immunocytochemical detection 63–4
  incorporation, detection by HO 258 64–6
buffers
  acetate buffer 260
  extract dialysis 204
  extraction buffer (EB) 180, 268
  GE buffer 5
  hypotonic 204
  phosphate-buffered saline 204
  phosphate-citrate buffer 147
  Pipes buffer 184
  regrowth buffer 279
  for replication-competent extract preparation (SV40) 204
  for T-antigen purification 202
  TAE buffer 224
  TBE buffer 224

calcofluor staining, *S. pombe (illus.)* 106, 108
cdc protein kinases
  assay, histone H1 267–9
  p34$^{cdc25}$ gene product 287
  p34$^{cdc28}$ protein kinase
    *S. cerevisiae* 80
    precipitation with p13$^{suc1}$ beads 81
  p34$^{cdc2}$ protein kinases 104–6, 257–8, 273–6, 287
    assay 302
    consensus target sequence 303
    production *in vitro* 303–5
    regulation 273–7
  *see also* cyclins; maturation-promoting factor
cdc-like proteins 301
cdk proteins 301
cDNA libraries
  isolation of higher eukaryotic gene homologues 85
  *S. pombe* 120
  yeast expression vectors 86

# Index

centrifugal elutriation, separation of cell-cycle phases 20–2
centrosomes, CHO, isolation 242–5
CHO cells
  isolation of centrosomes 242–5
  isolation of mitotic chromosomes 246
  mimosine arrest 5–6
  see also mammalian cells
chromatography, affinity, purification of MPF 299–300
co-segregation test, heterologous genes, S. cerevisiae 88–9
Colcemid, mitotic arrest 15
complementation, cloning by, Schizosaccharomyces pombe 120–2
creatine phosphate-diTris salt 206–7
crystal violet stock solution 13
cyclin A-p34$^{cdc2}$ kinase
  activity during fertilization 257
  assay and properties 302–3
  chimeric, production 304
  inactivation, M phase 301
cyclin-dependent protein kinases 285–309
  activation 273–6
  assay of cyclin degradation 305–7
  assay and properties 302–3
  family described 300-7
  maturation-promoting factor 287–309
  p34$^{cdc2}$/cyclin B kinase, DNA topoisomerase II inhibitors 9
  production in vitro 303–5
  see also maturation-promoting factor
cyclins
  A, B1, and B2, expression 271–2
  activation of cdc protein kinases 275–6
  activity
    analysis 273–7
    cyclins as fusion proteins 270–2
  assay, Xenopus egg extracts 270–7
  CLN$_{1-3}$ cyclins, yeast 301
  cyclin B1, activation 265
  degradation, Xenopus egg extracts, (illus.) 306
cycloheximide, cell cycle blocking 255–7
  S. pombe 111
cytosine arabinoside, lethal arrest 6
cytostatic factor (CSF) 292

DAPI see 4′,6-diamino-2-phenylindole
dejellying, Xenopus eggs 178, 180, 261
density substitution, BrdUTP 188
deoxyadenosine, block-release, in yeasts 36
4′,6-diamino-2-phenylindole (DAPI)
  and calcofluor staining, S. pombe (illus.) 106
  immunofluorescence staining, Drosophila 157

staining 78
Aspergillus nidulans 130
digestion solution, preparation 132
diphenylamine reagent 98
disinfection, agents 33
DMSO-freezing medium, mammalian mitotic cells 14
DNA, lambda DNA, preparation 184
DNA content
  and cell age 51
  coefficient of variation (CV), DNA-associated fluorescence 51
  measurement 51–3
DNA denaturation, and differential staining 57
DNA replication
  analysis of origins and directions 213–34
  in vitro SV40 DNA replication 197–210
  Xenopus egg extracts 177–95
DNA topoisomerase II inhibitors 9
Hoechst 33342 9–10
Drosophila
  brain preparations 163–9
    immunostaining larval neuroblasts 166–8
    in situ hybridization 168
  cell-cycle control
    during development 143–4
    gametogenesis 143
  cell division mutants
    generation 146–7
    mapping 147–53
    phenotypes 144–5
  embryos
    collection and dechorionation 158–9
    introducing reagents 159
    microinjection 159
    permeabilization 160
    preparation for immunostaining 160–3
  immunolocalization studies 156–8
  localization of transcripts 153–5
  mutations affecting male meiosis 169–73
    assays for non-disjunction 173
    cytological studies (illus.) 169–73
  tissue preparation 153–5
DTT (dithiothreitol) 206–7
Dynabeads 194

Edinburgh Minimal Medium, Scizosaccharomyces pombe 28
EGS, ethylene glycol bis-(succinic acid N-hydroxysuccinimide ester) 190
EGTA, ethylene glycol bis-($\beta$-aminoethyl ether) $N,N,N',N'$-tetraacetic acid 259–60, 296
electroporation, cloning by complementation 113–14

# Index

elutriation
  centrifugal
    separation of cell-cycle phases 20–2
  yeasts 32
enhancer traps 146–7
equipment
  addresses 311–15
*Escherichia coli*
  *lac-Z* expression 146–7
  TB1 strain 271
ethidium bromide, staining 107
extraction buffer (EB) 180, 268
extraction solution, preparation 132

FACS analysis *see* fluorescence activated cell sorter
fibroblasts
  human diploid, quiescence 2–4
  mouse, cell crisis, $G_0$ cells 3
  *Xenopus* (WAK cells), preparation 19
fission yeast *see Scizosaccharomyces pombe*
flow cytometry 50–60
  advantages 45
  BrdU incorporation 45
    detection by HO 258 64–6
    immunocytochemical detection 63–4
  kinetic methods, analysis, mammalian cell cycle 60–6
  multiparameter analysis
    DNA vs chromatin changes during cell cycle 56–7
    DNA vs proliferation-associated antigens 57–60
    DNA vs RNA content 53–5
  *S. pombe (illus.)* 101–4
  univariate cellular DNA content distribution 51–3
fluorescence activated cell sorter 75–8
5-fluoro-acetic acid (5-FOA) 89
fluorodeoxyuridine, lethal arrest 6
fraction of labelled mitoses (FLM) analysis
  mammalian cell cycle
    autoradiographic methods 47–9
    cell-cycle parameters *(illus.)* 48
freezing of mammalian mitotic cells 13–15
  medium 14
frog egg extracts *see Xenopus*

$G_0$ cells, non-cycling, generation 2–4
$G_0/G_1$ transition, production 4
$G_1$ populations
  late arrest, mimosine or HTCDT 5–6
  production 4–6
$G_2$
  prophase extracts *Xenopus* 257
  synchronization of mammalian cells 9–11

GE buffer 5
gel electrophoresis
  DNA preparation 220–33
    enrichment for replicating molecules 221–3
    first dimension electrophoresis 224–5
    incubation with restriction enzyme 220
    modification of N/N procedure 230–3
    second dimension electrophoresis 225–30
  DNA replication analysis of origins and directions 213–34
  gel replicon mapping methods 214–19
    advantages/disadvantages 219
    N/A mapping technique *(illus.)* 217–19
    N/N mapping technique *(illus.)* 215–17
    N/N and N/A techniques 214
    specific probes 233
  *Xenopus* MBP-cyclin $B_1$ fusion protein in bacteria *(illus.)* 274
genomic libraries in yeast expression vectors 113–14
  contraindications 120
  mammalian genes 85–6
germinal vesicle breakdown (GVBD), *Xenopus* 288, 291–2, 299
glutathione-S-transferase (GST) 270–1
glycerol cushions, preparation 192, 240
glycosaminoglycans, acridine orange staining 55
growth fraction (GF) estimation 49
growth media
  *S. cerevisiae*
    MYGP medium for feed–starve synchrony 30
    YPD (YEPD) broth 30, 72
  *S. pombe* 28, 30
guanosine-5'-O-(3-thiotriphosphate) (GTP-γ) 296
GVBD (germinal vesicle breakdown), *Xenopus* 288, 291–2, 299

[$^3$H]thymidine (TdR) labelling index, mammalian cell cycle, autoradiographic methods 46–7
[6-$^3$H]uracil, uptake, *S. pombe* 99
heat shock, synchronization method 38
HeLa cells
  isolation of mitotic chromosomes 248
  replication-competent extracts 204
  *see also* mammalian cells
hexylene glycol, isolation of mitotic organelles 239
histone H1 kinase assays
  calculation of activity 269
  *S. cerevisiae*, 80–1
Hoechst 33258 (HO 258) 64–6
  fluorescence *(illus.)* 66
  immunofluorescence staining 157

# Index

Hoechst 33342, DNA topoisomerase II inhibitor 9–10
Hoechst 768159 (HTCDT), late arrest of $G_1$ populations 5–6
*Holothuria forskali*, induction of meiotic maturation 295–6
human chorionic gonadotrophin (HCG) 258
humid chamber 192
hybridization, cloning by 121–2
hydroxyurea
  arrest of replication 6
  block-release
    in *Aspergillus* 140
    in yeasts 36
  DNA precursor inhibition, *S. pombe* 110
  preparation of arrested cells, *S. cerevisiae* 73–4

IgG exclusion assay 193
immunofluorescence staining
  *Drosophila* 156
  *S. pombe* 109–10
  *Xenopus* egg extracts 191–2
integrative mapping, *S. pombe* 117
interphase extract, *Xenopus* 255, 263–4
isolation of mitotic organelles 235–51
isoleucine-depletion arrest point 4–5
isoleucine-free growth medium 4–5

jump-starter 146

keratins, acridine orange staining 55

labelling protocols
  biotin-11-dUTP labelling 190
  [$^{32}$P]dCTP labelling 186–7
[$^{3}$H]leucine, incorporation into yeast cultures 34
lipofectin-addition, *S. pombe* transformation 113–14
lithium acetate, *S. pombe* transformation 113–14
lithium chloride, *S. pombe* transformation 113–14
lysis method, for mitotic chromosome 247

maltose binding protein (MBP) 270
mammalian cell cycle
  analysis 45–68
    autoradiographic methods 46–50
    flow cytometric kinetic methods 60–6
    flow cytometric methods 50–60
  arrest and alterations in control 19
  cyclin A loss 19
  DNA denaturation *(illus.)* 57
  DNA vs RNA content *(illus.)* 54
  estimation of duration of phases 53
  position in cell cycle *(illus.)* 51
  synchronization 1–23
mammalian cells
  BrdU pulse labelling *(illus.)* 64
  mitotic shake-off 12, 16, 237
    choice 11, 15
  replication-competent extracts (SV40) 204
  stathmokinetic methods 60–2
*Marthasterias see* starfish oocytes
maturation-promoting factor 287–309
  all-or-none law 299
  defined 286
  detection of activity
    species other than *Xenopus* 292–6
    *Xenopus* cell-free egg extracts 296–8
    *Xenopus* oocytes 287–96
  inactivation, M phase 301
  MPF amplification 298
  nature of activity in cell-free extracts 298–9
  phosphorylation state of proteins 286
  purification 299–300
MBC (methylbenzimadazol-2-yl carbamate) 111
metaphase chromosomes, CHO, isolation 246–50
1-methyladenine (1-MeAde) 285, 297
microcystin-LR 276
microinjection, volume-controlled system 292–6
microtubules
  hyperstabilization 241
  methanol fixation 162
  morphology, *S. cerevisiae* 78
  *Xenopus*, *(illus.)* 279
mimosine, late arrest of $G_1$ populations 5–6
mitochondrial DNA 77
mitotic cells
  checkpoint control
    classes of cells 15
    peculiarities 15
  collection by arrest and shake-off 12, 16, 237
  detachment and collection 11–14
    checkpoint, peculiarities 15
    freezing and recovery 13–14
    without drug use 11–13
mitotic organelles, isolation 235–51
mitotic shake-off 12, 16, 237
  choice of cells 11, 15
mitotic spindles
  isolation *(illus.)* 235–6
  spindle index, *Aspergillus* 130
  structure and components 238–42
Mowiol, preparation 280

# Index

MPF *see* maturation-promoting factor
MYGP medium for feed–starve synchrony, *S. cerevisiae* 30

N115 cells, isolation of centrosomes 244
nicotinamide, MPF in starfish 296
*nim1* and *wee1* gene products 287
nitrous oxide
   automatic delivery 18
   mitotic arrest 15
   synchrony in mitosis 16–19
*nmt* promoter, inducing overexpression 119
nocodazole
   mitotic arrest 15
   preparation of arrested cells, *S. cerevisiae* 73–4
Novozym 234, preparation 132
nuclear assembly, phase-contrast microscopy 191

octane 160
okadaic acid 276, 299

[$^{32}$P]deoxycytosine triphosphate (dCTP), labelling, *Xenopus* cell-free egg extracts 186–7
p13$^{suc1}$ beads 81
p13$^{suc1}$ protein, recombinant, in chromatography 300
p34$^{cdc2}$ protein kinases *see* cdc protein kinases
P-element promoter 146–7
PAGE *see* gel electrophoresis
paraformaldehyde solution, preparation 132
PCR, cloning by 121–2
PEM, preparation 133
phase-contrast microscopy, nuclear assembly 191
phosphate-buffered saline buffer 204
phosphate-citrate buffer 147
Pipes buffer 184
plastid loss test, heterologous genes. *S. cerevisiae* 88–9
pMAL$^{TM}$ vector 271
pregnant mare serum gonadotrophin (PMSG) 258
progesterone 285
propidium iodide, fluorescence activated cell sorting 75–8, 102, 107
protease inhibitors 202
*protocols*
   $\varrho^0$ mutants, generation 77
   assays
      Cln-associated histone H1 kinase activity, *S. cerevisiae* 82
      histone H1 kinase activity assay 268

mitotic cdc2 kinases 302
SV40 replication assay 208
autoradiography, single labelled cells, *S. pombe* 100
brain preparations
   *Drosophila* 164, 165, 166, 168, 169
   HO 258 staining 165
BrdU, immunocytochemical detection of 63
BrdUTP density substitution 188
cdc kinase activity, *S. pombe* 104
centrosomes
   isolation from CHO cells 242
   isolation from N115 cells 244
CHO cells
   arrest and synchronization, in $G_2$ 10–11
   freezing and recovery of mitotic cells 13
chromosome preparation, *Drosophila*
   acid orcein stained 164
   unstained 164
column preparation, SV40 T-antigen purification 200
conditional mutations, generation, *alc*A promoter, *Aspergillus* 136
CsCl gradient centrifugation 188
cyclin A-cdc2 kinase, chimeric, production 304
cyclin-dependent protein kinase activation, *Xenopus* egg extracts 276
cyclins, MPB-*Xenopus* cyclins 271
DAPI staining
   *Aspergillus* 130
   nuclear DNA, *S. pombe* 107
   *S. cerevisiae* 78
demembranated sperm heads, *Xenopus* 183
density substitution 188
DNA, lambda DNA, preparation 184
DNA content, determination of bulk DNA content, *S. pombe* 98
DNA replication
   density substitution, BrdUTP 188
   enrichment for replicating DNA 221
   monitoring via [6-$^3$H]uracil uptake, *S. pombe* 99
   SV40 208
   *Xenopus* egg extracts, labelling 187
egg extract preparation, *Xenopus* 180, 260, 263, 266, 296
FACS analysis, *S. cerevisiae* 76
fixation of *Drosophila* embryos 161
flow cytometry
   BrdU detection 63
   BrdU incorporation, detection by Hoechst 33258 65
   DNA vs chromatin changes during cell cycle 56
   DNA vs proliferation-associated antigens 59
   RNA vs DNA content 54
   *S. pombe* 101–4

# Index

flow cytometry (*cont.*):
fraction of labelled mitoses (FLM) analysis 48
gel electrophoresis
alkaline second dimension 229
first dimension 225
neutral second dimension 227
restriction enzyme digestion of DNA, first dimension gel 232
[$^3$H]thymidine (TdR) labelling index 46
heterokaryon gene disruption, *Aspergillus* 139
hormonal stimulation, *Xenopus* 258
immunodepletion with Dynabeads 194
immunofluorescence staining
indirect, *Drosophila* 157
mitotic spindles, *Aspergillus* 132
*Xenopus* cell-free egg extracts 192
*in situ* hybridization, localization of transcripts in *Drosophila* 154
inducing overexpression, *nmt* promoter *S. pombe* 119
integrative mapping, *S. pombe* 117
interphase extracts, *Xenopus*, preparation 305
large-scale isolation of cell-cycle staged cultures, *nim*T23 mutation, *Aspergillus* 141
mammalian cells
arrest and synchronization
in early S phase 7
in $G_0$ 3
in $G_2$ 10
using high-pressure nitrous oxide 17–18
mitotic selection from monolayers 12
separation of cell-cycle phases by centrifugal elutriation 21
microinjection
cytoplasmic transfer, *Xenopus* oocytes 294
needle preparation 294
microtubules, *Xenopus*, fixation and observation 279
mitotic chromosomes
isolation from CHO cells 246
isolation from HeLa cells 248
mitotic organelles, CHO, isolation
hexylene glycol method 239
taxol method 240
nmt promoter, inducing overexpression, *S. pombe* 119
nuclei, human, quiescent, isolation 185
permeabilization of *Drosophila* embryos with octane 160
plasmid and phenotype, cosegregation, *S. pombe* 114
plasmid recovery, *S. pombe* 116

polytene chromosomes in *Drosophila*
banding 148
detection of *in situ* hybridization, fluorescence 152
*in situ* hybridization 149, 150
staining with peroxidase 151
replication assay SV40 208
replication-competent extract preparation 205
rhodamine-phalloidin staining of F-actin, *S. pombe* 108
soluble extracts, *Xenopus* 266
spheroplast transformation, *S. cerevisiae* 86
staining, *lacZ* expression, *Drosophila* 147
starfish, cell-free egg extracts 297
stathmokinetic methods, mammalian cells 60–2
T-antigen purification 202
testes preparations, *Drosophila* 171, 172
transcripts, *Drosophila*, *in situ* hybridization 154
transformation methods *Aspergillus* 135
tumour growth fraction estimation 49
univariate cellular DNA content distribution 52
vitelline membrane removal, *Drosophila* embryos 162
*Xenopus* cell-free egg extracts
CsCl gradient centrifugation 188
density substitution 188
labelling with [$^{32}$P]dCTP
continuous 187
pulse 187
processing 188
preparation
for MPF 296
interphase extracts 263
mitotic extracts 260
replication competent extracts 180
soluble extracts 266
processing labelled samples 188
*Xenopus* oocytes, preparation 289
yeasts, arrest and synchronization
by $\alpha_1$-mating factor, *S. cerevisiae* 39
by feed–starve method, *Saccharomyces cerevisiae* 40
by gradient separation 29
by heat shock, *S. pombe* 38
of *S. pombe*, *cdc* mutants 37
of *S. pombe* using zonal rotor 35
using elutriation rotor 32

reciprocal shifts 79, 113
experiments *(illus.)* 80
regrowth buffer 279
replica plating, *S. pombe* 117
replication *see* DNA replication

**322**

# Index

replication-competent extract preparation buffers 204
rhodamine–phalloidin staining, F-actin 108
Ringer's solution, modified 289

S phase
  arresting agents 7
  early arrest, mammalian cell lines 8
  problems of extended inhibition 8
  synchronization of mammalian cells 6–9
*Saccharomyces cerevisiae*
  analysis of cell cycle 69–92
  budding
    assay 74
    bud formation 25–6
    clumpiness 75
    quantitation 75
    sick cells 75
  Cdc28 protein kinase 80–4
    precipitation with $p13^{suc1}$ agarose beads 81
  *cdc* mutants 38
  cell volume (equation) 74
  Cks1 protein, homologue of *S. pombe* Suc1 protein 81
  $CLN_{1-3}$ cyclins 301
  division profile 26–8
  DNA content and FACS analysis 75–8
  DNA staining with DAPI 78
  growth media 30, 72
  heterologous genes
    cloning of homologues by functional complementation 85–9
    co-segregation or plastid loss test 88–9
    testing for ability to complement mutations in yeast 89–90
    transformation 86–9
    without homologues 90
  histone H1 kinase assays 80–1
  isolation and retesting of complementing plasmids 89
  *MATAa* strain, $\alpha_1$-mating factor 39
  microtubules, morphology 78
  normal cell cycle *(illus.)* 70–1
    dependency map *(illus.)* 71
    monitoring cell cycle 71–84
  preparation of arrested cells
    cell growth 72
    drug arrests 73–4
    measurement of cell size 74
    temperature shifts *cdc* mutants 73
  preparation of synchronous cultures 25–44
    feed–starve method 40–2
  reciprocal shift experiments 79
  *see also* synchronous cultures, yeasts
*Schizosaccharomyces pombe*
  binary fission 26

calcofluor staining 106, 108
*cdc* mutants 37–8
cell cycle 93–125
  division profile 28–9
  generation time vs temperature 95
  measuring DNA content 97–9
  monitoring DNA replication 99–101
  normal life cycle *(illus.)* 93–5, 97
cell cycle genetics 111–22
  characterization of new mutants 112
  ordering gene functions by reciprocal shifts 113
  transition point determination 112–13
cell cycle inhibitors
  benomyl and related compounds 111
  cycloheximide 111
  hydroxyurea 110
cell length and septation (cell plate) index *(illus.)* 97
cell number determination 96–7
cell staining 106–10
  actin structures 108
  cell wall and septum 110
  electron microscopy 110
  immunofluorescence microscopy 109–10
  nuclear staining 106–17
cloning of cell cycle genes 113–18
  by complementation in *S. cerevisiae* 118
  cloning by complementation 113–15
  delimiting the functional region 118
  integrative mapping 117
  manipulation of cloned genes 118–19
  recovering plasmids from yeast transformants 115–16
  transformation methods, *S. pombe* 114
DAPI staining *(illus.)* 106–7
Edinburgh Minimal Medium 28
flow cytometry 101–4
genetic nomenclature 93
growing methods 95–6
growth determination, $OD_{595}$ 96–7
growth media 28, 30
integrative mapping 117
isolation of higher eukaryotic homologues
  cDNA libraries 120
  cloning by complementation 120–1
  cloning by physical methods 121–2
  manipulation of cloned genes
    cloning mutant alleles 119–20
    deletion or disruption 119
    overexpression 118–19
new mutants, transition point 112
nmt promoter, inducing overexpression 119
nuclear staining, DAPI *(illus.)* 106–7
$p13^{suc1}$ protein 81 299
$p34^{cdc2}$ H1 kinase activity 104–6
preparation of synchronous cultures 25–44, 110

# Index

*Schizosaccharomyces pombe (cont.)*:
  transition point, new mutants 112
  *see also* synchronous cultures, yeasts
sea urchin, cyclins, activation of cdc protein kinases 275
sheep secondary antibodies, Dynabeads 194
SLO, bacterial exotoxin 184
spectrophotometry, optical density readings 97
sperm, demembranated sperm heads, *Xenopus* 183
spheroplasts, transformation, *S. cerevisiae* 86
spindle index *see* mitotic spindle index
starfish oocytes
  maturation 285–6
  microinjection 295–6
stathmokinetic methods, mammalian cells 60–2
Suc1 protein 81, 299, 300
sucrose gradient, for mitotic chromosomes 247
SUNaSP solution 183
SV40 DNA replication 197–210
  extract preparation 204–5
  replication reaction
    assay 206–8
    reagents 206
  T-antigen purification 199–203
    column preparation 199–200
    immunoaffinity purification 201–2
synchronous cultures, mammalian cells 1–23
  apparatus *(illus.)* 17
  non-cycling $G_0$ cells 2–4
synchronous cultures, yeasts 25–44
  age fractionation 42
  density separation 31
  division profiles 26–9
  gradient separation in tubes 29–31
  induction methods of synchronization
    α-factor 39–40
    block and release 36–42
    *cdc* mutants 37–8
    feed–starve method 40–1
    heat shock 38–9
  preparation by elutriation 31–4
  selection synchrony 29–36


T-antigen purification 202
TAE buffer 224
taxol, isolation of mitotic organelles 240
TBE buffer 224
teniposide (VM26) 9
[$^3$H]thymidine (TdR) labelling index 46
thiabendazole 111
thymidine, arrest of replication 6
tissue culture cells, mitotic arrest 236–8
topoisomerase II inhibitors 9–10

transition point, defined 112–13
trichloracetic acid (TCA)–carrier DNA 207
trypsin, detachment and replating 3
tubulins, staining, anti-yeast α-tubulin monoclonal antibody 78
tumour growth fraction estimation, mammalian cell cycle, autoradiographic methods 49–50
two-dimensional gel electrophoresis *see* gel electrophoresis

[6-$^3$H]uracil, uptake, *S. pombe* 99

vertebrate oocytes, maturation 285–6
vinca alkaloids, mitotic arrest 15
VM26 (teniposide) 9

X-gal solution 147
*Xenopus*
  dejellying solution 178, 180
  egg activation 255
  egg laying 178, 258–9
    seasonal variations 177, 259
  fertilization process 255
  first cell cycle *(illus.)* 256
  GVBD (germinal vesicle breakdown) 288, 291–2, 299
  MPF activation, time course 292
  handling oocytes 287–8
  microinjection *(illus.)* 288, 287–8
  interphase extracts, preparation 305
  maintaining a colony 177–8
  WAK fibroblasts, arrest and synchronization 19
*Xenopus* cell-free egg extracts
  assays
    cyclin activity 270–7
    Histone H1 kinase 267–8
    other regulatory molecules 277
  choice of DNA template 182–6
    demembranated sperm heads 182–3
    double-stranded DNA plasmids 183–4
    somatic cell nuclei 184–5
  DNA replication 177–95
  extract preparation 178–80, 254–69
    Cdc2 H1 kinase activity as marker 267
    first embryonic cell cycle 264–5
    $G_2$ prophase extracts 277
    Histone H1 kinase assay, principle 267–9
    interphase extracts 255, 263–4
    mitotic extracts 260–3
    one-cycle extracts 257
    soluble extracts 257, 266
    types of extracts 255–8
      fresh vs frozen 181–2

## Index

labelling protocols 186–91
  biotin-11-dUTP labelling 190–1
  [$^{32}$P]dCTP labelling 186–7
microscopic analysis of nuclear assembly
    191–3
  immunofluorescence 191–2
  phase-contrast microscopy 191
microtubules, dynamics 277–81

somatic cell nuclei 184
study of mitosis 253–83

YPD (YEPD) medium, *S. cerevisiae* 30, 72

zonal rotor, synchronization of S. pombe 35

# ORDER OTHER TITLES OF INTEREST TODAY

## Price list for: UK, Europe, Rest of World (excluding US and Canada)

| | | | | |
|---|---|---|---|---|
| 138. | **Plasmids (2/e)** Hardy, K.G. (Ed) | | | |
| ...... | Spiralbound hardback | 0-19-963445-9 | **£30.00** | |
| ...... | Paperback | 0-19-963444-0 | **£19.50** | |
| 136. | **RNA Processing: Vol. II** Higgins, S.J. & Hames, B.D. (Eds) | | | |
| ...... | Spiralbound hardback | 0-19-963471-8 | **£30.00** | |
| ...... | Paperback | 0-19-963470-X | **£19.50** | |
| 135. | **RNA Processing: Vol. I** Higgins, S.J. & Hames, B.D. (Eds) | | | |
| ...... | Spiralbound hardback | 0-19-963344-4 | **£30.00** | |
| ...... | Paperback | 0-19-963343-6 | **£19.50** | |
| 134. | **NMR of Macromolecules** Roberts, G.C.K. (Ed) | | | |
| ...... | Spiralbound hardback | 0-19-963225-1 | **£32.50** | |
| ...... | Paperback | 0-19-963224-3 | **£22.50** | |
| 133. | **Gas Chromatography** Baugh, P. (Ed) | | | |
| ...... | Spiralbound hardback | 0-19-963272-3 | **£40.00** | |
| ...... | Paperback | 0-19-963271-5 | **£27.50** | |
| 132. | **Essential Developmental Biology** Stern, C.D. & Holland, P.W.H. (Eds) | | | |
| ...... | Spiralbound hardback | 0-19-963423-8 | **£30.00** | |
| ...... | Paperback | 0-19-963422-X | **£19.50** | |
| 131. | **Cellular Interactions in Development** Hartley, D.A. (Ed) | | | |
| ...... | Spiralbound hardback | 0-19-963391-6 | **£30.00** | |
| ...... | Paperback | 0-19-963390-8 | **£18.50** | |
| 129 | **Behavioural Neuroscience: Volume II** Sahgal, A. (Ed) | | | |
| ...... | Spiralbound hardback | 0-19-963458-0 | **£32.50** | |
| ...... | Paperback | 0-19-963457-2 | **£22.50** | |
| 128 | **Behavioural Neuroscience: Volume I** Sahgal, A. (Ed) | | | |
| ...... | Spiralbound hardback | 0-19-963368-1 | **£32.50** | |
| ...... | Paperback | 0-19-963367-3 | **£22.50** | |
| 127. | **Molecular Virology** Davison, A.J. & Elliott, R.M. (Eds) | | | |
| ...... | Spiralbound hardback | 0-19-963358-4 | **£35.00** | |
| ...... | Paperback | 0-19-963357-6 | **£25.00** | |
| 126. | **Gene Targeting** Joyner, A.L. (Ed) | | | |
| ...... | Spiralbound hardback | 0-19-963407-6 | **£30.00** | |
| ...... | Paperback | 0-19-963406-8 | **£19.50** | |
| 125. | **Glycobiology** Fukuda, M. & Kobata, A. (Eds) | | | |
| ...... | Spiralbound hardback | 0-19-963372-X | **£32.50** | |
| ...... | Paperback | 0-19-963371-1 | **£22.50** | |
| 124. | **Human Genetic Disease Analysis (2/e)** Davies, K.E. (Ed) | | | |
| ...... | Spiralbound hardback | 0-19-963309-6 | **£30.00** | |
| ...... | Paperback | 0-19-963308-8 | **£18.50** | |
| 122. | **Immunocytochemistry** Beesley, J. (Ed) | | | |
| ...... | Spiralbound hardback | 0-19-963270-7 | **£35.00** | |
| ...... | Paperback | 0-19-963269-3 | **£22.50** | |
| 123. | **Protein Phosphorylation** Hardie, D.G. (Ed) | | | |
| ...... | Spiralbound hardback | 0-19-963306-1 | **£32.50** | |
| ...... | Paperback | 0-19-963305-3 | **£22.50** | |
| 121. | **Tumour Immunobiology** Gallagher, G., Rees, R.C. & others (Eds) | | | |
| ...... | Spiralbound hardback | 0-19-963370-3 | **£40.00** | |
| ...... | Paperback | 0-19-963369-X | **£27.50** | |
| 120. | **Transcription Factors** Latchman, D.S. (Ed) | | | |
| ...... | Spiralbound hardback | 0-19-963342-8 | **£30.00** | |
| ...... | Paperback | 0-19-963341-X | **£19.50** | |
| 119. | **Growth Factors** McKay, I. & Leigh, I. (Eds) | | | |
| ...... | Spiralbound hardback | 0-19-963360-6 | **£30.00** | |
| ...... | Paperback | 0-19-963359-2 | **£19.50** | |
| 118. | **Histocompatibility Testing** Dyer, P. & Middleton, D. (Eds) | | | |
| ...... | Spiralbound hardback | 0-19-963364-9 | **£32.50** | |
| ...... | Paperback | 0-19-963363-0 | **£22.50** | |
| 117. | **Gene Transcription** Hames, B.D. & Higgins, S.J. (Eds) | | | |
| ...... | Spiralbound hardback | 0-19-963292-8 | **£35.00** | |
| ...... | Paperback | 0-19-963291-X | **£25.00** | |
| 116. | **Electrophysiology** Wallis, D.I. (Ed) | | | |
| ...... | Spiralbound hardback | 0-19-963348-7 | **£32.50** | |
| ...... | Paperback | 0-19-963347-9 | **£22.50** | |
| 115. | **Biological Data Analysis** Fry, J.C. (Ed) | | | |
| ...... | Spiralbound hardback | 0-19-963340-1 | **£50.00** | |
| ...... | Paperback | 0-19-963339-8 | **£27.50** | |
| 114. | **Experimental Neuroanatomy** Bolam, J.P. (Ed) | | | |
| ...... | Spiralbound hardback | 0-19-963326-6 | **£32.50** | |
| ...... | Paperback | 0-19-963325-8 | **£22.50** | |
| 113. | **Preparative Centrifugation** Rickwood, D. (Ed) | | | |
| ...... | Spiralbound hardback | 0-19-963208-1 | **£45.00** | |
| ...... | Paperback | 0-19-963211-1 | **£25.00** | |
| ...... | Paperback | 0-19-963099-2 | **£25.00** | |
| 112. | **Lipid Analysis** Hamilton, R.J. & Hamilton, Shiela (Eds) | | | |
| ...... | Spiralbound hardback | 0-19-963098-4 | **£35.00** | |
| ...... | Paperback | 0-19-963099-2 | **£25.00** | |
| 111. | **Haemopoiesis** Testa, N.G. & Molineux, G. (Eds) | | | |
| ...... | Spiralbound hardback | 0-19-963366-5 | **£32.50** | |
| ...... | Paperback | 0-19-963365-7 | **£22.50** | |
| 110. | **Pollination Ecology** Dafni, A. | | | |
| ...... | Spiralbound hardback | 0-19-963299-5 | **£32.50** | |
| ...... | Paperback | 0-19-963298-7 | **£22.50** | |
| 109. | **In Situ Hybridization** Wilkinson, D.G. (Ed) | | | |
| ...... | Spiralbound hardback | 0-19-963328-2 | **£30.00** | |
| ...... | Paperback | 0-19-963327-4 | **£18.50** | |
| 108. | **Protein Engineering** Rees, A.R., Sternberg, M.J.E. & others (Eds) | | | |
| ...... | Spiralbound hardback | 0-19-963139-5 | **£35.00** | |
| ...... | Paperback | 0-19-963138-7 | **£25.00** | |
| 107. | **Cell-Cell Interactions** Stevenson, B.R., Gallin, W.J. & others (Eds) | | | |
| ...... | Spiralbound hardback | 0-19-963319-3 | **£32.50** | |
| ...... | Paperback | 0-19-963318-5 | **£22.50** | |
| 106. | **Diagnostic Molecular Pathology: Volume I** Herrington, C.S. & McGee, J. O'D. (Eds) | | | |
| ...... | Spiralbound hardback | 0-19-963237-5 | **£30.00** | |
| ...... | Paperback | 0-19-963236-7 | **£19.50** | |
| 105. | **Biomechanics-Materials** Vincent, J.F.V. (Ed) | | | |
| ...... | Spiralbound hardback | 0-19-963223-5 | **£35.00** | |
| ...... | Paperback | 0-19-963222-7 | **£25.00** | |
| 104. | **Animal Cell Culture (2/e)** Freshney, R.I. (Ed) | | | |
| ...... | Spiralbound hardback | 0-19-963212-X | **£30.00** | |
| ...... | Paperback | 0-19-963213-8 | **£19.50** | |
| 103. | **Molecular Plant Pathology: Volume II** Gurr, S.J., McPherson, M.J. & others (Eds) | | | |
| ...... | Spiralbound hardback | 0-19-963352-5 | **£32.50** | |
| ...... | Paperback | 0-19-963351-7 | **£22.50** | |
| 102. | **Signal Transduction** Milligan, G. (Ed) | | | |
| ...... | Spiralbound hardback | 0-19-963296-0 | **£30.00** | |
| ...... | Paperback | 0-19-963295-2 | **£18.50** | |
| 101. | **Protein Targeting** Magee, A.I. & Wileman, T. (Eds) | | | |
| ...... | Spiralbound hardback | 0-19-963206-5 | **£32.50** | |
| ...... | Paperback | 0-19-963210-3 | **£22.50** | |
| 100. | **Diagnostic Molecular Pathology: Volume II: Cell and Tissue Genotyping** Herrington, C.S. & McGee, J.O'D. (Eds) | | | |
| ...... | Spiralbound hardback | 0-19-963239-1 | **£30.00** | |
| ...... | Paperback | 0-19-963238-3 | **£19.50** | |
| 99. | **Neuronal Cell Lines** Wood, J.N. (Ed) | | | |
| ...... | Spiralbound hardback | 0-19-963346-0 | **£32.50** | |
| ...... | Paperback | 0-19-963345-2 | **£22.50** | |

98. **Neural Transplantation** Dunnett, S.B. & Björklund, A. (Eds)
...... Spiralbound hardback 0-19-963286-3 £30.00
...... Paperback 0-19-963285-5 £19.50

97. **Human Cytogenetics: Volume II: Malignancy and Acquired Abnormalities (2/e)** Rooney, D.E. & Czepulkowski, B.H. (Eds)
...... Spiralbound hardback 0-19-963290-1 £30.00
...... Paperback 0-19-963289-8 £22.50

96. **Human Cytogenetics: Volume I: Constitutional Analysis (2/e)** Rooney, D.E. & Czepulkowski, B.H. (Eds)
...... Spiralbound hardback 0-19-963288-X £30.00
...... Paperback 0-19-963287-1 £22.50

95. **Lipid Modification of Proteins** Hooper, N.M. & Turner, A.J. (Eds)
...... Spiralbound hardback 0-19-963274-X £32.50
...... Paperback 0-19-963273-1 £22.50

94. **Biomechanics-Structures and Systems** Biewener, A.A. (Ed)
...... Spiralbound hardback 0-19-963268-5 £42.50
...... Paperback 0-19-963267-7 £25.00

93. **Lipoprotein Analysis** Converse, C.A. & Skinner, E.R. (Eds)
...... Spiralbound hardback 0-19-963192-1 £30.00
...... Paperback 0-19-963231-6 £19.50

92. **Receptor-Ligand Interactions** Hulme, E.C. (Ed)
...... Spiralbound hardback 0-19-963090-9 £35.00
...... Paperback 0-19-963091-7 £27.50

91. **Molecular Genetic Analysis of Populations** Hoelzel, A.R. (Ed)
...... Spiralbound hardback 0-19-963278-2 £32.50
...... Paperback 0-19-963277-4 £22.50

90. **Enzyme Assays** Eisenthal, R. & Danson, M.J. (Eds)
...... Spiralbound hardback 0-19-963142-5 £35.00
...... Paperback 0-19-963143-3 £25.00

89. **Microcomputers in Biochemistry** Bryce, C.F.A. (Ed)
...... Spiralbound hardback 0-19-963253-7 £30.00
...... Paperback 0-19-963252-9 £19.50

88. **The Cytoskeleton** Carraway, K.L. & Carraway, C.A.C. (Eds)
...... Spiralbound hardback 0-19-963257-X £30.00
...... Paperback 0-19-963256-1 £19.50

87. **Monitoring Neuronal Activity** Stamford, J.A. (Ed)
...... Spiralbound hardback 0-19-963244-8 £30.00
...... Paperback 0-19-963243-X £19.50

86. **Crystallization of Nucleic Acids and Proteins** Ducruix, A. & Giegé, R. (Eds)
...... Spiralbound hardback 0-19-963245-6 £35.00
...... Paperback 0-19-963246-4 £25.00

85. **Molecular Plant Pathology: Volume I** Gurr, S.J., McPherson, M.J. & others (Eds)
...... Spiralbound hardback 0-19-963103-4 £30.00
...... Paperback 0-19-963102-6 £19.50

84. **Anaerobic Microbiology** Levett, P.N. (Ed)
...... Spiralbound hardback 0-19-963204-9 £32.50
...... Paperback 0-19-963262-6 £22.50

83. **Oligonucleotides and Analogues** Eckstein, F. (Ed)
...... Spiralbound hardback 0-19-963280-4 £32.50
...... Paperback 0-19-963279-0 £22.50

82. **Electron Microscopy in Biology** Harris, R. (Ed)
...... Spiralbound hardback 0-19-963219-7 £32.50
...... Paperback 0-19-963215-4 £22.50

81. **Essential Molecular Biology: Volume II** Brown, T.A. (Ed)
...... Spiralbound hardback 0-19-963112-3 £32.50
...... Paperback 0-19-963113-1 £22.50

80. **Cellular Calcium** McCormack, J.G. & Cobbold, P.H. (Eds)
...... Spiralbound hardback 0-19-963131-X £35.00
...... Paperback 0-19-963130-1 £25.00

79. **Protein Architecture** Lesk, A.M.
...... Spiralbound hardback 0-19-963054-2 £32.50
...... Paperback 0-19-963055-0 £22.50

78. **Cellular Neurobiology** Chad, J. & Wheal, H. (Eds)
...... Spiralbound hardback 0-19-963106-9 £32.50
...... Paperback 0-19-963107-7 £22.50

77. **PCR** McPherson, M.J., Quirke, P. & others (Eds)
...... Spiralbound hardback 0-19-963226-X £30.00
...... Paperback 0-19-963196-4 £19.50

76. **Mammalian Cell Biotechnology** Butler, M. (Ed)
...... Spiralbound hardback 0-19-963207-3 £30.00
...... Paperback 0-19-963209-X £19.50

75. **Cytokines** Balkwill, F.R. (Ed)
...... Spiralbound hardback 0-19-963218-9 £35.00
...... Paperback 0-19-963214-6 £25.00

74. **Molecular Neurobiology** Chad, J. & Wheal, H. (Eds)
...... Spiralbound hardback 0-19-963108-5 £30.00
...... Paperback 0-19-963109-3 £19.50

73. **Directed Mutagenesis** McPherson, M.J. (Ed)
...... Spiralbound hardback 0-19-963141-7 £30.00
...... Paperback 0-19-963140-9 £19.50

72. **Essential Molecular Biology: Volume I** Brown, T.A. (Ed)
...... Spiralbound hardback 0-19-963110-7 £32.50
...... Paperback 0-19-963111-5 £22.50

71. **Peptide Hormone Action** Siddle, K. & Hutton, J.C.
...... Spiralbound hardback 0-19-963070-4 £32.50
...... Paperback 0-19-963071-2 £22.50

70. **Peptide Hormone Secretion** Hutton, J.C. & Siddle, K. (Eds)
...... Spiralbound hardback 0-19-963068-2 £35.00
...... Paperback 0-19-963069-0 £25.00

69. **Postimplantation Mammalian Embryos** Copp, A.J. & Cockroft, D.L. (Eds)
...... Spiralbound hardback 0-19-963088-7 £15.00
...... Paperback 0-19-963089-5 £12.50

68. **Receptor-Effector Coupling** Hulme, E.C. (Ed)
...... Spiralbound hardback 0-19-963094-1 £30.00
...... Paperback 0-19-963095-X £19.50

67. **Gel Electrophoresis of Proteins (2/e)** Hames, B.D. & Rickwood, D. (Eds)
...... Spiralbound hardback 0-19-963074-7 £35.00
...... Paperback 0-19-963075-5 £25.00

66. **Clinical Immunology** Gooi, H.C. & Chapel, H. (Eds)
...... Spiralbound hardback 0-19-963086-0 £32.50
...... Paperback 0-19-963087-9 £22.50

65. **Receptor Biochemistry** Hulme, E.C. (Ed)
...... Paperback 0-19-963093-3 £25.00

64. **Gel Electrophoresis of Nucleic Acids (2/e)** Rickwood, D. & Hames, B.D. (Eds)
...... Spiralbound hardback 0-19-963082-8 £32.50
...... Paperback 0-19-963083-6 £22.50

63. **Animal Virus Pathogenesis** Oldstone, M.B.A. (Ed)
...... Spiralbound hardback 0-19-963100-X £15.00
...... Paperback 0-19-963101-8 £12.50

62. **Flow Cytometry** Ormerod, M.G. (Ed)
...... Paperback 0-19-963053-4 £22.50

61. **Radioisotopes in Biology** Slater, R.J. (Ed)
...... Spiralbound hardback 0-19-963080-1 £32.50
...... Paperback 0-19-963081-X £22.50

60. **Biosensors** Cass, A.E.G. (Ed)
...... Spiralbound hardback 0-19-963046-1 £30.00
...... Paperback 0-19-963047-X £19.50

59. **Ribosomes and Protein Synthesis** Spedding, G. (Ed)
...... Spiralbound hardback 0-19-963104-2 £15.00
...... Paperback 0-19-963105-0 £12.50

58. **Liposomes** New, R.R.C. (Ed)
...... Spiralbound hardback 0-19-963076-3 £35.00
...... Paperback 0-19-963077-1 £22.50

57. **Fermentation** McNeil, B. & Harvey, L.M. (Eds)
...... Spiralbound hardback 0-19-963044-5 £30.00
...... Paperback 0-19-963045-3 £19.50

56. **Protein Purification Applications** Harris, E.L.V. & Angal, S. (Eds)
...... Spiralbound hardback 0-19-963022-4 £30.00
...... Paperback 0-19-963023-2 £18.50

55. **Nucleic Acids Sequencing** Howe, C.J. & Ward, E.S. (Eds)
...... Spiralbound hardback 0-19-963056-9 £30.00
...... Paperback 0-19-963057-7 £19.50

54. **Protein Purification Methods** Harris, E.L.V. & Angal, S. (Eds)
...... Spiralbound hardback 0-19-963002-X £30.00
...... Paperback 0-19-963003-8 £22.50

53. **Solid Phase Peptide Synthesis** Atherton, E. & Sheppard, R.C.
...... Spiralbound hardback 0-19-963066-6 £15.00
...... Paperback 0-19-963067-4 £12.50

52. **Medical Bacteriology** Hawkey, P.M. & Lewis, D.A. (Eds)
...... Paperback 0-19-963009-7 £25.00

51. **Proteolytic Enzymes** Beynon, R.J. & Bond, J.S. (Eds)
...... Spiralbound hardback 0-19-963058-5 £30.00
...... Paperback 0-19-963059-3 £19.50

50. **Medical Mycology** Evans, E.G.V. & Richardson, M.D. (Eds)
...... Spiralbound hardback 0-19-963010-0 £37.50
...... Paperback 0-19-963011-9 £25.00

49. **Computers in Microbiology** Bryant, T.N. & Wimpenny, J.W.T. (Eds)
...... Paperback 0-19-963015-1 £12.50

| No. | Title | Editor/Author | Format | ISBN | Price |
|---|---|---|---|---|---|
| 48. | Protein Sequencing | Findlay, J.B.C. & Geisow, M.J. (Eds) | | | |
| | | | Spiralbound hardback | 0-19-963012-7 | £15.00 |
| | | | Paperback | 0-19-963013-5 | £12.50 |
| 47. | Cell Growth and Division | Baserga, R. (Ed) | | | |
| | | | Spiralbound hardback | 0-19-963026-7 | £15.00 |
| | | | Paperback | 0-19-963027-5 | £12.50 |
| 46. | Protein Function | Creighton, T.E. (Ed) | | | |
| | | | Spiralbound hardback | 0-19-963006-2 | £32.50 |
| | | | Paperback | 0-19-963007-0 | £22.50 |
| 45. | Protein Structure | Creighton, T.E. (Ed) | | | |
| | | | Spiralbound hardback | 0-19-963000-3 | £32.50 |
| | | | Paperback | 0-19-963001-1 | £22.50 |
| 44. | Antibodies: Volume II | Catty, D. (Ed) | | | |
| | | | Spiralbound hardback | 0-19-963018-6 | £30.00 |
| | | | Paperback | 0-19-963019-4 | £19.50 |
| 43. | HPLC of Macromolecules | Oliver, R.W.A. (Ed) | | | |
| | | | Spiralbound hardback | 0-19-963020-8 | £30.00 |
| | | | Paperback | 0-19-963021-6 | £19.50 |
| 42. | Light Microscopy in Biology | Lacey, A.J. (Ed) | | | |
| | | | Spiralbound hardback | 0-19-963036-4 | £30.00 |
| | | | Paperback | 0-19-963037-2 | £19.50 |
| 41. | Plant Molecular Biology | Shaw, C.H. (Ed) | | | |
| | | | Paperback | 1-85221-056-7 | £12.50 |
| 40. | Microcomputers in Physiology | Fraser, P.J. (Ed) | | | |
| | | | Spiralbound hardback | 1-85221-129-6 | £15.00 |
| | | | Paperback | 1-85221-130-X | £12.50 |
| 39. | Genome Analysis | Davies, K.E. (Ed) | | | |
| | | | Spiralbound hardback | 1-85221-109-1 | £30.00 |
| | | | Paperback | 1-85221-110-5 | £18.50 |
| 38. | Antibodies: Volume I | Catty, D. (Ed) | | | |
| | | | Paperback | 0-947946-85-3 | £19.50 |
| 37. | Yeast | Campbell, I. & Duffus, J.H. (Eds) | | | |
| | | | Paperback | 0-947946-79-9 | £12.50 |
| 36. | Mammalian Development | Monk, M. (Ed) | | | |
| | | | Hardback | 1-85221-030-3 | £15.00 |
| | | | Paperback | 1-85221-029-X | £12.50 |
| 35. | Lymphocytes | Klaus, G.G.B. (Ed) | | | |
| | | | Hardback | 1-85221-018-4 | £30.00 |
| 34. | Lymphokines and Interferons | Clemens, M.J., Morris, A.G. & others (Eds) | | | |
| | | | Paperback | 1-85221-035-4 | £12.50 |
| 33. | Mitochondria | Darley-Usmar, V.M., Rickwood, D. & others (Eds) | | | |
| | | | Hardback | 1-85221-034-6 | £32.50 |
| | | | Paperback | 1-85221-033-8 | £22.50 |
| 32. | Prostaglandins and Related Substances | Benedetto, C., McDonald-Gibson, R.G. & others (Eds) | | | |
| | | | Hardback | 1-85221-032-X | £15.00 |
| | | | Paperback | 1-85221-031-1 | £12.50 |
| 31. | DNA Cloning: Volume III | Glover, D.M. (Ed) | | | |
| | | | Hardback | 1-85221-049-4 | £15.00 |
| | | | Paperback | 1-85221-048-6 | £12.50 |
| 30. | Steroid Hormones | Green, B. & Leake, R.E. (Eds) | | | |
| | | | Paperback | 0-947946-53-5 | £19.50 |
| 29. | Neurochemistry | Turner, A.J. & Bachelard, H.S. (Eds) | | | |
| | | | Hardback | 1-85221-028-1 | £15.00 |
| | | | Paperback | 1-85221-027-3 | £12.50 |
| 28. | Biological Membranes | Findlay, J.B.C. & Evans, W.H. (Eds) | | | |
| | | | Hardback | 0-947946-84-5 | £15.00 |
| | | | Paperback | 0-947946-83-7 | £12.50 |
| 27. | Nucleic Acid and Protein Sequence Analysis | Bishop, M.J. & Rawlings, C.J. (Eds) | | | |
| | | | Hardback | 1-85221-007-9 | £35.00 |
| | | | Paperback | 1-85221-006-0 | £25.00 |
| 26. | Electron Microscopy in Molecular Biology | Sommerville, J. & Scheer, U. (Eds) | | | |
| | | | Hardback | 0-947946-64-0 | £15.00 |
| | | | Paperback | 0-947946-54-3 | £12.50 |
| 25. | Teratocarcinomas and Embryonic Stem Cells | Robertson, E.J. (Ed) | | | |
| | | | Paperback | 1-85221-004-4 | £19.50 |
| 24. | Spectrophotometry and Spectrofluorimetry | Harris, D.A. & Bashford, C.L. (Eds) | | | |
| | | | Hardback | 0-947946-69-1 | £15.00 |
| | | | Paperback | 0-947946-46-2 | £12.50 |
| 23. | Plasmids | Hardy, K.G. (Ed) | | | |
| | | | Paperback | 0-947946-81-0 | £12.50 |
| 22. | Biochemical Toxicology | Snell, K. & Mullock, B. (Eds) | | | |
| | | | Paperback | 0-947946-52-7 | £12.50 |
| 19. | Drosophila | Roberts, D.B. (Ed) | | | |
| | | | Hardback | 0-947946-66-7 | £32.50 |
| | | | Paperback | 0-947946-45-4 | £22.50 |
| 17. | Photosynthesis: Energy Transduction | Hipkins, M.F. & Baker, N.R. (Eds) | | | |
| | | | Hardback | 0-947946-63-2 | £15.00 |
| | | | Paperback | 0-947946-51-9 | £12.50 |
| 16. | Human Genetic Diseases | Davies, K.E. (Ed) | | | |
| | | | Hardback | 0-947946-76-4 | £15.00 |
| | | | Paperback | 0-947946-75-6 | £12.50 |
| 14. | Nucleic Acid Hybridisation | Hames, B.D. & Higgins, S.J. (Eds) | | | |
| | | | Hardback | 0-947946-61-6 | £15.00 |
| | | | Paperback | 0-947946-23-3 | £12.50 |
| 13. | Immobilised Cells and Enzymes | Woodward, J. (Ed) | | | |
| | | | Hardback | 0-947946-60-8 | £15.00 |
| 12. | Plant Cell Culture | Dixon, R.A. (Ed) | | | |
| | | | Paperback | 0-947946-22-5 | £19.50 |
| 11a. | DNA Cloning: Volume I | Glover, D.M. (Ed) | | | |
| | | | Paperback | 0-947946-18-7 | £12.50 |
| 11b. | DNA Cloning: Volume II | Glover, D.M. (Ed) | | | |
| | | | Paperback | 0-947946-19-5 | £12.50 |
| 10. | Virology | Mahy, B.W.J. (Ed) | | | |
| | | | Paperback | 0-904147-78-9 | £19.50 |
| 9. | Affinity Chromatography | Dean, P.D.G., Johnson, W.S. & others (Eds) | | | |
| | | | Paperback | 0-904147-71-1 | £19.50 |
| 7. | Microcomputers in Biology | Ireland, C.R. & Long, S.P. (Eds) | | | |
| | | | Paperback | 0-904147-57-6 | £18.00 |
| 6. | Oligonucleotide Synthesis | Gait, M.J. (Ed) | | | |
| | | | Paperback | 0-904147-74-6 | £18.50 |
| 5. | Transcription and Translation | Hames, B.D. & Higgins, S.J. (Eds) | | | |
| | | | Paperback | 0-904147-52-5 | £12.50 |
| 3. | Iodinated Density Gradient Media | Rickwood, D. (Ed) | | | |
| | | | Paperback | 0-904147-51-7 | £12.50 |

## Sets

| Title | Editor | Format | ISBN | Price |
|---|---|---|---|---|
| Essential Molecular Biology: 2 vol set | Brown, T.A. (Ed) | | | |
| | | Spiralbound hardback | 0-19-963114-X | £58.00 |
| | | Paperback | 0-19-963115-8 | £40.00 |
| Antibodies: 2 vol set | Catty, D. (Ed) | | | |
| | | | 0-19-963063-1 | £33.00 |
| Cellular and Molecular Neurobiology: 2 vol set | Chad, J. & Wheal, H. (Eds) | | | |
| | | Spiralbound hardback | 0-19-963255-3 | £56.00 |
| | | Paperback | 0-19-963254-5 | £38.00 |
| Protein Structure and Protein Function: 2 vol set | Creighton, T.E. (Ed) | | | |
| | | Spiralbound hardback | 0-19-963064-X | £55.00 |
| | | Paperback | 0-19-963065-8 | £38.00 |
| DNA Cloning: 2 vol set | Glover, D.M. (Ed) | | | |
| | | Paperback | 1-85221-069-9 | £30.00 |
| Molecular Plant Pathology: 2 vol set | Gurr, S.J., McPherson, M.J. & others (Eds) | | | |
| | | Spiralbound hardback | 0-19-963354-1 | £56.00 |
| | | Paperback | 0-19-963353-3 | £37.00 |
| Protein Purification Methods, and Protein Purification Applications: 2 vol set | Harris, E.L.V. & Angal, S. (Eds) | | | |
| | | Spiralbound hardback | 0-19-963048-8 | £48.00 |
| | | Paperback | 0-19-963049-6 | £32.00 |
| Diagnostic Molecular Pathology: 2 vol set | Herrington, C.S. & McGee, J. O'D. (Eds) | | | |
| | | Spiralbound hardback | 0-19-963241-3 | £54.00 |
| | | Paperback | 0-19-963240-5 | £35.00 |
| RNA Processing: 2 vol set | Higgins, S.J. & Hames, B.D. (Eds) | | | |
| | | Spiralbound hardback | 0-19-963473-4 | £54.00 |
| | | Paperback | 0-19-963472-6 | £35.00 |
| Receptor Biochemistry; Receptor-Effector Coupling; Receptor-Ligand Interactions: 3 vol set | Hulme, E.C. (Ed) | | | |
| | | Paperback | 0-19-963097-6 | £62.50 |
| Human Cytogenetics: 2 vol set (2/e) | Rooney, D.E. & Czepulkowski, B.H. (Eds) | | | |
| | | Hardback | 0-19-963314-2 | £58.50 |
| | | Paperback | 0-19-963313-4 | £40.50 |
| Behavioural Neuroscience: 2 vol set | Sahgal, A. (Ed) | | | |
| | | Spiralbound hardback | 0-19-963460-2 | £58.00 |
| | | Paperback | 0-19-963459-9 | £40.00 |
| Peptide Hormone Secretion/Peptide Hormone Action: 2 vol set | Siddle, K. & Hutton, J.C. (Eds) | | | |
| | | Spiralbound hardback | 0-19-963072-0 | £55.00 |
| | | Paperback | 0-19-963073-9 | £38.00 |

# ORDER FORM for UK, Europe and Rest of World

## (Excluding USA and Canada)

| Qty | ISBN | Author | Title | Amount |
|-----|------|--------|-------|--------|
|     |      |        |       |        |
|     |      |        |       |        |
|     |      |        |       |        |
|     |      |        |       |        |
|     |      |        | P&P   |        |
|     |      |        | *VAT  |        |
|     |      |        | TOTAL |        |

Please add postage and packing: £1.75 for UK orders under £20; £2.75 for UK orders over £20; overseas orders add 10% of total.
* EC customers please note that VAT must be added (excludes UK customers)

Name ............................................................................

Address ........................................................................

....................................................................................

.................................................... Post code ....................

[ ] Please charge £ ..................... to my credit card
Access/VISA/Eurocard/AMEX/Diners Club (circle appropriate card)

Card No ............................... Expiry date ........................

Signature .....................................................................

Credit card account address if different from above:

....................................................................................

.................................................... Postcode ....................

[ ] I enclose a cheque for £.....................

Please return this form to: OUP Distribution Services, Saxon Way West, Corby, Northants NN18 9ES, UK

OR ORDER BY CREDIT CARD HOTLINE: Tel +44-(0)536-741519 or Fax +44-(0)536-746337

# ORDER OTHER TITLES OF INTEREST TODAY

## Price list for: USA and Canada

| No. | Title | Authors | Format | ISBN | Price |
|---|---|---|---|---|---|
| 128. | Behavioural Neuroscience: Volume I | Sahgal, A. (Ed) | | | |
| | | | Spiralbound hardback | 0-19-963368-1 | $57.00 |
| | | | Paperback | 0-19-963367-3 | $37.00 |
| 127. | Molecular Virology | Davison, A.J. & Elliott, R.M. (Eds) | | | |
| | | | Spiralbound hardback | 0-19-963358-4 | $49.00 |
| | | | Paperback | 0-19-963357-6 | $32.00 |
| 126. | Gene Targeting | Joyner, A.L. (Ed) | | | |
| | | | Spiralbound hardback | 0-19-963407-6 | $49.00 |
| | | | Paperback | 0-19-9634036-8 | $34.00 |
| 124. | Human Genetic Disease Analysis (2/e) | Davies, K.E. (Ed) | | | |
| | | | Spiralbound hardback | 0-19-963309-6 | $54.00 |
| | | | Paperback | 0-19-963308-8 | $33.00 |
| 123. | Protein Phosphorylation | Hardie, D.G. (Ed) | | | |
| | | | Spiralbound hardback | 0-19-963306-1 | $65.00 |
| | | | Paperback | 0-19-963305-3 | $45.00 |
| 122. | Immunocytochemistry | Beesley, J. (Ed) | | | |
| | | | Spiralbound hardback | 0-19-963270-7 | $62.00 |
| | | | Paperback | 0-19-963269-3 | $42.00 |
| 121. | Tumour Immunobiology | Gallagher, G., Rees, R.C. & others (Eds) | | | |
| | | | Spiralbound hardback | 0-19-963370-3 | $72.00 |
| | | | Paperback | 0-19-963369-X | $50.00 |
| 120. | Transcription Factors | Latchman, D.S. (Ed) | | | |
| | | | Spiralbound hardback | 0-19-963342-8 | $48.00 |
| | | | Paperback | 0-19-963341-X | $31.00 |
| 119. | Growth Factors | McKay, I. & Leigh, I. (Eds) | | | |
| | | | Spiralbound hardback | 0-19-963360-6 | $48.00 |
| | | | Paperback | 0-19-963359-2 | $31.00 |
| 118. | Histocompatibility Testing | Dyer, P. & Middleton, D. (Eds) | | | |
| | | | Spiralbound hardback | 0-19-963364-9 | $60.00 |
| | | | Paperback | 0-19-963363-0 | $41.00 |
| 117. | Gene Transcription | Hames, B.D. & Higgins, S.J. (Eds) | | | |
| | | | Spiralbound hardback | 0-19-963292-8 | $72.00 |
| | | | Paperback | 0-19-963291-X | $50.00 |
| 116. | Electrophysiology | Wallis, D.I. (Ed) | | | |
| | | | Spiralbound hardback | 0-19-963348-7 | $56.00 |
| | | | Paperback | 0-19-963347-9 | $39.00 |
| 115. | Biological Data Analysis | Fry, J.C. (Ed) | | | |
| | | | Spiralbound hardback | 0-19-963340-1 | $80.00 |
| | | | Paperback | 0-19-963339-8 | $60.00 |
| 114. | Experimental Neuroanatomy | Bolam, J.P. (Ed) | | | |
| | | | Spiralbound hardback | 0-19-963326-6 | $59.00 |
| | | | Paperback | 0-19-963325-8 | $39.00 |
| 113. | Preparative Centrifugation | Rickwood, D. (Ed) | | | |
| | | | Spiralbound hardback | 0-19-963208-1 | $78.00 |
| | | | Paperback | 0-19-963211-1 | $44.00 |
| 111. | Haemopoiesis | Testa, N.G. & Molineux, G. (Eds) | | | |
| | | | Spiralbound hardback | 0-19-963366-5 | $59.00 |
| | | | Paperback | 0-19-963365-7 | $39.00 |
| 110. | Pollination Ecology | Dafni, A. | | | |
| | | | Spiralbound hardback | 0-19-963299-5 | $56.95 |
| | | | Paperback | 0-19-963298-7 | $39.95 |
| 109. | In Situ Hybridization | Wilkinson, D.G. (Ed) | | | |
| | | | Spiralbound hardback | 0-19-963328-2 | $58.00 |
| | | | Paperback | 0-19-963327-4 | $36.00 |
| 108. | Protein Engineering | Rees, A.R., Sternberg, M.J.E. & others (Eds) | | | |
| | | | Spiralbound hardback | 0-19-963139-5 | $64.00 |
| | | | Paperback | 0-19-963138-7 | $44.00 |
| 107. | Cell-Cell Interactions | Stevenson, B.R., Gallin, W.J. & others (Eds) | | | |
| | | | Spiralbound hardback | 0-19-963319-3 | $55.00 |
| | | | Paperback | 0-19-963318-5 | $38.00 |
| 106. | Diagnostic Molecular Pathology: Volume I | Herrington, C.S. & McGee, J. O'D. (Eds) | | | |
| | | | Spiralbound hardback | 0-19-963237-5 | $50.00 |
| | | | Paperback | 0-19-963236-7 | $33.00 |
| 105. | Biomechanics-Materials | Vincent, J.F.V. (Ed) | | | |
| | | | Spiralbound hardback | 0-19-963223-5 | $70.00 |
| | | | Paperback | 0-19-963222-7 | $50.00 |
| 104. | Animal Cell Culture (2/e) | Freshney, R.I. (Ed) | | | |
| | | | Spiralbound hardback | 0-19-963212-X | $55.00 |
| | | | Paperback | 0-19-963213-8 | $35.00 |
| 103. | Molecular Plant Pathology: Volume II | Gurr, S.J., McPherson, M.J. & others (Eds) | | | |
| | | | Spiralbound hardback | 0-19-963352-5 | $65.00 |
| | | | Paperback | 0-19-963351-7 | $45.00 |
| 102. | Signal Transduction | Milligan, G. (Ed) | | | |
| | | | Spiralbound hardback | 0-19-963296-0 | $60.00 |
| | | | Paperback | 0-19-963295-2 | $38.00 |
| 101. | Protein Targeting | Magee, A.I. & Wileman, T. (Eds) | | | |
| | | | Spiralbound hardback | 0-19-963206-5 | $75.00 |
| | | | Paperback | 0-19-963210-3 | $50.00 |
| 100. | Diagnostic Molecular Pathology: Volume II: Cell and Tissue Genotyping | Herrington, C.S. & McGee, J.O'D. (Eds) | | | |
| | | | Spiralbound hardback | 0-19-963239-1 | $60.00 |
| | | | Paperback | 0-19-963238-3 | $39.00 |
| 99. | Neuronal Cell Lines | Wood, J.N. (Ed) | | | |
| | | | Spiralbound hardback | 0-19-963346-0 | $68.00 |
| | | | Paperback | 0-19-963345-2 | $48.00 |
| 98. | Neural Transplantation | Dunnett, S.B. & Bj‹148›rklund, A. (Eds) | | | |
| | | | Spiralbound hardback | 0-19-963286-3 | $69.00 |
| | | | Paperback | 0-19-963285-5 | $42.00 |
| 97. | Human Cytogenetics: Volume II: Malignancy and Acquired Abnormalities (2/e) | Rooney, D.E. & Czepulkowski, B.H. (Eds) | | | |
| | | | Spiralbound hardback | 0-19-963290-1 | $75.00 |
| | | | Paperback | 0-19-963289-8 | $50.00 |
| 96. | Human Cytogenetics: Volume I: Constitutional Analysis (2/e) | Rooney, D.E. & Czepulkowski, B.H. (Eds) | | | |
| | | | Spiralbound hardback | 0-19-963288-X | $75.00 |
| | | | Paperback | 0-19-963287-1 | $50.00 |
| 95. | Lipid Modification of Proteins | Hooper, N.M. & Turner, A.J. (Eds) | | | |
| | | | Spiralbound hardback | 0-19-963274-X | $75.00 |
| | | | Paperback | 0-19-963273-1 | $50.00 |
| 94. | Biomechanics-Structures and Systems | Biewener, A.A. (Ed) | | | |
| | | | Spiralbound hardback | 0-19-963268-5 | $85.00 |
| | | | Paperback | 0-19-963267-7 | $50.00 |
| 93. | Lipoprotein Analysis | Converse, C.A. & Skinner, E.R. (Eds) | | | |
| | | | Spiralbound hardback | 0-19-963192-1 | $65.00 |
| | | | Paperback | 0-19-963231-6 | $42.00 |
| 92. | Receptor-Ligand Interactions | Hulme, E.C. (Ed) | | | |
| | | | Spiralbound hardback | 0-19-963090-9 | $75.00 |
| | | | Paperback | 0-19-963091-7 | $50.00 |
| 91. | Molecular Genetic Analysis of Populations | Hoelzel, A.R. (Ed) | | | |
| | | | Spiralbound hardback | 0-19-963278-2 | $65.00 |
| | | | Paperback | 0-19-963277-4 | $45.00 |

| 90. | Enzyme Assays Eisenthal, R. & Danson, M.J. (Eds) | | |
|---|---|---|---|
| | Spiralbound hardback | 0-19-963142-5 | **$68.00** |
| | Paperback | 0-19-963143-3 | **$48.00** |
| 89. | Microcomputers in Biochemistry Bryce, C.F.A. (Ed) | | |
| | Spiralbound hardback | 0-19-963253-7 | **$60.00** |
| | Paperback | 0-19-963252-9 | **$40.00** |
| 88. | The Cytoskeleton Carraway, K.L. & Carraway, C.A.C. (Eds) | | |
| | Spiralbound hardback | 0-19-963257-X | **$60.00** |
| | Paperback | 0-19-963256-1 | **$40.00** |
| 87. | Monitoring Neuronal Activity Stamford, J.A. (Ed) | | |
| | Spiralbound hardback | 0-19-963244-8 | **$60.00** |
| | Paperback | 0-19-963243-X | **$40.00** |
| 86. | Crystallization of Nucleic Acids and Proteins Ducruix, A. & Giegé, R. (Eds) | | |
| | Spiralbound hardback | 0-19-963245-6 | **$60.00** |
| | Paperback | 0-19-963246-4 | **$50.00** |
| 85. | Molecular Plant Pathology: Volume I Gurr, S.J., McPherson, M.J. & others (Eds) | | |
| | Spiralbound hardback | 0-19-963103-4 | **$60.00** |
| | Paperback | 0-19-963102-6 | **$40.00** |
| 84. | Anaerobic Microbiology Levett, P.N. (Ed) | | |
| | Spiralbound hardback | 0-19-963204-9 | **$75.00** |
| | Paperback | 0-19-963262-6 | **$45.00** |
| 83. | Oligonucleotides and Analogues Eckstein, F. (Ed) | | |
| | Spiralbound hardback | 0-19-963280-4 | **$65.00** |
| | Paperback | 0-19-963279-0 | **$45.00** |
| 82. | Electron Microscopy in Biology Harris, R. (Ed) | | |
| | Spiralbound hardback | 0-19-963219-7 | **$65.00** |
| | Paperback | 0-19-963215-4 | **$45.00** |
| 81. | Essential Molecular Biology: Volume II Brown, T.A. (Ed) | | |
| | Spiralbound hardback | 0-19-963112-3 | **$65.00** |
| | Paperback | 0-19-963113-1 | **$45.00** |
| 80. | Cellular Calcium McCormack, J.G. & Cobbold, P.H. (Eds) | | |
| | Spiralbound hardback | 0-19-963131-X | **$75.00** |
| | Paperback | 0-19-963130-1 | **$50.00** |
| 79. | Protein Architecture Lesk, A.M. | | |
| | Spiralbound hardback | 0-19-963054-2 | **$65.00** |
| | Paperback | 0-19-963055-0 | **$45.00** |
| 78. | Cellular Neurobiology Chad, J. & Wheal, H. (Eds) | | |
| | Spiralbound hardback | 0-19-963106-9 | **$73.00** |
| | Paperback | 0-19-963107-7 | **$43.00** |
| 77. | PCR McPherson, M.J., Quirke, P. & others (Eds) | | |
| | Spiralbound hardback | 0-19-963226-X | **$55.00** |
| | Paperback | 0-19-963196-4 | **$40.00** |
| 76. | Mammalian Cell Biotechnology Butler, M. (Ed) | | |
| | Spiralbound hardback | 0-19-963207-3 | **$60.00** |
| | Paperback | 0-19-963209-X | **$40.00** |
| 75. | Cytokines Balkwill, F.R. (Ed) | | |
| | Spiralbound hardback | 0-19-963218-9 | **$64.00** |
| | Paperback | 0-19-963214-6 | **$44.00** |
| 74. | Molecular Neurobiology Chad, J. & Wheal, H. (Eds) | | |
| | Spiralbound hardback | 0-19-963108-5 | **$56.00** |
| | Paperback | 0-19-963109-3 | **$36.00** |
| 73. | Directed Mutagenesis McPherson, M.J. (Ed) | | |
| | Spiralbound hardback | 0-19-963141-7 | **$55.00** |
| | Paperback | 0-19-963140-9 | **$35.00** |
| 72. | Essential Molecular Biology: Volume I Brown, T.A. (Ed) | | |
| | Spiralbound hardback | 0-19-963110-7 | **$65.00** |
| | Paperback | 0-19-963111-5 | **$45.00** |
| 71. | Peptide Hormone Action Siddle, K. & Hutton, J.C. | | |
| | Spiralbound hardback | 0-19-963070-4 | **$70.00** |
| | Paperback | 0-19-963071-2 | **$50.00** |
| 70. | Peptide Hormone Secretion Hutton, J.C. & Siddle, K. (Eds) | | |
| | Spiralbound hardback | 0-19-963068-2 | **$70.00** |
| | Paperback | 0-19-963069-0 | **$50.00** |
| 69. | Postimplantation Mammalian Embryos Copp, A.J. & Cockroft, D.L. (Eds) | | |
| | Spiralbound hardback | 0-19-963088-7 | **$70.00** |
| | Paperback | 0-19-963089-5 | **$50.00** |
| 68. | Receptor-Effector Coupling Hulme, E.C. (Ed) | | |
| | Spiralbound hardback | 0-19-963094-1 | **$70.00** |
| | Paperback | 0-19-963095-X | **$45.00** |
| 67. | Gel Electrophoresis of Proteins (2/e) Hames, B.D. & Rickwood, D. (Eds) | | |
| | Spiralbound hardback | 0-19-963074-7 | **$75.00** |
| | Paperback | 0-19-963075-5 | **$50.00** |
| 66. | Clinical Immunology Gooi, H.C. & Chapel, H. (Eds) | | |
| | Spiralbound hardback | 0-19-963086-0 | **$69.95** |
| | Paperback | 0-19-963087-9 | **$50.00** |
| 65. | Receptor Biochemistry Hulme, E.C. (Ed) | | |
| | Paperback | 0-19-963093-3 | **$50.00** |
| 64. | Gel Electrophoresis of Nucleic Acids (2/e) Rickwood, D. & Hames, B.D. (Eds) | | |
| | Spiralbound hardback | 0-19-963082-8 | **$75.00** |
| | Paperback | 0-19-963083-6 | **$50.00** |
| 63. | Animal Virus Pathogenesis Oldstone, M.B.A. (Ed) | | |
| | Spiralbound hardback | 0-19-963100-X | **$68.00** |
| | Paperback | 0-19-963101-8 | **$40.00** |
| 62. | Flow Cytometry Ormerod, M.G. (Ed) | | |
| | Paperback | 0-19-963053-4 | **$50.00** |
| 61. | Radioisotopes in Biology Slater, R.J. (Ed) | | |
| | Spiralbound hardback | 0-19-963080-1 | **$75.00** |
| | Paperback | 0-19-963081-X | **$45.00** |
| 60. | Biosensors Cass, A.E.G. (Ed) | | |
| | Spiralbound hardback | 0-19-963046-1 | **$65.00** |
| | Paperback | 0-19-963047-X | **$43.00** |
| 59. | Ribosomes and Protein Synthesis Spedding, G. (Ed) | | |
| | Spiralbound hardback | 0-19-963104-2 | **$75.00** |
| | Paperback | 0-19-963105-0 | **$45.00** |
| 58. | Liposomes New, R.R.C. (Ed) | | |
| | Spiralbound hardback | 0-19-963076-3 | **$70.00** |
| | Paperback | 0-19-963077-1 | **$45.00** |
| 57. | Fermentation McNeil, B. & Harvey, L.M. (Eds) | | |
| | Spiralbound hardback | 0-19-963044-5 | **$65.00** |
| | Paperback | 0-19-963045-3 | **$39.00** |
| 56. | Protein Purification Applications Harris, E.L.V. & Angal, S. (Eds) | | |
| | Spiralbound hardback | 0-19-963022-4 | **$54.00** |
| | Paperback | 0-19-963023-2 | **$36.00** |
| 55. | Nucleic Acids Sequencing Howe, C.J. & Ward, E.S. (Eds) | | |
| | Spiralbound hardback | 0-19-963056-9 | **$59.00** |
| | Paperback | 0-19-963057-7 | **$38.00** |
| 54. | Protein Purification Methods Harris, E.L.V. & Angal, S. (Eds) | | |
| | Spiralbound hardback | 0-19-963002-X | **$60.00** |
| | Paperback | 0-19-963003-8 | **$40.00** |
| 53. | Solid Phase Peptide Synthesis Atherton, E. & Sheppard, R.C. | | |
| | Spiralbound hardback | 0-19-963066-6 | **$58.00** |
| | Paperback | 0-19-963067-4 | **$39.95** |
| 52. | Medical Bacteriology Hawkey, P.M. & Lewis, D.A. (Eds) | | |
| | Paperback | 0-19-963009-7 | **$50.00** |
| 51. | Proteolytic Enzymes Beynon, R.J. & Bond, J.S. (Eds) | | |
| | Spiralbound hardback | 0-19-963058-5 | **$60.00** |
| | Paperback | 0-19-963059-3 | **$39.00** |
| 50. | Medical Mycology Evans, E.G.V. & Richardson, M.D. (Eds) | | |
| | Spiralbound hardback | 0-19-963010-0 | **$69.95** |
| | Paperback | 0-19-963011-9 | **$50.00** |
| 49. | Computers in Microbiology Bryant, T.N. & Wimpenny, J.W.T. (Eds) | | |
| | Paperback | 0-19-963015-1 | **$40.00** |
| 48. | Protein Sequencing Findlay, J.B.C. & Geisow, M.J. (Eds) | | |
| | Spiralbound hardback | 0-19-963012-7 | **$56.00** |
| | Paperback | 0-19-963013-5 | **$38.00** |
| 47. | Cell Growth and Division Baserga, R. (Ed) | | |
| | Spiralbound hardback | 0-19-963026-7 | **$62.00** |
| | Paperback | 0-19-963027-5 | **$38.00** |
| 46. | Protein Function Creighton, T.E. (Ed) | | |
| | Spiralbound hardback | 0-19-963006-2 | **$65.00** |
| | Paperback | 0-19-963007-0 | **$45.00** |
| 45. | Protein Structure Creighton, T.E. (Ed) | | |
| | Spiralbound hardback | 0-19-963000-3 | **$65.00** |
| | Paperback | 0-19-963001-1 | **$45.00** |
| 44. | Antibodies: Volume II Catty, D. (Ed) | | |
| | Spiralbound hardback | 0-19-963018-6 | **$58.00** |
| | Paperback | 0-19-963019-4 | **$39.00** |
| 43. | HPLC of Macromolecules Oliver, R.W.A. (Ed) | | |
| | Spiralbound hardback | 0-19-963020-8 | **$54.00** |
| | Paperback | 0-19-963021-6 | **$45.00** |
| 42. | Light Microscopy in Biology Lacey, A.J. (Ed) | | |
| | Spiralbound hardback | 0-19-963036-4 | **$62.00** |
| | Paperback | 0-19-963037-2 | **$38.00** |
| 41. | Plant Molecular Biology Shaw, C.H. (Ed) | | |
| | Paperback | 1-85221-056-7 | **$38.00** |
| 40. | Microcomputers in Physiology Fraser, P.J. (Ed) | | |
| | Spiralbound hardback | 1-85221-129-6 | **$54.00** |
| | Paperback | 1-85221-130-X | **$36.00** |
| 39. | Genome Analysis Davies, K.E. (Ed) | | |
| | Spiralbound hardback | 1-85221-109-1 | **$54.00** |
| | Paperback | 1-85221-110-5 | **$36.00** |
| 38. | Antibodies: Volume I Catty, D. (Ed) | | |
| | Paperback | 0-947946-85-3 | **$38.00** |
| 37. | Yeast Campbell, I. & Duffus, J.H. (Eds) | | |
| | Paperback | 0-947946-79-9 | **$36.00** |

| 36. | Mammalian Development | Monk, M. (Ed) | |
|---|---|---|---|
| ...... | Hardback | 1-85221-030-3 | **$60.00** |
| ...... | Paperback | 1-85221-029-X | **$45.00** |
| 35. | **Lymphocytes** | Klaus, G.G.B. (Ed) | |
| ...... | Hardback | 1-85221-018-4 | **$54.00** |
| 34. | **Lymphokines and Interferons** Clemens, M.J., Morris, A.G. & others (Eds) | | |
| ...... | Paperback | 1-85221-035-4 | **$44.00** |
| 33. | **Mitochondria** Darley-Usmar, V.M., Rickwood, D. & others (Eds) | | |
| ...... | Hardback | 1-85221-034-6 | **$65.00** |
| ...... | Paperback | 1-85221-033-8 | **$45.00** |
| 32. | **Prostaglandins and Related Substances** Benedetto, C., McDonald-Gibson, R.G. & others (Eds) | | |
| ...... | Hardback | 1-85221-032-X | **$58.00** |
| ...... | Paperback | 1-85221-031-1 | **$38.00** |
| 31. | **DNA Cloning: Volume III** Glover, D.M. (Ed) | | |
| ...... | Hardback | 1-85221-049-4 | **$56.00** |
| ...... | Paperback | 1-85221-048-6 | **$36.00** |
| 30. | **Steroid Hormones** Green, B. & Leake, R.E. (Eds) | | |
| ...... | Paperback | 0-947946-53-5 | **$40.00** |
| 29. | **Neurochemistry** Turner, A.J. & Bachelard, H.S. (Eds) | | |
| ...... | Hardback | 1-85221-028-1 | **$56.00** |
| ...... | Paperback | 1-85221-027-3 | **$36.00** |
| 28. | **Biological Membranes** Findlay, J.B.C. & Evans, W.H. (Eds) | | |
| ...... | Hardback | 0-947946-84-5 | **$54.00** |
| ...... | Paperback | 0-947946-83-7 | **$36.00** |
| 27. | **Nucleic Acid and Protein Sequence Analysis** Bishop, M.J. & Rawlings, C.J. (Eds) | | |
| ...... | Hardback | 1-85221-007-9 | **$66.00** |
| ...... | Paperback | 1-85221-006-0 | **$44.00** |
| 26. | **Electron Microscopy in Molecular Biology** Sommerville, J. & Scheer, U. (Eds) | | |
| ...... | Hardback | 0-947946-64-0 | **$54.00** |
| ...... | Paperback | 0-947946-54-3 | **$40.00** |
| 24. | **Spectrophotometry and Spectrofluorimetry** Harris, D.A. & Bashford, C.L. (Eds) | | |
| ...... | Hardback | 0-947946-69-1 | **$56.00** |
| ...... | Paperback | 0-947946-46-2 | **$39.95** |
| 23. | **Plasmids** Hardy, K.G. (Ed) | | |
| ...... | Paperback | 0-947946-81-0 | **$36.00** |
| 22. | **Biochemical Toxicology** Snell, K. & Mullock, B. (Eds) | | |
| ...... | Paperback | 0-947946-52-7 | **$40.00** |
| 19. | **Drosophila** Roberts, D.B. (Ed) | | |
| ...... | Hardback | 0-947946-66-7 | **$67.50** |
| ...... | Paperback | 0-947946-45-4 | **$46.00** |
| 17. | **Photosynthesis: Energy Transduction** Hipkins, M.F. & Baker, N.R. (Eds) | | |
| ...... | Hardback | 0-947946-63-2 | **$54.00** |
| ...... | Paperback | 0-947946-51-9 | **$36.00** |
| 16. | **Human Genetic Diseases** Davies, K.E. (Ed) | | |
| ...... | Hardback | 0-947946-76-4 | **$60.00** |
| ...... | Paperback | 0-947946-75-6 | **$34.00** |
| 14. | **Nucleic Acid Hybridisation** Hames, B.D. & Higgins, S.J. (Eds) | | |
| ...... | Hardback | 0-947946-61-6 | **$60.00** |
| ...... | Paperback | 0-947946-23-3 | **$36.00** |
| 12. | **Plant Cell Culture** Dixon, R.A. (Ed) | | |
| ...... | Paperback | 0-947946-22-5 | **$36.00** |
| 11a. | **DNA Cloning: Volume I** Glover, D.M. (Ed) | | |
| ...... | Paperback | 0-947946-18-7 | **$36.00** |
| 11b. | **DNA Cloning: Volume II** Glover, D.M. (Ed) | | |
| ...... | Paperback | 0-947946-19-5 | **$36.00** |
| 10. | **Virology** Mahy, B.W.J. (Ed) | | |
| ...... | Paperback | 0-904147-78-9 | **$40.00** |
| 9. | **Affinity Chromatography** Dean, P.D.G., Johnson, W.S. & others (Eds) | | |
| ...... | Paperback | 0-904147-71-1 | **$36.00** |
| 7. | **Microcomputers in Biology** Ireland, C.R. & Long, S.P. (Eds) | | |
| ...... | Paperback | 0-904147-57-6 | **$36.00** |
| 6. | **Oligonucleotide Synthesis** Gait, M.J. (Ed) | | |
| ...... | Paperback | 0-904147-74-6 | **$38.00** |
| 5. | **Transcription and Translation** Hames, B.D. & Higgins, S.J. (Eds) | | |
| ...... | Paperback | 0-904147-52-5 | **$38.00** |
| 3. | **Iodinated Density Gradient Media** Rickwood, D. (Ed) | | |
| ...... | Paperback | 0-904147-51-7 | **$36.00** |

## Sets

| | **Essential Molecular Biology: 2 vol set** Brown, T.A. (Ed) | | |
|---|---|---|---|
| ...... | Spiralbound hardback | 0-19-963114-X | **$118.00** |
| ...... | Paperback | 0-19-963115-8 | **$78.00** |
| | **Antibodies: 2 vol set** Catty, D. (Ed) | | |
| ...... | Paperback | 0-19-963063-1 | **$70.00** |
| | **Cellular and Molecular Neurobiology: 2 vol set** Chad, J. & Wheal, H. (Eds) | | |
| ...... | Spiralbound hardback | 0-19-963255-3 | **$133.00** |
| ...... | Paperback | 0-19-963254-5 | **$79.00** |
| | **Protein Structure and Protein Function: 2 vol set** Creighton, T.E. (Ed) | | |
| ...... | Spiralbound hardback | 0-19-963064-X | **$114.00** |
| ...... | Paperback | 0-19-963065-8 | **$80.00** |
| | **DNA Cloning: 2 vol set** Glover, D.M. (Ed) | | |
| ...... | Paperback | 1-85221-069-9 | **$92.00** |
| | **Molecular Plant Pathology: 2 vol set** Gurr, S.J., McPherson, M.J. & others (Eds) | | |
| ...... | Spiralbound hardback | 0-19-963354-1 | **$110.00** |
| ...... | Paperback | 0-19-963353-3 | **$75.00** |
| | **Protein Purification Methods, and Protein Purification Applications: 2 vol set** Harris, E.L.V. & Angal, S. (Eds) | | |
| ...... | Spiralbound hardback | 0-19-963048-8 | **$98.00** |
| ...... | Paperback | 0-19-963049-6 | **$68.00** |
| | **Diagnostic Molecular Pathology: 2 vol set** Herrington, C.S. & McGee, J. O'D. (Eds) | | |
| ...... | Spiralbound hardback | 0-19-963241-3 | **$105.00** |
| ...... | Paperback | 0-19-963240-5 | **$69.00** |
| | **Receptor Biochemistry; Receptor-Effector Coupling; Receptor-Ligand Interactions: 3 vol set** Hulme, E.C. (Ed) | | |
| ...... | Paperback | 0-19-963097-6 | **$130.00** |
| | **Human Cytogenetics: (2/e): 2 vol set** Rooney, D.E. & Czepulkowski, B.H. (Eds) | | |
| ...... | Hardback | 0-19-963314-2 | **$130.00** |
| ...... | Paperback | 0-19-963313-4 | **$90.00** |
| | **Peptide Hormone Secretion/Peptide Hormone Action: 2 vol set** Siddle, K. & Hutton, J.C. (Eds) | | |
| ...... | Spiralbound hardback | 0-19-963072-0 | **$135.00** |
| ...... | Paperback | 0-19-963073-9 | **$90.00** |

# ORDER FORM for USA and Canada

| Qty | ISBN | Author | Title | Amount |
|---|---|---|---|---|
|  |  |  |  |  |
|  |  |  |  |  |
|  |  |  |  |  |
|  |  |  |  |  |
|  |  |  | S&H |  |
|  | CA and NC residents add appropriate sales tax ||| |
|  |  |  | TOTAL |  |

Please add shipping and handling: US $2.50 for first book, (US $1.00 each book thereafter)

Name ...................................................................................

Address ................................................................................

................................................................................................

............................................................. Zip ............................

[ ] Please charge $ ...................... to my credit card
Mastercard/VISA/American Express (circle appropriate card)

Acct. ..................................... Expiry date ............................

Signature ..............................................................................

Credit card account address if different from above:

................................................................................................

............................................................. Zip ............................

[ ] I enclose a cheque for US $............

Mail orders to: Order Dept. Oxford University Press, 2001 Evans Road, Cary, NC 27513

QH
605
.C423
1993